T0257841

Encyclopedia of Soybean: Pest Resistance
Volume X

Edited by **Albert Marinelli and
Kiara Woods**

New York

Published by Callisto Reference,
106 Park Avenue, Suite 200,
New York, NY 10016, USA
www.callistoreference.com

Encyclopedia of Soybean: Pest Resistance
Volume X
Edited by Albert Marinelli and Kiara Woods

© 2015 Callisto Reference

International Standard Book Number: 978-1-63239-305-0 (Hardback)

This book contains information obtained from authentic and highly regarded sources. Copyright for all individual chapters remain with the respective authors as indicated. A wide variety of references are listed. Permission and sources are indicated; for detailed attributions, please refer to the permissions page. Reasonable efforts have been made to publish reliable data and information, but the authors, editors and publisher cannot assume any responsibility for the validity of all materials or the consequences of their use.

The publisher's policy is to use permanent paper from mills that operate a sustainable forestry policy. Furthermore, the publisher ensures that the text paper and cover boards used have met acceptable environmental accreditation standards.

Trademark Notice: Registered trademark of products or corporate names are used only for explanation and identification without intent to infringe.

Printed in the United States of America.

Contents

Preface

In my initial years as a student, I used to run to the library at every possible instance to grab a book and learn something new. Books were my primary source of knowledge and I would not have come such a long way without all that I learnt from them. Thus, when I was approached to edit this book; I became understandably nostalgic. It was an absolute honor to be considered worthy of guiding the current generation as well as those to come. I put all my knowledge and hard work into making this book most beneficial for its readers.

This book comprehensively discusses the pest resistance characteristics of soybean. Legumes are an essential part of a healthy diet for a major part of the world's population. They are an excellent source of carbohydrates, protein, vitamins and minerals. Soybeans are of great significance in the overall agriculture and trade as well as in their contribution to food supply. They are highly rich in protein content and have no cholesterol in comparison with animal food sources and conventional legume. Additionally, soybean is a cheap source of food as well as medicinal value because of its photochemical, genistein, isoflavones content. Studies have reflected that soybean is extremely helpful in fighting numerous medical conditions including diabetes, heart disease and cancer. Presently, soybean protein and calories are being employed for the prevention of body wasting mostly linked with HIV. Soybean's nutritional value is of extreme importance in places where medication facilities are not readily available. It holds great economical potential for extensive use in industrial purposes. This quality of soybean can be hugely beneficial for small-scale soybean producers.

I wish to thank my publisher for supporting me at every step. I would also like to thank all the authors who have contributed their researches in this book. I hope this book will be a valuable contribution to the progress of the field.

Editor

Developing Host-Plant Resistance for Hemipteran Soybean Pests: Lessons from Soybean Aphid and Stink Bugs

Raman Bansal, Tae-Hwan Jun, M. A. R. Mian and
Andy P. Michel

Additional information is available at the end of the chapter

1. Introduction

Soybean is world's leading agricultural crop with multiple uses including human food, animal feed, edible oil, biofuel, industrial products, cosmetics, etc. In soybean production, United States (US) is the leading producer with 33% of the world's total production of 251.5 million MT, amounting to $38.5 billion in production value [1,2]. In North-America, there has been an exponential increase of soybean acreage during the second half of last century, but there is a continuous threat of pests attacking this crop. Soybean yield is impacted by various kinds of pests such as fungi, bacteria, and insects [3]. Indeed, the strategies and input costs for pest management in soybean have changed dramatically with time [3-5]. For example, there has been a 130-fold increase in insecticide use across the North-Central US states since 2001 [4].

In regards to insects, soybean has been traditionally attacked by foliage-feeding Lepidopteran and Coleopteran pests such as soybean looper, velvet bean caterpillar, beet armyworm, bean leaf beetle, stem borer, Mexican bean beetle, and soybean leaf miner [6]. However, during the last decade, the invasion of soybean aphid [*Aphis glycines* Matsumura], brown-marmorated stink bug (BMSB) [*Halyomorpha halys* (Stål)], and (although technically not a stink bug) kudzu bug [*Megacopta cribraria* (F.)] in north-central, eastern, and southeastern US, respectively, and the emergence of red-banded stink bug [*Piezodorus guildinii* (Westwood)] as major pest in southern US have drastically changed the pest complex in soybean [4,7,8]. The threat posed by soybean aphid and stink bugs has the potential to rapidly increase as these insects continue to expand their geographical range. For example, in less than 10 years since its initial detection in Wisconsin, soybean aphid had spread across 30 US states and 3 Canadian provinces by 2009

[4], and the BMSB has already been detected in 38 US states since first being seen in Pennsylvania in 1996 [7].

Both soybean aphid and stink bug belong to order Hemiptera which also includes other economic pests such as whiteflies and leafhoppers. To minimize the damage by Hemipteran pests, host-plant resistance in soybean cultivars should constitute an integral part of an integrated pest management (IPM) program. In the current chapter, we attempt to review the recent research advances made on soybean resistance to Hemipteran pests. In the light of various challenges to manage Hemipteran pests, we have proposed strategies for successful and sustainable use of host plant resistance (HPR) in soybean against these pests.

2. Hemipteran pests of soybean

The soybean aphid, various stink bug species and kudzu bug are the major Hemipteran pests of soybean (Figure 1). Although soybean aphid and stink bugs share basic features of Hemipteran insects, there is much evolutionary divergence between them, their suborders having diverged more than 250 million years ago [9]. Aphids belong to the suborder Homoptera which have uniform, membranous forewings and hindwings. In homopteran insects, wings are held roof-like over their abdomen. Stink bugs belong to suborder Heteroptera having forewings that are leathery basally and membranous distally, in contrast to membranous hindwings. In Heteroptera, wings are folded flat over the abdomen [10]. Stink bugs can also be identified by five-segmented antennae and a conspicuous scutellum [10].

2.1. Soybean aphid

The soybean aphid is a recent invasive species in North-America [4,5,11]. This species was first detected during the summer of 2000 and is believed to have been introduced from its native Asian range [11,12]. Soybean aphid is a pest of significant economic importance as it can cause up to 58% yield losses in soybean [13]. Losses due to yield have been estimated to be $2.4 billion annually [14-15].

The biology of soybean aphid in North-America has been reviewed recently [4,5]. In general, it is a typical holocyclic (asexual and sexual reproduction) heterecious (alternates between 2 hosts) aphid species. In autumn, sexual reproduction occurs on its primary host, buckthorn (*Rhamnus* spp.), and the resulting eggs undergo overwintering. The following spring, the eggs hatch, and the fundatrices (i.e. stem mothers) begin to produce female clones. After 2-3 asexual generations, winged females are formed that disperse to emerging soybean, where about 15 additional asexual generations occur, and when damage to soybean is most severe. Upon soybean maturity, sexual forms are formed and migrate back to buckthorn.

2.2. Stink bugs

In Hemiptera, stink bugs are in the family Pentatomidae. The name stink bug is attributed to the malodorous substance they emit for defense [16]. In the past, three species [southern green

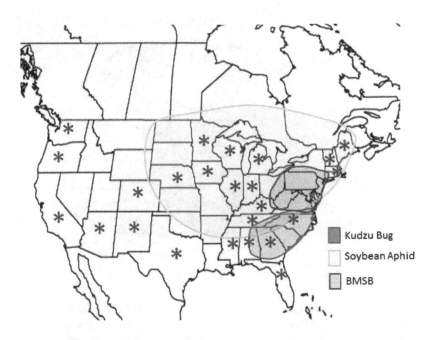

Figure 1. Approximate distribution of the main, invasive Hemipteran pests of soybean. For BMSB, shaded range is
where damage is heaviest, * represent states where detection has been observed.

stink bug, *Nezara viridula (L.)*; green stink bug, *Acrosternum hilare* (Say); brown stink bug,
Euschistus servus (Say)] constituted the stink bug complex that attacked soybean crop in
southern US [6,17]. Amongst these, *N. viridula* is the most abundant [18] which has caused the
most severe damage to the field crop [17]. In more northern latitudes, the relative abundance
of *A. hilare* is higher compared to that of *N. viridula* [6]. During the last decade, redbanded stink
bug, *Piezodorus guildinii* (Westwood), and brown marmorated stink bug (BMSB), *Halyomorpha
halys* (Stål) have established themselves as important members of stink bug complex that attack
soybean in the US [7, 19,20]. From 1960 onwards, the redbanded stink bug has been detected
in US soybean but without causing any economic damage. However during last decade,
redbanded stink bug infestations have reached above the threshold levels. By 2009, it was the
most serious stink bug species attacking soybean in southern US [21, 22]. The BMSB is native
to North and South Korea, Japan, China [23] and is invasive in North America and Europe. In
US, BMSB has been confirmed as pest of soybean crop with a high damaging potential [24,
25]. Although not in the Pentatomidae, the kudzu bug (Plataspididae: *Megacopta cribraria*) is
another recent invasive pest, first detected in Georgia in 2009 [26] and has now spread to 8
southeastern states of US [8]. It is known to feed on both kudzu and soybean, and damage on
soybean can be quite severe [27].

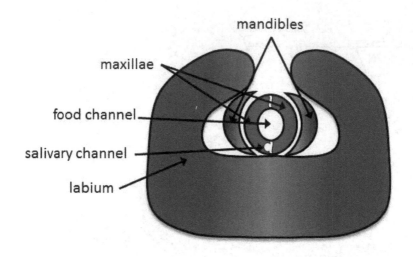

Figure 2. Generalized mouthpart structure of Hemiptera. Redrawn from [29.]

The biology of various stink bug species has been described in detail [10]. In general, stink bugs pass through five nymphal instars and an adult stage during their life cycle [10]. These insects overwinter as adults beneath the leaf litter of various host plants including grasses, shrubs and trees. Several species can also overwinter in homes and these infestations can be severe as seen with the BMSB [7]. In early summer (April-June), overwintered and first generation adults feed on crops like tomato, okra, crucifers, and legumes, but for BMSB, may also feed on woody trees like *Paulownia* or *Ailanthus altissima* (tree of heaven) early in the spring before moving into crops [28]. However, they will feed on soybean if early planted crop is available. In most cases, stink bugs will move into soybean to feed on the developing pods and seeds. In North America, the number of generations per year is largely dependent upon environmental factors but usually varies between 1 in the north to 5 in the south.

2.3. Hemiptera feeding and damage to soybean

Hemipteran pests inflict the damage on soybean by feeding on plant juices. These insects possess piercing and sucking mouthparts, the most characteristic feature of Hemiptera which are highly adapted for extracting the liquid contents of plants. The mouthparts' structures are held in a grove present on the anterior side of the insect's lower tip i.e. labium (also called as rostrum) (Figure 2, [29]). On the either side of maxillae, two mandibles are present. The mandibles, which are often barbed at the tip, form the main piercing structure called the stylet. Two opposing maxillae which are held together by a system of tongues and groves, form two canals: a food canal and a salivary canal. The food canal is used for uptake of plant liquids whereas salivary canal is for egestion of saliva into the plant structures. Because of the segmented structure, labium can fold itself when stylets penetrate into the plant surface.

Both the soybean aphid and stink bugs can feed on various above-ground plant parts like leaves, stems, flowers, and pods. Soybean aphid prefers to feed on the undersides of leaves [15] whereas stink bugs prefer to feed on pod and seeds [6]. During the early seedling stages of soybean, soybean aphids are mostly found on freshly growing trifoliate leaves or the stems [30]. Later in the season, soybean aphids are more likely to be found lower in the canopy, on leaves that are attached to nodes further away from the terminal bud. During feeding, soybean aphid withdraws sap from soybean leaves which results in loss of photosynthates. Heavy infestations by soybean aphid can result in yellow and wrinkled leaves, reduction in plant height, reduced pod set and lesser number of seeds within pods [15,31]. Infested leaves may turn black due to sooty mold growing on the sugary excretions or "honeydew" produced by soybean aphid [5,15]. The severity of plant losses caused by soybean aphid is largely dependent on the physiological status of the soybean plant. Soybean aphid populations that reach their peak density during the early-vegetative or mid-reproductive stages (R3-R5) are more likely to cause serious damage compared to populations that peak during late reproductive stages (R6-R7) [32].

Except the first instar which is a non-feeding stage, all other developmental stages of stink bugs feed upon plants. In soybean, the most damage is caused by adults and/or fifth instars stages [10]. Due to their preference for pods and seeds, reproductive stages of soybean are the most susceptible to damage by stink bugs [17]. Further, stink bugs prefer to feed upon pods present on upper half of the plant. However in case of severe infestation, these insects may also feed on lower pods. Stink bugs cause injury to soybean seeds as they insert the stylets through the pod wall into the seed for feeding on plant juices. In immature seeds, discolored necrotic areas may occur around these punctures [10]. Mature seeds show puncturing marks, discoloration, and internal irregular white spots which may have a chalky appearance [33]. Heavy feeding on mature seeds may result in smaller size, irregular shape, including wrinkled areas around punctures (Figure 3) [33, 34]. Stink bug damage in soybean results in decreased pod number, fewer seed per pod, lower seed weight, changes in fatty acids composition, and lower seed quality [10]. The germination of soybean seeds may be prevented due to single puncture in radicle-hypocotyl axis of seeds [35]. On the other hand, several punctures in cotyledons may not prevent germination but affect the vigor. Heavy infestation of stink bugs can result in foliar retention, delayed maturation and abnormal growth of the soybean crop [10].

Hemipteran pests also cause indirect damage to crops by vectoring the transmission of microbial pathogens. The virus transmission by soybean aphid to various crops has been reviewed [4]. In soybean, soybean aphid has the potential to transmit *Soybean mosaic virus* [4] but so far, there is no report of significant damage. Stink bugs generally transmit fungal pathogens [36] but recently, they were also found to transmit bacterial pathogens [37]. Besides vectoring of fungal pathogens, feeding by stink bugs provides the entrance points for microbial pathogens [10].

3. Host Plant Resistance (HPR)

In agriculture, HPR represents the ability of a certain plant variety to produce a larger yield of better quality compared to other varieties of the same crop at the same level of insect

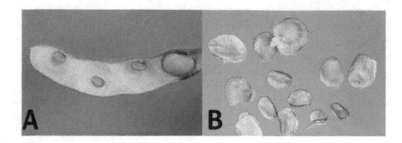

Figure 3. Picture of BMSB damage to soybean in Columbus, Ohio. A) Damaged seeds found in the pods compared to undamaged seed (far right); B) Sample of shriveled seeds collected from BMSB infested soybean. (Picture courtesy of R. B. Hammond)

infestation [38]. R. H. Painter, a pioneer researcher on HPR described it as the relative amount of heritable qualities possessed by the plant that influence the ultimate degree of damage done by the insect [39]. Painter's definition was extended to emphasize the relative nature of HPR [40]. HPR refers to the "...sum of the constitutive, genetically inherited qualities that result in a plant of one cultivar or species being less damaged than a susceptible plant lacking these qualities." [40]. Thus, plant resistance to insects should be measured on a relative scale, by comparing the damage to susceptible control plants.

Plant resistance to Hemipteran insects have been found and utilized in many crop plants [40-41]. From historical perspective, HPR to manage Hemipteran insect-pests of soybean has been highly successful. The resistance to potato leaf hopper (*Empoasca fabae*), a serious pest of soybean in the past, occurs due to the presence of pubescence on plant leaves [42-43]. The subsequent incorporation of pubescence trait into commercial varieties of soybean consigned the potato leaf hopper to a non-pest status [44]. Specifically for aphids, resistant genes have been identified in several crops, including cereals, vegetables, fruits and forages [40, 41, 44]. Identification and deployment of aphid resistance genes against Russian wheat aphid (*Diuraphis noxia*) in wheat [46, 47] and in barley [48] are very good examples of host plant resistance to Hemipertan pests. The Russian wheat aphid resistant wheat cultivars are commercially grown throughout USA and South Africa.

Plant resistance to important pests and pathogens is an integral component of soybean IPM and development of soybean cultivars with resistance to insects has been a long time priority for soybean breeders. However, one major drawback of employing HPR as a control measure against insect pests is the potential lack of its durability. Principally, the HPR based on major genes has not proved to be a long term solution for pest management because insect populations emerge which can overcome plant defenses resulting from those genes [40, 45], often referred to as insect biotypes. Smith [45] defines biotypes as "...populations within an arthropod species that differ in their ability to utilize a particular trait in a particular plant genotype". The success of HPR against Hemipteran pests is particularly compromised by the presence and emergence of insect biotypes. Variable number of biotypes have been reported in many Hemipteran insects like brown plant hopper (*Nilaparvata lugens*), White fly (*Bemisia tabaci*), and in a number of aphid species. A total of 14 aphid species with known biotypes are listed in [45], and this does not include the soybean aphid. Managing the emergence and increase of insect biotypes will be critically important to extend the durability of HPR. The topic of biotypes will be discussed in detail in the subsequent sections of this chapter.

3.1. Soybean HPR to hemipteran pests

3.1.1. Soybean aphid

A limited number of studies were conducted in China on soybean host plant resistance to the soybean aphid before its invasion into North America and few aphid resistant soybean lines were identified [49-51]. None of the aphid resistant genes in these lines were reported to be genetically characterized. Several soybean lines with resistance to the soybean aphid have been recently identified by researchers in US [52-56]. Most of these studies used locally collected soybean aphid biotypes for greenhouse and field screenings of soybeans in early growth stages and estimating the number of soybean aphids on each seedling after 2 to 4 weeks of infestation with a known number of aphids. However, there are currently 3 soybean aphid biotypes recognized, biotype 1, 2, and 3 [57]. Biotype 1 is unable to survive (*i.e.* avirulent) on any known HPR soybean, whereas biotype 2 and biotype 3 can survive (*i.e. virulent*) on *Rag1* and *Rag2*, respectively. More recent soybean aphid HPR characterizations have included some or all biotypes.

Three lines with resistance to three soybean lines – Dowling, Jackson and a plant introduction (PI) 71506 were reported in [52]. These lines have resistance to biotype 1 but not to biotype 2 of the soybean aphid. The resistance in Dowling, Jackson, and PI71506 exhibited both antixenosis [58,59] and antibiosis [52]. Three PIs – PI243540, PI567301B and PI567324 – were identified to have resistance to soybean aphid biotypes 1 and 2 [56]. PI243540 showed antibiosis whereas PI567301B and PI567324 showed antixenosis type of resistance in no-choice tests. A total of 2147 soybean germplasm accessions were evaluated in choice tests and four PIs - PI 567598B, PI 567541B, PI567543C, and PI 567597C- with resistance to soybean aphid were identified [55]. The soybean aphids used in this study were collected from Michigan fields that comprised of unknown biotypes. In a no-choice test, PI 567598B and PI567541B were found to possess antibiosis resistance and PI567543C, and PI 567597C showed antixenosis resistance. Diaz- Montano [58] reported 11 soybean genotypes with resistance to soybean aphid of unknown biotype. Among the 11 genotypes, nine showed moderate antibiotic effects and the other two showed not only a strong antibiotic effect but also exhibited antixenosis as a category of resistance to the aphid. Pierson et al. [60] documented moderate resistance to the soybean aphid in soybean lines KS4202, K1639 and K1621 during the reproductive stages of development.

3.1.2. Stink bugs

Based on field and laboratory trials, Jones and Sullivan [61] found that 3 Mexican bean beetle-resistant PIs i.e. PI 171451, PI 227687, and PI 229358 [62] were also effective against southern green stink bug. Subsequently these PIs were found to possess resistance to multiple pests including Lepidoptera, and thus were used as donor parents for major breeding programs across US [44]. These PIs exhibited antixenosis and antibiosis type of resistance against different pests. Amongst these, PI229358 appeared to be most resistant to stink bugs. Gilman et al. [63] evaluated 894 PIs and 26 cultivars (maturity groups V-VIII) for their resistance to the southern green stink bug. The resistance to stink bugs was, in general, associated with plant maturity as early maturing genotypes showed lesser damage compared to late maturing ones. They identified PI 171444 (MG VI) to be the highly resistant against southern green stink bug. The stink bug resistance in PI 171444 is exhibited as antixenosis, antibiosis and temporal separation [64].

Most research on soybean HPR against stink bugs has been conducted in Brazil, where the damage is often the heaviest [65]. The cultivar IAC-100 having PI 229358 and PI 274454 in its genealogy was officially released in Brazil, and it carries resistance to stink bug complex [66, 67]. In order to develop soybean lines adaptable to southeastern US, McPherson et al. [68] developed 65 breeding lines carrying IAC-100 in their genealogy. During 5 years of field testing, all these lines showed variable amount of resistance against stink bugs primarily the southern green stink bug. Among these, four breeding lines with either Hutcheson x IAC-100 or IAC-100 x V71-370 in their genealogy were identified as promising material for future development of stink bug resistant soybean. Recently, Campos et al [69] screened 16 genotypes for resistance against southern green stink bug by caging adult bugs on pods. Based on number of feeding punctures, three genotypes V00-0742, V00-0842, and V99-1685 were identified as resistant. Based on reduced seed weight loss, two genotypes (PI 558040 and V00-0870) were further identified as resistant.

3.2. Genetics of soybean resistance against Hemipterans

3.2.1. Soybean aphid

Inheritance of several major soybean genes (R-genes or single gene) for resistance to soybean aphids has been reported [53, 54, 70]. The aphid resistance in each of the two soybean cultivars Dowling and Jackson is controlled by a single dominant gene. The gene in Dowling was designated as *Rag1* and the *Rag* gene in Jackson remained unnamed. The aphid resistance in PI 243540 is controlled by a single dominant gene [70]. In contrast, the aphid resistance in PI 567541B is controlled by quantitative trait loci (QTL) and resistance in PI 567598B is controlled by two recessive genes [71, 72] A total of six major genes for resistance to the soybean aphids have been reported by 2012 (Table 1).

Li et al. [78] have mapped *Rag1* and the unnamed *Rag* gene from Jackson to the same genomic region on soybean chromosome 7 (LG M), indicating that these two resistance genes may be allelic. PI 243540 was resistant to both biotypes 1 and 2 and the resistance in this soybean accession from Japan was controlled by a single dominant gene [70]. Mian et al. [75] mapped the *Rag2* gene on soybean chromosome 13 (LG F) between SSR markers Satt334 and Sct_033. Zhang et al. [79] mapped two recessive loci in PI 567541B controlling soybean aphid resistance.

Gene name	Source PI (s)	Resistance category	Biotype response*		
			B1	B2	B3
Rag1	PI 548663 [53]; PI 548657[54]	Antibiosis [52,73] and antixenosis [58,59]	A	V	A/V†
Rag2	PI 200538 [74], PI 543540 [70-75]	Antibiosis [56,74]	A	A	V
Rag3	PI 567543C [72]	Antixenosis [72]	A	A	V
rag4	PI 567541B [72]	Antibiosis [72]	A	A	V
Rag5	PI 567301B [76]	Antixenosis [56]	A	A	V

*A-Avirulent; V-Virulent

†Virulent in choice tests, Avirulent in no-choice tests [77]

Table 1. Major HPR genes identified in soybean response to soybean aphid biotypes

One locus was mapped on chromosome 13 (LG F) and the other loci on chromosome 7 (LG M). The major locus on chromosome 7 in this study was only about 3 cM from Satt435, which was the closely linked marker to the Rag1 locus and designated as rag1_provisional [80]. Thus, the major locus identified in this study is located in the same genomic region as Rag1, which indicates that they are either allelic at the same locus or two different loci tightly linked to each other. The second aphid resistance locus in this PI was mapped on chromosome 13 nearly 50 cM away from the Rag2 locus and this recessive gene was designated as rag4. Zhang et al. [80] mapped single dominant locus in PI567543C for soybean aphid resistance on chromosome 16 (LG J). This locus provided a near complete resistance to the soybean aphid indicating a single gene resistance and the gene has been named as Rag3. Jun et al. [76] mapped a major soybean aphid resistance gene on chromosome 13 near the Rag2 locus in PI567301B. However, the resistance in PI 567301B is antixenosis type while Rag2 is a locus for antibiosis resistance and thus the locus in PI567301B has been tentatively named as Rag5, pending approval from the Soybean Genetics Committee. Jun et al. [82] recently mapped three QTL for oligogenic resistance to the soybean aphids in PI567324 (in review). The inheritance and genetic mapping studies on tolerant lines have not been conducted yet.

3.2.2. Stink bug

Research on genetics of soybean resistance to stink bugs has been limited and confined to only a few PIs. Multiple studies have confirmed that the pest resistance (including against stink bugs) in 4 PIs i.e. PI 171451, PI 227687, PI 229358, and PI 229321 is a quantitative trait and is controlled by two or three major genes (reviewed in [44]). Because of quantitative nature of both resistance and yield in these PIs, conventional breeding strategies were not successful in introgressing these traits into the locally adapted varieties. The stink bug resistance in tolerant cultivars like IAC-100, IAC-74 2832, IAC-78 2318 is a complex polygenic trait that is exhibited as additive, dominant and epistatic effects of multiple genes [82, 83].

3.3. Mechanisms of soybean resistance against hemipterans

3.3.1. Soybean aphid

Until now, a limited amount of information is available on the mechanism of resistance provided by *Rag* genes to soybean aphid. Using cDNA microarrays, the transcript profiles of cultivars Dowling (*Rag1*, soybean aphid resistant) and Williams 82 (soybean aphid susceptible) were compared after aphid infestation [84]. Out of ~18,000 soybean genes tested, only 140 showed differential expression between resistant and susceptible cultivars after 6 and 12hrs of aphid feeding. In the resistant cultivar, genes involved in the salicylic acid (SA) and jasmonic acid (JA) pathways were upregulated compared to their expression in susceptible cultivar. Both SA and JA are signaling molecules that mediate the stress response in a resistant plant upon being attacked by an insect. In the downstream of defense signaling pathway, both SA and JA lead to the production of defensive allelochemicals that are deterrent or lethal to the insect. In addition, SA signaling may result in the production of reactive oxygen species that kill the insect due to oxidative injury [41].

3.3.2. Stink bugs

Most of the research into the mechanism of soybean HPR to stink bugs has been in the distant past. As mentioned earlier, 3 PIs PI 171451, PI 227687, PI 229358 show resistance to multiple pests (including stink bugs) through antixenosis and antibiosis. The mechanism of resistance in these PIs has been elucidated in various studies that involved various lepidopteran and coleopteran pests of soybean but not stink bugs (reviewed in [44]). In general, antixenosis and antibiosis resistance in these PIs is manifested through plant allelochemicals, which are isoflavones (plaseol, aformosin), phenolic acids, phytoalexins. However, the mechanism of resistance in stink bug-tolerant cultivars (IAC-100) is better understood. It includes pest evasion by shorter pod filling period, rejection of young damaged pods and replacement with new pods, normal leaf senescence under stress, and higher number of seeds in pods [82].

3.4. Soybean traits as selection criteria for breeding

In breeding for insect resistance, selection is the key step. Since soybean aphids build up in huge numbers due to asexual reproduction, susceptible germplasm of soybean is not able to withstand early stage infestation. General vigor, chlorosis, curling, infestation levels (*e.g.* insect counts) will allow for selection of cultivars showing antixenosis and antibiosis. Since tolerant cultivars continue to withstand high infestation levels of soybean aphid, yield and seed quality are the most important selection criteria for evaluation of this trait. During vegetative stages, plant health and rate of growth of infested plants compared to the uninfested plants can be used as criteria of tolerance. While on mature plants, measurement of agronomic traits, including plant height, maturity, lodging, seed yield and quality (discoloration and wrinkling of seed coats, shriveled seeds) are some of the traits that can be measured for determination of soybean tolerance to the aphid. Evaluations at both vegetative and mature plant stages will be more desirable. For selection of stink bug resistance germplasm, insect counts may not be the best indicators because insects may evade monitoring. Again the best indicators of tolerance are the seed yield and seed quality of infested plants in reference to the uninfested plants. As stink bugs mostly feed on developing and developed pods and seeds of soybean, there is no reliable way of evaluation of soybean tolerance to these insects during the vegetative stages.

3.5. Sustainable HPR against soybean aphid biotypes

Smith [40] has reviewed the occurrence of biotypes in various insect species. Among insects, 'biotypes' is the most abundant phenomenon in aphids. Smith [40] argued that aphids will continue to produce biotypes because of their parthenogenic reproduction, high reproductive potential and clonal diversity. There is no clear evidence to suggest the cause behind occurrence of biotypes in insects. In some insects like Hessian fly and Russian wheat aphid, biotypes emerged as these were exposed to resistant cultivars. These biotypes developed probably due to selection pressure placed by agronomic production, recombination or mutation to overcome the defense due to resistance genes. However in insects like soybean aphid and green bug, virulent biotypes have been discovered in field populations *before* the deployment of resistant cultivars. Three biotypes of soybean aphid are known so far that can defeat resistance genes identified in several PI's (Table 1). Thus, the success of resistance genes has been greatly hindered by the occurrence of virulent biotypes of soybean aphid. Based on successful examples in other plant-insect systems and resistance-management approaches, here we discuss strategies and questions that need to be answered to sustain the success of HPR in soybean against soybean aphid.

1.   Gene deployment based on biotype distribution:

The knowledge on biotype distribution is extremely important for the success of resistance gene deployment. To characterize soybean aphid populations from various geographical locations, there is a need to perform regular and systematic sampling to monitor the soybean aphid populations from various geographical locations. Aphids that are collected from the field can be tested and characterized under laboratory conditions to investigate levels of virulence. There are two possible ways to detect for the presence of novel biotypes. First, collected populations can be exposed to a set of plant differentials containing known major genes for resistance, and their response can be compared with the known biotypes. Second, PCR based strategy using biotype-specific markers can be helpful (see [85] for detection of *Orseolia oryzae* biotypes). Recent work has focused on expanding the molecular resources for the soybean aphid [86-88] but to date, no biotype diagnostic marker for field populations exist. Both of these methods will provide data to develop biotype distribution maps. This could be extremely important for resistant gene deployment in that growers may avoid planting a resistant variety if it is unlikely to control soybean aphid populations. A geographic based approach will also help to avoid growing of a particular resistance cultivar over a wide area, which has hastened the development of virulent biotypes in insects such as in case of early wheat cultivars and Hessian fly biotype adaptation [40]. However, more research is needed in understanding soybean aphid migration and how virulent biotypes may spread. While the overwintering host of buckthorn is restricted to more northern latitudes (above 40°N, [15]), dispersal of winged aphids across much of the US soybean growing region occurs late in the growing season [89]. Not only does this spread clonal and genetic diversity immediately prior to sexual reproduction, it may also allow virulent biotypes to rapidly move across the environment.

2.   Gene pyramiding:

This involves the release of a variety containing more than one major resistance gene. Pyramided varieties are likely to have extended durability as survival and subsequent multiplication of virulent individuals appearing in a susceptible pest population are highly reduced because of multiple resistance genes. Another advantage is that the pyramided varieties may yield more compared to single gene varieties due to higher reduction in pest population. Cultivars with both *Rag1* and *Rag2* had less aphid numbers and less yield reduction than soybeans with only one resistant gene [90].

3.   Variety mixture:

Using seed mixtures of resistant and susceptible plants may extend durability of soybean aphid-HPR; this system would be analogous to the refuge requirements for transgenic insect resistance [91-93]. In most cases, 80-95% of a field would be resistant, with 5-20% of the plants being susceptible to provide a population of insects unexposed to the resistance genes. For maize, modeling predicts that using refuge plants extends durability, particularly if resistant plants have multiple genes (*i.e.* pyramids) [94]. However, for any HPR strategy that involves insect resistance management (IRM), several questions remain regarding soybean aphid biology that differs for models developed for corn pests. For example, mating, and therefore the transfer of genetic variation for virulence, is dependent on the overwintering host buckthorn, which, as stated previously, is not randomly distributed across the environment. In most corn pests, mating occurs in the field and therefore may allow a more random assortment of virulent (*i.e.* resistant) and avirulent individuals. Additionally, the inheritance of resistance is still unknown. Most importantly, soybean aphids asexually reproduce in the presence of the HPR selection pressure. In these cases of asexual reproduction, resistance can be delayed when 1) refuges are large, 2) resistance genotypes are low in frequency, 3) resistant individuals are less fit than susceptible individuals on refuge plants (i.e. fitness among biotypes, fitness costs), and 4) resistant individuals are less fit on resistant plants than susceptible plants (fitness of virulent biotypes on different plants, i.e. incomplete resistance) [95]. Fitness costs can be due to both physiological mechanisms and direct competition among susceptible and resistant individuals on the same plant. Few studies have investigated the impact of differing reproductive strategies such as parthenogenesis, but in the most basic sense, the genotype (virulent or avirulent) with the highest fitness has the highest reproductive output and becomes the most common [95]. Through simulation modeling, Crowder and Carrière [95] determined that the key parameters for delaying the evolution of resistance in parthenogenetic organisms were the presence of fitness costs and incomplete resistance. Fitness costs had the least effect, but "...incomplete resistance delayed resistance evolution more than fitness costs, and in some cases, resistance was prevented with incomplete resistance and fitness costs." [95]. While recessive resistance can delay resistance, "...such delays [in resistance] are not possible in haplodiploid or parthenogenetic pests without additional factors such as fitness costs and incomplete resistance." [95]. These parameters have not been experimentally estimated for soybean aphid in an IRM framework, but must be understood to develop and evaluate appropriate IRM strategies for soybean aphids

3.6. Integration of soybean HPR with other IPM tactics

IPM is loosely defined as the integration of multiple tactics to control insect pests [96]. These tactics include chemical, biological and cultural (agronomic), and are decided based on economic, environmental and societal impacts [96]. Much has changed in the past 15 years regarding IPM in soybean [97]. In the past, less than <1% of soybean acreage in the Midwestern USA was treated with insecticides [5], and insect pests of interest centered on defoliators such as bean leaf beetle (*Cerotoma trifurcata* (Forster)), and Mexican bean beetle (*Epilachna varivestis* Mulsant), as well as various Lepidopteran and Gastropodan (*e.g.* slugs) pests [6, 98]. Apart from various native stink bugs such as the brown and green stink bugs, very little attention was given to Hemipteran pests of soybean. Within the past 10 years, focus of insect pests of soybean has shifted toward the invasive Hemipterans such as the soybean aphid and BMSB. In fact, Ragsdale et al. [4] report that in less than ten years, insecticide use in soybean has increased 130-fold, in large part due to soybean aphid infestations. As these invasive Hemipterans expand their distribution, soybean researchers and producers will look to implement various methods of control.

3.6.1. Chemical control

The use of chemical insecticides remains the most widely used option for control of Hemipteran soybean pests because of mainly 2 reasons. First, most exotic pests invade new environments lacking natural enemies, resistant plants, or even basic biological information; often research is published in different languages and can be difficult or time-consuming to translate. Additionally, as invasive pests adapt to their new environment, previous data from native environments may lose relevancy. Secondly, most commercial insecticides are widely available and broad-spectrum which can act quickly and effectively to control economic populations. Therefore, in most cases, insecticides are the only short-term options.

There are several insecticide classes that have proven effective against Hemipteran soybean pests [5, 99-101]. While the wide availability of effective insecticides provides soybean producers with choices, responsible use of these chemicals requires timely applications based on field scouting and economic analysis. Economic injury levels (EIL, the pest population at which plant injury occurs) and economic or action thresholds (ET, the pest population at which treatment is recommended to prevent plant injury) have been estimated for a few Hemipteran soybean pests such as the soybean aphid, brown and green stink bug [5,32, 102]. Although general recommendations based on field observations are known for the BMSB and kudzu bug, more research is necessary to determine the EIL and the ET. Finally, the use of neonicinotoid seed treatments appears to be increasing, however these chemicals are only active for 30-40 days. These seed treatments will control early season soybean aphids, they do not prevent late season aphid infestations [5]. Likewise, most stink bug feeding occurs in the reproductive stages when the activity of the seed treatment has decreased dramatically.

3.6.2. Cultural control

There are very few options for adapting crop production methods for controlling Hemipteran soybean pests. One of the most common practices is to alter planting dates. The heaviest damage from Hemipteran feeding occurs late in the season—by planting early, most of the yield potential has been already made. In practice, planting dates are often at the mercy of ideal weather conditions rather than based on managing insect pressure. Additionally, early emerged fields may act as trap crops for other soybean insect pests such as bean leaf beetle. Studies on variable planting dates with the soybean aphid have been unclear [5] but likely vary across geography based on the reliance of its overwintering host buckthorn (Rhamnus spp.). The presence of buckthorn has been shown to be the key predictor of aphid infestation through ecological modeling and was supported by population genetic evidence [89,103]. Virtually no research has been performed with the BMSB and kudzu bug on soybean planting date, but [104] reported that planting date impacted the presence of native stink bugs.

Cultural control also involves manipulation of the environment. As most Hemipteran pests migrate into soybean fields during the season, controlling these source populations could limit pest damage. For example, research has been directed on the impact of buckthorn on aphid movement and dispersal [103], including a citizen science project to map the distribution of buckthorn and detect the presence of aphids [105[. Similarly, the kudzu bug is also known to feed on kudzu before moving to soybean [27]. The impact of removing these host plants in preventing pest outbreaks is unknown; however, removing buckthorn and kudzu may be beneficial regardless due to their devastating impacts on ecosystems [106]. For BMSB, early observations from soybean suggest most damage is restricted to field edges, particularly along edges close to forest patches. Control may be achieved by either restricting spraying to these edges, or keeping the most susceptible soybean varieties away from edges.

3.6.3. Biological control

The importance of natural enemies for Hemipteran soybean control has been comprehensively documented in several studies, although to date, most research has focused on the soybean aphid (see [4] for a review of biological control for soybean aphid). Natural enemies of Hemipteran pests include parasitoids, predators and diseases [4, 15, 107-109]. General predators such as lady beetles, insidious flower bug and ground beetles are probably the most important natural enemies [4, 108] as most of invasive Hemipteran pests lack specific natural enemies that provide control in native regions [4, 110]. The indigenous parasitoids for most of the invasive Hemipteran soybean pests are either poorly adapted or are just beginning to attack these new hosts [4, 27]. Foreign exploration for natural enemies has resulted in candidates for classical biological control, with at least one species, Binodoxys communis, undergoing field evaluations for persistent control of soybean aphid. Preliminary exploration for parasitoids of BMSB has revealed several candidates of Scelionid wasps (Trissolcus and Telenomus) [111], and at least 1 egg parasitoid for kudzu bug [112]. While the role of natural enemies has extensively been researched for soybean aphid, more research is needed for their role in controlling other invasive Hemipterans such as BMSB and kudzu bug.

3.6.4. Integrating HPR for hemipteran pest control

Host-plant resistance offers many benefits to soybean producers in controlling insect pests, but also must fit within production practices. HPR varieties need to be in the proper maturity group, high-yielding, and without any increased susceptibility to other pests and diseases. For example, an interaction among soybean aphids, soybean cyst nematode, and brown stem rot was noticed [113]. Indeed, as other invasive Hemipterans spread into new areas, HPR with resistance to multiple pests would be desirable.

Similar to the rise of insecticide resistance, insects have also shown adaptation potential to overcome HPR. If HPR is to be a successful component in IPM of Hemipteran pests, more research is needed to develop strategies that preserve the utility of these traits. For example, the durability of HPR could be predicted through ecological modeling, similar to research for transgenic maize [91-94]. In addition to modeling, more basic biological research is needed for all Hemipteran pests including migration and gene flow estimates, virulent biotype frequencies, and competitive interactions between biotypes. Results from this research would also help to estimate the accuracy of modeling and improve any strategy for managing insect virulence and preserving HPR traits. Early research [90] with soybean aphid suggests that stacking resistance genes (e.g. *Rag1/Rag2* stack) offers better and more sustainable protection from soybean aphid than single gene resistance. In addition, more research needs to be studied in terms of how using HPR may alter efficiency of natural enemies [114-117]). The cues to which aphid or stink bug parasitoids use to locate prey are unknown, and might be either from the prey or plant host. Any breeding for host-plant resistance should also be careful not to disrupt volatile signaling [117]. Plant nutrients and resistance influence growth rate and size of herbivores which, in turn, could influence natural enemy biology [45]. Indeed, the parasitoid *Binodoxys communis* had lower fitness when attacking aphids on *Rag1* plants compared to aphids on susceptible plants [117].

In the US and Canada, soybean HPR for soybean aphid have been available since 2010, and are often combined with insecticidal seed treatments. Combining seed treatments with HPR may allow a greater opportunity for natural enemies to maintain populations below the economic threshold and therefore prevent a chemical application. However, the full benefit of seed treatments may not be fulfilled where soybean aphid infestations mainly occur late in the season, and may unnecessarily increase selection pressure for resistance.

4. Future directions in breeding for Hemipteran resistance in soybean

While much of current research has focused on traditional and classical breeding and screening methods for host-plant resistance, research is emerging which incorporates newer genomic and molecular technology. Likely, a combined approach will be necessary to improve the durability and decrease development time of Hemiptera resistant soybean. Here we list several considerations for the future of HPR in soybean:

1. Tolerant cultivars must be emphasized in breeding programs. As mentioned earlier, tolerant cultivars have the ability to withstand or recover from damage caused by insect

populations equal to those on susceptible cultivars. Thus, unlike antixenosis and antibiosis, tolerance comprises of plant features which are not involved in plant-insect interaction. Though breeding for tolerant cultivars is more difficult and time consuming (as the crop has to be grown till maturity in the infested condition) tolerance based resistant cultivars nonetheless offer two major advantages. 1) New virulent biotypes are less likely to emerge in a cropping system based on tolerant cultivars. While feeding upon tolerant plants, infesting insect populations are not reduced as they are on plants exhibiting antibiosis and antixenosis. As a result, there is no selection pressure and frequency of novel virulence trait remains lower. This is directly in contrast to what is observed in insects feeding on plants showing antixenosis and antibiosis, as various physical and chemical factors in these plants allows for selection of virulent individuals. Thus, the chances for development of biotypes that can overcome resistance genes are significantly reduced through the use of tolerant cultivars. Russian wheat aphid populations were not able to overcome tolerant plants but can break antibiosis based resistance in wheat [46]. 2) Tolerant cultivars are highly compatible with biological control measures, thus can be combined in IPM program. The allelochemical based toxins in plants exhibiting antibiosis and antixenosis could be detrimental for natural enemies of insect pests [45]. On the other hand, tolerance based cultivars do not have any known adverse impact on beneficial insects. In North-America, biological control employing natural enemies makes up an important component for IPM of soybean aphid. Thus, tolerant cultivars provide an excellent opportunity for integrating HPR and biological control in IPM.

2. Marker assisted breeding will facilitate faster and more efficient development of resistant cultivars. Markers will also be useful for pyramiding major genes as well as quantitative trait loci (QTL) for multigeneic defense against the insect as found in tolerant soybeans. Development of closely linked markers for known resistant genes in soybean will enhance the selection efficiency.

3. Exploration for new sources of HPR: In South America especially Brazil, extensive research on HPR against stink bugs has been conducted. But in US especially, the southern states, lepidopteran foliage feeders have been the focus of HPR research [68]. Further, due to the preference for insecticide based control, soybean germplasm has not been extensively explored for resistance against stink bugs. No cultivar of soybean showing resistance against stink bugs has been released so far in US. There is a need to identify native sources of resistance against stink bugs. Though the selection for resistance against stink bugs is relatively time consuming and labor intensive, novel sources that offer wider pool of traits such as pest resistance, yield, etc. should be incorporated into breeding programs.

Traditionally, lepidopteran foliage feeders and 3 species of stink bugs (*N. viridula, A. hilare, E servus*) were major insect pests of soybean in southern us. However, as mentioned earlier, recent expansions of red-banded stink bug in southern US, brown-marmorated stink bugs in eastern, central and southern US, and soybean aphid in north central us have significantly changed the pest scenario in these regions. Thus, for effective HPR based IPM program, there is a need to identify novel germplasm sources that are resistant against more than one insect. Further, the known resistant sources against a single pest can be explored for their response to other insect pests e.g. in breeding programs for pest management in north central US, soybean aphid resistant (*Rag* containing soybean) PIs can be screened for response against brown marmorated stink bug. Previously, several soybean PIs have shown resistance against multiple insect pests e.g. PI171444 which originally identified for resistance against stink bug complex also showed resistance against bean leaf beetle and banded cucumber beetle [44,118]. Thus, there is strong potential for soybean PIs having multiple pest resistance and to be incorporated into the breeding programs.

4.   Bt soybean potential against Hemipterans: The development of transgenic (e.g. Bt) resistance against the Hemipteran insects has not succeeded so far. Pyramiding of Bt genes with HPR genes may be an useful strategy. In Lepidoteran insects, the highest level of resistance to Lepidopteran insects obtained through MAS using the native soybean genes was 70% reduction in feeding [119]. However, when the soybean insect resistance loci were pyramided with a *cry1Ac* transgene from *Bacillus thuringiensis* (Bt) the level of feeding damage was reduced to 90% compared to susceptible checks [119]. Such native crop gene and transgene pyramids may be useful in several aspects of insect resistance. First, the Cry protein from a single Bt transgene may only protect the host plant from one or at best two classes of insects. For instance, the Cry1Ac toxin provides resistance against many Lepidopteran pests, but not to Coleopteran pests. A combination of native insect resistance gene with resistance to beetle (e.g, insect resistance loci on chromosome 7) with the Bt transgene could broaden resistance of plants to include Coleopteran pests that are insensitive to Cry1Ac toxins. Second, several insect pests have demonstrated the ability to develop resistance to Cry toxins, so effective strategies are needed to manage resistance to Bt [120]. Some populations of the diamondback moth [*Plutella xylostella* (L.)] have developed resistance to Bt toxins in different parts of the world where Bt are routinely used on cruciferous crops [121]. Soybean lines carrying the PI 229358 allele at the insect resistance locus on chromosome 7 in addition to a Cry1Ac transgene were more protected against defoliation by corn earworm and soybean looper than related transgenic lines lacking the PI 229358 allele [119]. Studies to investigate weight gain of tobacco budworm larvae from Cry1Ac-resistant and Cry1Ac-sensitive strains demonstrated that larvae fed on leaves of plants with both a Cry1Ac transgene and the native insect resistance allele on chromosome 7 gained weight more slowly than larvae fed on leaves from transgenic plants lacking the native resistance allele [119].

5.   RNAi and other genomic approaches: Given issues with ineffectiveness of Bt on Hemipterans, RNA interference (RNAi)- mediated control presents an attractive avenue for management of these pests. RNAi results in sequence specific knockdown of gene expression at the post-transcriptional level as introduced dsRNA causes the degradation of identical mRNAs [122]. Crops based on RNAi-mediated pest protection are expected

to achieve the same level of success as Bt-based transgenic crops [123]. Though there are various categories of insect genes that could be silenced through RNAi to achieve the desired results, targeting of genes encoding for effector proteins in salivary glands of Hemipteran insects has been promising. At the start of feeding, Hemipteran insects inject the saliva produced by salivary glands into plant tissues. Hemipteran saliva contains various chemical substances such as digestive enzymes that facilitate feeding. Importantly, the saliva also contains the effector proteins that are determinants of virulence for these insects. RNAi knockdown of *coo2*, which is an effector protein of pea aphid secreted into the fava beans leaves during feeding, significantly reduced the survival of this insect [124,125]. In addition to pea aphid, successful RNAi studies in Hemipterans insects like peach aphid (*Myzus persicae*), Brown plant hopper (*Nilaparvata lugens*) have also been reported [126-129]. To develop soybean employing RNAi-based management of Hemipteran pests, there is a need to generate significant amount of molecular resources for these insects. To date, a whole genome sequence is only known for 1 Hemipteran insect, the pea aphid, *Acyrthosiphon pisum* [130]. Besides RNAi, there are other novel approaches such as the transgenic plant resistance against Hemipteran pests. The management of Hemipteran pests by use of transgenic plants expressing lectins and protease inhibitors has been recently reviewed [131], thus not discussed in this chapter.

5. Conclusions

Although there has been some success with HPR for Hemipteran pests, for example the glandular hairs for potato leafhopper, there are many opportunities for expanding this important pest management tool. Research has already resulted in the commerical availability of HPR against the soybean aphid, with many more varieties to come. However, HPR research for the other major Hemipteran pests of soybean continues to lag behind. More molecular and genomic techniques increase the feasibility of finding HPR loci and improve the ability to combine both traditional HPR approaches and newer RNAi methodologies. This includes not only developing resistance to multiple insect pests, but potentially other pathogens that they may interact with to impact soybean [113]. However, these new varieties will need to be studies and balanced in terms of the other aspects of integrated pest management (i.e. chemical and biological control) to both limit non-target impacts and extend durability in the face of insect adaptation.

Acknowledgements

We would like to thank the members of the Michel and Mian Laboratories for assistance in Hemipteran soybean pest research, specifically L. Orantes, J. Wenger, C. Wallace, R. Mian, W. Zhang, J. Todd, T. Mendiola, K. Freewalt. Funding for this work was provided by the Ohio Soybean Council, the North Central Soybean Research Project, OARDC, The Ohio State University and USDA-ARS.

Author details

Raman Bansal[1], Tae-Hwan Jun[1,2], M. A. R. Mian[2,3] and Andy P. Michel[1]

1 Department of Entomology, Ohio Agricultural Research and Development Center, The
Ohio State University, USA

2 Department of Horticulture and Crop Sciences, Ohio Agricultural Research and Development Center, The Ohio State University, USA

3 USDA-ARS Corn and Soybean Research Unit, Madison Ave., Wooster, OH, USA

References

[1] Soystats: A reference guide to important soybean facts and figures. American Soybean
Association. http://www.soystats.com/ (accessed 20 June 2012).

[2] USDA-National Agriculture Statistical Service. http://www.nass.usda.gov/ (accessed
20 June 2012).

[3] Hartman GL, West ED, Herman TK. Crops that feed the World 2. Soybean—worldwide
production, use, and constraints caused by pathogens and pests. Food Sec 2011;3: 5-17.

[4] Ragsdale DW, Landis DA, Brodeur J, Heimpel GE, Desneux N. Ecology and management of the soybean aphid in North America. Annu Rev Entomol 2011;56: 375-399.

[5] Hodgson EW, McCornack BP, Tilmon K, Knodel JJ. Management Recommendations
for Soybean Aphid (Hemiptera: Aphididae) in the United States. J Integrated Pest
Management 2012;3: E1-E10.

[6] Higley LG, Boethel DJ., editors. Handbook of Soybean Insect Pests. Entomological
Society of America; 1994.

[7] Leskey TC, Hamilton GC, Nielsen AL, Polk DF, Rodriguez-Saona C, Bergh JC, Herbert
DA, Kuhar TP, Pfeiffer D, G. Dively F, et al. Pest status of the brown marmorated stink
bug, Halyomorpha halys, in the USA. Outlooks Pest Manag 2012;23(5): 218-226.

[8] The University of Georgia - Center for Invasive Species and Ecosystem Health: Kudzu
Bug. http://www.kudzubug.org/distribution_map.cfm (accessed 7 December 2012).

[9] Li M, Tian Y, Zhao Y, Bu W. Higher Level Phylogeny and the First Divergence Time
Estimation of Heteroptera (Insecta: Hemiptera) Based on Multiple Genes. PLoS ONE
2012;7(2): e32152.

[10] McPherson JE., McPherson RM. General introduction to stink bugs. In: Stink Bugs of
Economic Importance in America North of Mexico. Boca Raton: CRC Press; 2000a. p1-6.

[11] Ragsdale DW, Voegtlin DJ, O'Neil RJ. Soybean aphid biology in North America. Ann Entomol Soc Am 2004;97: 204-208.

[12] Venette RC, Ragsdale DW. Assessing the invasion by soybean aphid (Homoptera: Aphididae): where will it end? Ann Entomol Soc Am 2004;97: 219-226.

[13] Wang XB, Fang YH, Lin SZ, zhang LR, Wang HD. A study on the damage and economic threshold of the soybean aphid at the seedling stage. Plant Prot 1994;20: 12-13.

[14] Song F, Swinton SM, DiFonzo C, O'Neal M, Ragsdale DW. Profitability analysis of soybean aphid control treatments in three northcentral states. Michigan State University Department of Agricultural Economics: Staff Paper; 2006-24.

[15] Tilmon KJ, Hodgson EW, O'Neal ME, Ragsdale DW. Biology of the Soybean Aphid, Aphis glycines (Hemiptera: Aphididae) in the United States. J Integ Pest Mngmt 2011;2(2): e1-e7.

[16] Drake CJ. The southern green stink-bug in Florida. Florida State Plant Board Quarterly Bull 1920;4: 41-94.

[17] McPherson JE., McPherson RM. Nezara viridula L. In: Stink Bugs of Economic Importance in America North of Mexico. Boca Raton: CRC Press; 2000b. p71-100.

[18] Todd JW., Herzog DC. Sampling phytophagous Pentatomidae on soybean. In: Kogan M., Herzog DC. (ed.) Sampling methods in soybean entomology. New York: Springer-Verlag; 1980. p438-478.

[19] Temple JH, Leonard BR, Davis J, Fontenot K. Insecticide efficacy against redbanded-stinkbug, Piezodorus guildinii (Westwood), a new stinkbug pest of Louisiana soybean. MidSouth Ent 2009;2: 68-69.

[20] Musser FR, Catchot AL, Gibson BK, Knighten KS. Economic injury levels for southern green stinkbugs (Hemiptera: Pentatomidae) in R7 growth stage soybeans. Crop Protection 2011;30: 63-69.

[21] Musser FR, Lorenz GM, Stewart SD, Catchot AL. 2009 Soybean insect losses for Mississippi, Tennessee, and Arkansas. Midsouth Entomol 2010;3: 48-54.

[22] Temple J, Davis JA, Hardke J, Price P, Micinski S, Cookson C, Richter A, Leonard BR. Seasonal abundance and occurrence of the redbanded stink bug in Louisiana soybeans. Louisiana Agric 2011;54: 20-22.

[23] Hoebeke ER, Carter ME. Halyomorpha halys (Stål) (Heteroptera: Pentatomidae): A polyphagous plant pest from Asia newly detected in North America. Proc Entomol Soc Washington 2003;105: 225-237.

[24] Herbert A. Brown Marmorated Stink Bug: A Confirmed New Pest of Soybean. Plant management network; 2011.

[25] Nielsen AL, Hamilton GC, Shearer PW. Seasonal Phenology and Monitoring of the Non-Native Halyomorpha halys (Hemiptera: Pentatomidae) in Soybean. Environmental entomology 2011;40(2): 231-238.

[26] Suiter DR, Eger JE, Gardner WA, Kemerait RC, All JN, Roberts PM, Greene JK, Ames LM, Buntin GD, Jenkins TM, Douce GK. Discovery and Distribution of Megacopta cribraria (Hemiptera: Heteroptera: Plataspidae) in Northeast Georgia. J of Integrated Pest Management 2010;1: F1-F5.

[27] Zhang Y, Hannula JL, Horn S. The Biology and Preliminary Host Range of Megacopta cribraria (Heteroptera: Plataspidae) and Its Impact on Kudzu Growth. Env Ent 2012;42:40-50.

[28] USDA-ARS. 2011 Update Activity of the Invasive Brown Marmorated Stink Bug, Halyomorpha halys (Stål), in Tree Fruit. Kearneysville: USDA-ARS Appalachian Fruit Research Station; 2011.

[29] Chapman RF. The Insects; Structure and Function. New York: Cambridge University Press; 1998.

[30] McCornack BP, Costamagna AC, Ragsdale DW. Within-plant distribution of soybean aphid (Hemiptera: Aphididae) and development of node-based sample units for estimating whole-plant densities in soybean. J Econ Entomol 2008;101: 1288-1500.

[31] Lin C, Li L, Wang Y, Xun Z, Zhang G, Li S. Effects of aphid density on the major economic characters of soybean. Soybean Science 1993;12: 252-254.

[32] Ragsdale DW, Mccornack BP, Venette RC, Potter BD, Macrae IV, Hodgson EW, O'neal ME, Johnson KD, O'neil RJ, Difonzo CD, Hunt TE, Glogoza PA, Cullen EM. Economic threshold for soybean aphid (Hemiptera: Aphididae). J Econ Entomol 2007;100: 1258-67.

[33] Miner FD. Stink bug damage to soybeans. Arkansas Agric Exp Stn Farm Res 1961;10: 12.

[34] Miner FD, Dumas B. Stored soybean and stink bug damage. Arkansas Farm Res 1980;29: 14.

[35] Jensen RL, Newsom LD. Effects of stink bug damaged soybean seed on germination, emergence, and yield. J Econ Entomol 1972;65: 261-264.

[36] Mitchell PL. Heteroptera as vectors of plant pathogens. Neotropical Entomology 2004;335(1): 519-545.

[37] Medrano EG, Esquivel JF, Bell AA. Transmission of cotton seed and boll rotting bacteria by the southern green stink bug (Nezara viridula L.). J Appl Microbiol 2007;103(2): 436-44.

[38] Dhaliwal GS, Arora R. Integrated pest Management: Concepts and Approaches. Ludhiana: Kalyani publishers; 2001.

[39] Painter RH. Insect resistance in crop plants. New York: Macmillan; 1951.

[40] Smith CM. Plant resistance to arthropods. Dordrecht: Kluwer Academic Publishers; 2005.

[41] Smith CM, Boyko EV. The molecular bases of plant resistance and defense responses to aphid feeding: current status. Entomol Exp Appl 2007;122(1): 1-16.

[42] Hollowell EA, Johnson HW. Correlation between rough-hairy pubescence in soybean and freedom from injury by Empoasca fabae. Phytopathology 1934;24: 12.

[43] Johnson HW, Hollowell EA. Pubescent and glabrous character of soybeans as related to injury by the potato leafhopper. J Agric Res 1935;51: 371-381.

[44] Boethel DJ. Assessment of soybean germplasm for multiple insect resistance. In: Clement SL., Quisenberry SS. (ed.) Global plant genetic resources for insect-resistant crops. Boca Raton: CRC; 1999. p101-129.

[45] van Emden H. Host-plant Resistance. In: van Emden H, Harrington R. (ed.) Aphids as crop pests. Oxfordshire: CABI; 2007. p447-468.

[46] Basky Z. Biotypic variation and pest status differences between Hungarian and South African populations of Russian wheat aphid, Diuraphis noxia (Kurdjumov) (Homoptera: Aphididae). Pest Manag Sci 2003;59: 1152-1158.

[47] Randolph TL, Peairs FB, Kroening MK, Armstrong JS, Hammon RW, Walker CB, Quick JS. Plant damage and yield response to the Russian wheat aphid (Homoptera: Aphididae) on susceptible and resistant winter wheats in Colorado. J Econ Entomol 2003;96(2): 352-360.

[48] Bregitzer P, Mornhinweg DW, Jones BL. Resistance to Russian Wheat Aphid Damage Derived from STARS 9301B Protects Agronomic Performance and Malting Quality When Transferred to Adapted Barley Germplasm. Crop Sci 2003;43(6): 2050-2057.

[49] He F, Liu X, Yan F, Wang Y. Soybean resistance to the soybean aphid. Liaoning Agric Sci 1995;4: 30-34.

[50] Hu Q, Zhao J, Cui D. Relationship between content of secondary catabolite-lignin-in soybean and soybean resistance to the soybean aphid. Plant Prot 1993;19: 8-9.

[51] Yu D, Guo S, Shan Y. Resistance of wild soybean Glycine soja to Aphis glycines. I. Screening for resistant varieties. Jilin Agric Sci 1989;3: 15-19.

[52] Hill CB, Li Y, Hartman GL. Resistance to the soybean aphid in soybean germplasm. Crop Sci 2004;44: 98-106.

[53] Hill CB, Li Y, Hartman GL. A single dominant gene for resistance to the soybean aphid in the soybean cultivar Dowling. Crop Sci 2006a;46: 1601-1605.

[54] Hill CB, Li Y, Hartman GL. Soybean aphid resistance in soybean Jackson is controlled by a single dominant gene. Crop Sci 2006b;46: 1606-1608.

[55] Mensah C, DiFonzo C, Nelson RL, Wang D. Resistance to soybean aphid in early maturing soybean germplasm. Crop Sci 2005;45: 2228-2233.

[56] Mian MAR, Hammond RB, St. Martin SK. New Plant Introductions with Resistance to the Soybean Aphid. Crop Sci 2008a;48(3): 1055-1061.

[57] Hill CB, Crull L, Herman TK, Voegtlin DJ, Hartman GL. A new soybean aphid (Hemiptera: Aphididae) biotype identified. J Econ Entomol 2010;103(2): 509-515.

[58] Diaz-Montano J, Reese JC, Schapaugh WT, Campbell LR. Characterization of antibiosis and antixenosis to the soybean aphid (Hemiptera: Aphididae) in several soybean genotypes. J Econ Entomol 2006;99(5): 1884-1889.

[59] Hesler LS, Dashiell KE. Antixenosis to Aphis glycines (Hemiptera: Aphididae) among soybean lines. The Open Entomology Journal 2011;5: 39-44.

[60] Pierson LM, Heng-Moss TM, Hunt TE, Reese JC. Categorizing the resistance of soybean genotypes to the soybean aphid (Hemiptera: Aphididae). J Econ Entomol 2010;103(4): 1405-1411.

[61] Jones WA Jr., Sullivan MJ. Overwintering habitats, spring emergence patterns, and winter mortality of some South Carolina Hemiptera. Environ. Entomol 1981;10: 409-414.

[62] Van Duyn JW, Turnipseed SG, Maxwell JD. Resistance in soybeans to the Mexican bean beetle: I. Sources of resistance. Crop Sci 1971;11: 572-573.

[63] Gilman DF, McPherson RM, Newsom LD, Herzog DC, Williams C. Resistance in Soybeans to the Southern Green Stink Bug. Crop Sci 1982;22(3): 573-576.

[64] Kester KM, Smith CM, Gilman DF. Mechanisms of resistance in soybean (Glycine max [L.] Merrill) genotype PI 171444 to the southern green stink bug., Nezara viridula (L.) (Hemiptera: Pentatomidae). Environmental Entomology 1984;13(5): 1208-1215.

[65] Borges M., Moraes MCB., Laumann RA., Pareja M., Silva CC., Michereff MFF., Paula DB. Chemical Ecology Studies in Soybean Crop in Brazil and Their Application to Pest Management. In: Ng T-B. (ed.) Soybean - Biochemistry, Chemistry and Physiology. Rijeka, Croatia: InTech Publishing; 2011. p31-66.

[66] Rosseto CJ. Breeding for resistance to stink bugs. In: Pascale AJ. (ed.) Proceedings of the World Soybean Research Conference IV. Buenos Aires, Argentina: Assoc Argentina de la Soja Press; 1989. p2046-2060.

[67] Carrao-Panizzi MC, Kitamura K. Isoflavone content in Brazilian soybean cultivars. Breed Sci 1995;45: 295-300.

[68] McPherson RM, Buss GR, Roberts PM. Assessing stink bug resistance in soybean breeding lines containing genes from germplasm IAC- 100. J Econ Entomol 2007;100: 1456-1463.

[69] Campos M, Knutson A, Heitholt J, Campos C. Resistance to Seed Feeding by Southern Green Stink Bug, Nezara viridula (Linnaeus), in Soybean, Glycine max (L.) Merrill. Southwestern Entomologist 2010;35(3): 233-239.

[70] Kang S, Mian MAR, Hammond RB. Soybean aphid resistance in PI 243540 is controlled by a single dominant gene. Crop Sci 2008;48: 1744-1748.

[71] Chen Y, Mensah C, DiFonzo C, Wang D. Identification of QTLs underlying soybean aphid resistance in PI 567541B. ASA-CSSA-SSSA Annual Meeting 2006: conference proceedings, November 10–14, 2006, Indianapolis, USA.

[72] Mensah C, DiFonzo C, Wang D. Inheritance of Soybean Aphid Resistance in PI 567541B and PI 567598B. Crop Sci 2008;48: 1759-1763.

[73] Li Y, Hill CB, Hartman GL. Effect of three resistant soybean genotypes on the fecundity, mortality, and maturation of soybean aphid (Homoptera: Aphididae). J Econ Entomol 2004;97: 1106-1111.

[74] Hill CB, Kim KS, Crull L, Diers BW, Hartman GL. Inheritance of resistance to the soybean aphid in soybean PI200538. Crop Sci 2009;49: 1193-1200.

[75] Mian MAR, Kang ST, Beil SE, Hammond RB. Genetic linkage mapping of the soybean aphid resistance gene in PI 243540. Theor Appl Genet 2008b;117(6): 955-962.

[76] Jun TH, Mian MAR, Michel AP. Genetic mapping revealed two loci for soybean aphid resistance in PI 567301B. Theor Appl Genet 2012;124(1): 13-22.

[77] Hill CB, Chirumamilla A, Hartman GL. Resistance and virulence in the soybean-Aphis glycines interaction. Euphytica 2012; DOI 10.1007/s10681-012-0695-z.

[78] Li Y, Hill C, Carlson S, Diers B, Hartman G. Soybean aphid resistance genes in the soybean cultivars Dowling and Jackson map to linkage group M. Mol Breed 2007;19(1): 25-34.

[79] Zhang G, Gu C, Wang D. Molecular mapping of soybean aphid resistance genes in PI 567541B. Theor Appl Genet 2009;118(3): 473-482.

[80] Zhang G, Gu C, Wang D. A novel locus for soybean aphid resistance. Theor Appl Genet 2010;120(6): 1183-1191.

[81] Jun TH, Mian, MAR, Michel AP. Genetic Mapping of Three Quantitative Trait Loci for Soybean Aphid Resistance in PI 567324. Heredity 2012; in review.

[82] Rossetto CJ, Gallo PB, Razera LF, Bortoletto N, Igue T, Medina PF, Tisselli Filho O, Aquilera V, Veiga RFA, Pinheiro JB. Mechanisms of resistance to stink bug complex in the soybean cultivar 'IAC-100'. An Soc Entomol Bras 1995;24: 517-522.

[83] Souza RF, Toledo JFF. Genetic analysis of soybean resistance to stinkbug. Braz J Genet 1995;18: 593-598.

[84] Li Y, Zou JJ, Zou J, Li M, Bilgin DD, Vodkin LO, Hartman GL, Clough SJ. Soybean defense responses to the soybean aphid. New Phytologist 2008;179(1): 185-195.

[85] Behura SK, Nair S, Sahu SC, Mohan M. An AFLP marker that differentiates biotypes of the Asian rice gall midge (Orseolia oryzae, Wood-Mason) is sex-linked and also linked to avirulence. Mol Genet Genomics 2000;263: 328-334.

[86] Bai X, Zhang W, Orantes L, Jun TH, Mittapalli O, Mian MAR, Michel AP. Combining next-generation sequencing strategies for rapid molecular resource development from an invasive aphid species, Aphis glycines. PLoS One 2010;5(6): e11370.

[87] Jun TH, Michel AP, Mian MAR. Development of soybean aphid genomic SSR markers using next generation sequencing. Genome 2011;54: 360-7.

[88] Jun TH, Michel AP, Mian MAR. Characterization of EST-based microsatellites from the soybean aphid, Aphis glycines. Journal of Applied Entomology 2012; DOI: 10.1111/j.1439-0418.2011.01697.x.

[89] Orantes LO, Zhang W, Mian MAR, Michel AP. Maintaining genetic diversity and population panmixia through dispersal and not gene flow in a holocyclic heteroecious aphid species. Heredity 2012; In Press.

[90] Wiarda SL, Fehr WR, O'Neal ME. Soybean Aphid (Hemiptera: Aphididae) Development on Soybean With Rag1 Alone, Rag2 Alone, and Both Genes Combined. J Econ Entomol 2012;105(1): 252-258.

[91] Tabashnik BE. 1990 Modeling and evaluation of resistance management tactics. In: Roush RT., Tabashnik BE. (ed.) Pesticide resistance in arthropods, New York: Chapman and Hall; p153-182.

[92] Tabashnik BE. Delaying insect adaptation to transgenic plants: seed mixtures and refugia reconsidered. Proc R Soc Lond B 1994;255: 7-12.

[93] Tabashnik BE, Gassmann AJ, Crowder DW, Carrière Y. Insect resistance to Bt crops: Evidence versus theory. Nat Biotechnol 2008;26: 199–202.

[94] Onstad DW, Meinke LJ. Modeling Evolution of Diabrotica virgifera virgifera (Coleoptera:Chrysomelidae) to Transgenic Corn With Two Insecticidal Traits. J Econ Entomol 2010;103: 849-860.

[95] Crowder DW, Carrière Y. Comparing the refuge strategy for managing the evolution of insect resistance under different reproductive strategies. J Theor Biol 2009;261: 423-430.

[96] Radcliffe EB, Hutchison WD, Cancelado RE., editor. Radcliffe's IPM World Textbook. St. Paul, MN: University of Minnesota; 2011. http://ipmworld.umn.edu/ (accessed 20 June 2012).

[97] O'Neal ME., Johnson KD. Insect pests of soybean and their management. In: Singh G. (ed.) The soybean - Botany, production and uses. Cambridge: CABI; 2010. p300–324.

[98] Hammond RB. Soybean Insect IPM. In: Radcliffe EB., Hutchison WD., Cancelado RE. (ed.) Radcliffe's IPM World Textbook. St. Paul, MN: University of Minnesota; 1996.

Available from http://ipmworld.umn.edu/chapters/Hammond.htm (accessed 20 June 2012).

[99]   Snodgrass GL, Adamczyk JJ, Gore J. Toxicity of Insecticides in a Glass-Vial Bioassay to Adult Brown, Green, and Southern Green Stink Bugs (Heteroptera: Pentatomidae). J Econ Entomol 2005;98: 177-181.

[100] Nielsen AL, Shearer PW, Hamilton GC. Toxicity of Insecticides to Halyomorpha halys (Hemiptera: Pentatomidae) Using Glass-Vial Bioassays. J Econ Entomol 2008;101: 1439-1442.

[101] Hodgson EW, VanNostrand G, O'Neal ME. 2010 yellow book: report of insecticide evaluation for soybean aphid. Department of Entomology, Iowa State University: Publication 287–10; 2010.

[102] Agricultural MU Guide, University of Missouri Extension: Soybean Pest Management: Stink Bugs. http://extension.missouri.edu/explorepdf/agguides/pests/g07151.pdf (accessed 20 June 2012).

[103] Bahlai CA, Sikkema S, Hallett RH, Newman J, Schaafsma AW. Modeling distribution and abundance of soybean aphid in soybean fields using measurements from the surrounding landscape. Env Entomol 2010;39: 50–56.

[104] Smith JF, Luttrell RG, Greene JK, Tingle C. Early-season soybean as a trap crop for stink bugs (Heteroptera: Pentatomidae) in Arkansas' changing system of soybean production. Environmental Ent 2009;8(2): 450-458.

[105] Gardiner MM, Prajzner SP, Landis DA, Michel AP, O'Neal ME, Woltz JM. Buckthorn Watch: Studying the Invasive Plant Common Buckthorn, Wooster, OH. Michigan State University Extension (Report No. E-3146); 2011.

[106] Heimpel GE, Frelich LE, Landis DA, Hopper KR, Hoelmer KA, Sezen Z, Asplen MK, Wu K. European buckthorn and Asian soybean aphid as components of an extensive invasional meltdown in North America. Biol Invas 2010;12: 2913–2931.

[107] Koppel AL, Herbert DA Jr, Kuhar TP, Kamminga K. Survey of stink bug (Hemiptera: Pentatomidae) egg parasitoids in wheat, soybean, and vegetable crops in southeast Virginia. Env Ent 2009;38: 375-379.

[108] Gardiner MM, Landis DA, Gratton C, Schmidt N, O'Neal M, Mueller E, Chacon J, Heimpel GE, DiFonzo CD. Landscape composition influences patterns of native and exotic lady beetle abundance. Diversity and Distributions 2009;15(4): 554-564.

[109] Gouli V, Gouli S, Skinner M, Hamilton G, Kim JS, Parker BL. Virulence of select entomopathogenic fungi to the brown marmorated stink bug, Halyomorpha halys (Stål) (Heteroptera: Pentatomidae). Pest Manag Sci 2012;68: 155-157.

[110] Rutledge CE, O'Neil RJO, Fox TB, Landis D. Soybean Aphid Predators and Their Use in Integrated Pest Management. Ann Entomol Soc Am 2004;97(2): 240-248.

[111] USDA-APHIS-PPQ. Qualitative analysis of the pest risk potential of the brown marmorated stink bug (BMSB), Halyomorpha halys (Stål), in the United States. 2010.

[112] UGA News Service. Tiny wasp may hold key to controlling kudzu bug. http://onlineathens.com/uga/2012-05-02/tiny-wasp-may-hold-key-controlling-kudzu-bug (accessed 20 June 2012).

[113] McCarville MT, O'Neal M, Tylka GL, Kanobe C, MacIntosh GC. A nematode, fungus, and aphid interact via a shared host plant: implications for soybean management. Entomologia Experimentalis et Applicata 2012;143: 55-66.

[114] van Emden HF., Way MJ. Host Plants in the Population Dynamics of Insects. In: van Emden HR. (ed.) Insect-Plant Relationships. London: Royal Entomological Society; 1972. p81-199.

[115] Price PW, Bouton CE, Gross P, McPheron BA, Thompson JN, Weis AE. Interactions among three trophic levels: influences of plants on interactions between insect herbivores and natural enemies. Ann Rev Ent 1980;11: 41.

[116] Boethel DJ, Eikenbary RD., editors. Interactions of Plant Resistance and Parasitoids and Predators of Insects. Ellis Horwood Ltd; 1986.

[117] Ghising K, Harmon JP, Beauzay PB, Prischmann-Voldseth DA, Helms TC, Ode PJ, Knodel JJ. Impact of Rag1 aphid resistant soybeans on Binodoxys communis (Hymenoptera: Braconidae), a parasitoid of soybean aphid (Hemiptera: Aphididae). Environ Entomol 2012;41: 282-288.

[118] Layton MB, Boethel DJ, Smith CM. Resistance to adult bean leaf beetle and banded cucumber beetle (Coleoptera: Chrysomelidae) in soybean. J Econ Entomol 1987;80: 151-155.

[119] Walker DR, Narvel JM, Boerma HR, All JN, Parrott WA. A QTL that enhances and broadens Bt insect resistance in soybean. Theor Appl Genet 2004;109: 1051-1057.

[120] Roush RT. Bt-transgenic crops: just another pretty insecticide or a chance for a new start in resistance management? Pest Manag Sci 1999;51: 328-334.

[121] Tabashnik BE, Liu Y-B, Finson N, Masson L, Heckel DG. One gene in diamondback moth confers resistance to four Bacillus thuringiensis toxins. Proc Natl Acad Sci USA 1997;94: 1640-1644.

[122] Hannon GJ. RNA interfenence. Nature 2002;418: 244-251.

[123] Gordon KH, Waterhouse PM. RNAi for insect-proof plants. Nat Biotechnol 2007;25(11): 1231-2.

[124] Mutti NS, Louis J, Pappan LK, Pappan K, Begum K, Chen M-S, Park Y, Dittmer N, Marshall J, Reese JC, Reeck GR. A protein from the salivary glands of the pea aphid, Acyrthosiphon pisum, is essential in feeding on a host plant. Proc Natl Acad Sci USA 2008;105: 9965-9969.

[125] Mutti NS, Park Y, Reese JC, Reeck GR. RNAi knockdown of a salivary transcript leading to lethality in the pea aphid, Acyrthosiphon pisum. Journal of Insect Science 2006;6: 1-7.

[126] Jaubert-Possamai S, Le Trionnaire G, Bonhomme J, Christophides GK, Rispe C, Tagu D. Gene knockdown by RNAi in the pea aphid Acyrthosiphon pisum. Bmc Biotechnology 2007;7: 8.

[127] Chen J, Zhang D, Yao Q, Zhang J, Dong X, Tian H, Chen J, Zhang W. Feeding-based RNA interference of a trehalose phosphate synthase gene in the brown planthopper, Nilaparvata lugens. Insect Mol Biol 2010;19: 777-786.

[128] Shakesby AJ, Wallace IS, Isaacs HV, Pritchard J, Roberts DM, Douglas AE. A water-specific aquaporin involved in aphid osmoregulation. Insect Biochemistry and Molecular Biology 2009;39: 1-10.

[129] Pitino M, Coleman AD, Maffei ME, Ridout CJ, Hogenhout SA. Silencing of Aphid Genes by dsRNA Feeding from Plants. PLoS ONE 2011;6(10): e25709.

[130] The International Aphid Genomics Consortium. Genome sequence of the pea aphid Acyrthosiphon pisum. PLoS Biol 2010;8(2): e1000313.

[131] Chougule NP, Bonning BC. Toxins for transgenic resistance to Hemipteran pests. Toxins 2012;4: 405-429.

Chapter 2

Weed Management in the Soybean Crop

Alexandre Ferreira da Silva, Leandro Galon,
Ignacio Aspiazú, Evander Alves Ferreira,
Germani Concenço,
Edison Ulisses Ramos Júnior and
Paulo Roberto Ribeiro Rocha

Additional information is available at the end of the chapter

1. Introduction

Inadequate weed control is one of the main factors related to decrease in soybean production. Weeds compete with crops by resources (water, light and nutrients). This competition is important mainly in the initial stages of crop development, due to possible losses in production that can be up to 80% or even, in extreme cases, hinders harvest operations [1].

Weeds have traits which confer them great aggressiveness even in adverse environments. High number of seeds, seed dormancy, discontinuous germination, effective dispersal mechanisms and population heterogeneity, are very important for weed establishment during crop development. During this phase, weeds may rapidly capture resources and occupy space; this is often linked to their competitive ability, because rapid growth requires the prompt and efficient conversion of resources into biomass. Thus, the yield is reduced and production costs increase, resulting in a decrease in farmer's income.

Besides reducing crop yield, weeds can cause other problems, like reduce grain quality, cause loss and difficulty during harvesting and serve as hosts of pests and diseases. The role of weeds as alternate hosts for soybean crop pests and diseases and their interference with cultivation operations resulting into higher costs of production must not be over looked. Weeds can also release toxins highly harmful to crop development. However, despite weeds show many negative aspects, they can also show advantages, like: providing food for the wildlife; potential source of germoplasm; recycling nutrients and preventing soil erosion.

Competition is defined as the condition that exists when requirements of one or more organisms living in a community cannot be obtained from available resources. Because competition involves many direct and indirect factors, it is often, preferable to consider it as interference of a plant community on another one, rather than competition. Interference is a natural phenomenon in a plant community where limited resources exist, and tends to be more harmful to competitors as more equal are the environmental demands and vegetative habit between them.

In agricultural ecosystems, weeds show competitive advantages over crop plants, because the aim of crop breeding is to increase the economic productivity, and this is almost always accompanied by a decrease in the competitive potential. Another important aspect in weed interference is the capacity of weeds in reducing or preventing cultivated plants to get access to resources. Thus, when those are limited, weeds almost always stand out, due to its higher efficiency in either capturing or using them. It is up to farmers and agronomists to use weed control methods and cultural practices in order to increase the chances of the crop overcoming weeds in the competition for resources.

Reduction in weed competition is perfectly achievable with the wide spectrum of tools and herbicides existing in the market, but weed management strategies are not related solely to the use of herbicides [2]. Weed control consists in suppressing the development and/or decreasing the number of weeds per area, until an acceptable levels for the coexistence between the species involved is reached, with minimum damages to both. In soybean crop, weed control can be achieved by using one or more control methods that are: preventive, mechanical, chemical, biological and cultural. Farmers can also use the integrated weed management (IWM), in which two or more of these methods are adopted.

The IWM approaches incorporate multiple tactics of prevention, avoidance, monitoring and suppression of weeds, undergirded by the knowledge of the agroecosystem biology [3]. The development of IWM was motivated by a desire to provide farmers with systematic approaches to reduce reliance upon herbicides [4] and, consequently, retard the selection of herbicide-resistant biotypes. The use of integrated control facilitates weed control during all crop cycle. The cultural practices, like soil tillage, fertilization, cultivar choice, sowing time, number of plants per area and crop rotation should be done in order to benefit crop development, and in some cases can reduce or eliminate the need of using other control methods.

The aim of this chapter is to summarize basic information about weed interference and weed management in the soybean crop, subsidizing technicians in the adoption of suitable positions regarding problems with weed control.

2. Competition between weeds and soybean by abiotic and biotic factors

Plants genetically improved by human action, aiming increases in productivity, lost part of their aggressive nature and therefore the ability to survive and compete against adversities imposed by the environment. Thus, most of the weeds show higher extraction capacity and utilization of environmental resources compared to cultivated species. The competition for limited resources or not, directly or indirectly, can be described as: *spatial* competition, which is generated by the physical dominance of a given species over another, simultaneously; a second classification that could be addressed is *temporal* competition, that results from competition over the time in which the crop is under development [5].

The various aspects of competition occurring between weeds and crops may also be named *ecological*, being classified as to their nature in biotic or abiotic [5]. The former are those from the live action elements of the ecosystem, such as predation, parasitism, commensalism, morphophysiological factors among others. The latter is a result of the action of non-living environmental factors, such as climatic and soil factors.

2.1. Competition between soybean and weeds for biotic factors

The biotic factors that determine the increased competitiveness of certain species over others are: plant size and architecture, growth rate, extension of root system, dry mass production, increased susceptibility to environmental elements (such as frost and dry spells), greater leaf area index and greater capacity for production and release of chemicals with allelopathic properties [6].

Morphophysiological traits of plants influence the competitive relationship between crop and weeds. Plant height and development cycle, for example, are features that have been positively associated with competitive ability in soybean; cultivars with higher cycle length and height reduce seeds production and size of weed species due to the increase in competitiveness of the crop [7].

Moreover, yield losses due to competition tend to be higher the more similar are the individuals, i.e. their morphophysiological traits, reaching maximum stress within the same species, because in this case neighboring plants compete for the same resources and occupy the same ecological niche [5].

2.1.1. Plant traits indicators of higher competitive ability

The competitive ability of crops can be expressed according to the crop ability to compete with weeds, reducing the production of seeds and dry mass accummulation by weeds, which is called *suppressive ability*. There is also the crop ability to *tolerate competition* with weeds, when under competition the crop is capable of maintaining yields almost unchanged [8, 9]. For Jordan [10], the suppressive ability should be preferred because it reduces seeds production by weeds and its benefits remain for subsequent growing seasons, while tolerance to weeds limits its benefits only to the current growing season. It is worth noting that in case crops do not have the ability to suppress weeds, the probability of yield reduction is increased, regardless of crop tolerance to competition.

Olofsdotter [11] remarks that several traits which confer competitive ability are genetically changeable, and can be manipulated by plant breeding, as they are elucidated by research. According to the author, it is necessary to identify one or more traits as well as their genetic variability in the crop. After demonstrating its variability, studies are needed to indicate the mechanisms involved and the environmental effect on the expression of these traits. Finally, it is necessary to involve geneticists and breeders in the identification of genes coding for the desired trait, as well as to evaluate the usefulness of indicators in the selection, i.e. if the character can be selected.

Differences in competitive ability between soybean cultivars with weeds have been reported by Bussan et al. [12]; Jannink et al. [9]; Lamego et al. [13]; Bianchi et al. [14] and Fleck et al. [15]. Suitable conditions for crop planting, such as moist soil, proper and uniform planting depth, close contact between seed and soil, as well as certified quality seeds, are essential to ensure competitive advantage to the crop by promoting the rapid emergence and establishment of uniform populations. In a study with soybeans, higher size of seeds resulted in seedlings with higher hypocotyl expansion rates, which may constitute a favorable feature in adverse conditions of emergence as in the case of soil crusting following heavy rainfalls [16].

2.1.2. Exploring competitive traits

The use of cultural methods for weed management can minimize weeds interference on soybean. Among the most efficient management practices for the suppression of weeds, the population density of the crop can be highlighted, as well as equal plants arrangement, development cycle and root growth of the crop.

2.1.3. Population density

In areas of agricultural production, the density of cultivated plants is kept constant along the field while weeds density varies with the degree of infestation, which is determined by the soil seed bank richness [17, 5]. According to these authors a variation occurs in the crop/weeds density ratio, making important to understand in competition studies not only the influence of density in the competition process – additive studies, but also the influence of the variation in the species proportion in the population - substitutive studies [5].

The duration of the period planting-emergence is also affected by seeding rate, temperature and soil moisture, planting depth and seed traits [18]. The duration of this period changes seedling height and subsequently, the intra-specific competitive ability. According to this author, the effects on the duration of this period are more evident under high plant densities.

2.1.4. Emergence speed

The use of high vigor seeds, which provide immediate plant emergence after planting, is important for the cultural management of weeds. In the dispute for limited environmental resources, the advantage is granted for plants that exhibit early establishment. A growing plant must quickly seize space and other resources, and its competitive success depends on the anticipated use of them. Plants stop growing when its area is restricted by competitors, so that the last individuals appear to grow very little due to shading. Thus, a fast emergence is often more important than the spatial arrangement of individuals in determining the competitiveness of the population [19].

Plants that have rapid and uniform emergence can compete more effectively for environmental resources [13]. These authors reported that the emergence rate is positively correlated with the ability of soybean cultivars to compete with weeds. In this sense, Fischer & Miles [19] formulated theoretical principles in which the greater the rate of development of a plant, the higher the shoot and edaphic volume explored. The first seedlings to emerge, probably present higher yields because they have priority in using water, light and nutrients, i.e., they occupy the niche early [20].

Plants that use resources earlier will shade the others, reducing the amount and quality of light available for the neighbors [20]. Weeds which establish before the crop, with big size and high number of seeds, will increase its frequence in the soil seed bank and keep infesting subsequent crops [6]. Another problem resulting from the establishment of crops later in relation to weeds is the need for increasing herbicide rates for their control [21].

Ecologically, weeds are less demanding in true growth factors in relation to crop plants, which confers great competitive ability for them [6]. In a study conducted by Carranza et al. [22], it was found that the relative intraspecific competition (yield loss per weed unit) decreased when weed population increased. According to the authors, plants that emerged earlier were 1.5 times more competitive than those who had delayed emergence.

Crop management practices such as use of high quality seeds, appropriate management of soil and planting at the recommended time and depth significantly increase chances of crop plants to be more competitive. The adoption of these practices, along with the use of cultivars with fast establishment, are key points to accelerate crop growth and focus on their success in competition with weeds.

2.1.5. Soybean plants arrangement and its relationship with weeds

The better arrangement of crop plants may be more important for those species with less potential for branching or tillering. The increase in grain yield of soybean with narrow row spacing has been demonstrated in several studies [23, 24, 16]. Positive results are obtained with this practice especially in wet years with the use of early maturing cultivars [16], in soil well supplied with nutrients [24], and also with late planting [25].

In the case of planting soybean after the recommended period, Board et al. [25] found that the reduced spacing resulted in higher dry mass of branches of plants at maturity (R8), which was highly correlated with grain yield. They also observed that the yield components of the branches, such as number, length and number of nodes in the branches, were higher in the smaller spacing, justifying the greater yield in reduced spacing system in late planting.

The dry weight of soybean can be used as a criterion to choose between wider or narrower spacings between rows. For Board & Harville [26], if plant dry mass of late-maturing cultivars, in the stage R8, is at least 800 g m^{-2} in wide spacing, probably no benefit will be obtained by reducing row spacing. However, this value should be used carefully because it may cause lower levels of total dry mass, for example, if planting is accomplished out of the indicated time interval (early or late planting).

The removal of weeds by using reduced spacing was evaluated by Legere & Schreiber [27]. These authors found that in the middle of the soybean growing season, the contribution of pigweed (*Amaranthus retroflexus*) for total dry mass per area was 43% in soybean row spacing of 76 cm, but was only 24% for rows spaced in 25 cm. In this work, in any situation of weed infestation, grain yield was always higher in the smaller row spacing.

In some situations, however, an adequate suppression of weeds may not occur. Burnside [28] found no difference in yields between spacings of 38 and 76 cm in the presence of weeds during different periods of coexistence. Also, Nice et al. [29] found no effect on the population of sicklepod (*Senna obtusifolia*) by reducing soybean row spacing, when there was no increase in the population of the crop. Under higher populations and smaller row spacings, a reduction in population and seed production of sicklepod was observed.

Another advantage of the smaller row spacing is the possibility of using lower doses of certain herbicides due to the effect of the additional shading of weeds by crop plants. Young et al. [30] observed that the reduction in row spacing from 76 cm to 38 cm, increased weed control after herbicide application. In contrast, glyphosate presented weed control superior to 90% at row spacing of 19 cm, controlling between 75% and 90% of the weeds when crop was planted at a spacing of 76 cm.

The set of morphological and physiological traits of cultivars defines its ability to compete with weeds for environmental resources [31, 14]. However, the competitive ability of cultivars can be altered by agronomic practices [32]. Weed population and its emergence delay in relation to the crop, often define the relationships of competition between species [33].

According to Lamego et al. [13] soybean cultivars with early emergence, fast leaf area expansion, high growth rate, and higher plant height in early stages, are more capable of competing with weeds. On the other hand, weed species with fast emergence, like the ones from Genus *Brachiaria, Digitaria, Euphorbia, Bidens* and *Raphanus*, among others, are able to compete earlier for environmental resources in relation to soybean [34, 13, 14, 35]. Rizzardi et al. [36], studying competition between soybean and *Euphorbia heterophylla, Ipomoea ramosissima, Bidens pilosa* or *Sida rhombifolia*, reported that several management practices can minimize their interference on soybeans, like the use of more competitive cultivars, correction of soil fertility and the adequacy of the arrangement of plants.

2.2. Competition between soybean and weeds for abiotic factors

2.2.1. Competition for water

Water is the most limiting factor essential for plant growth and production [37]. The rainfall and soil moisture strongly influence the growth of weeds, affecting, therefore, competition with crops [38]. Certain morphological and physiological traits determine the ability of plants to compete for soil water. In nature, species with C_3 metabolism predominate in temperate regions, while the C_4 are prevalent in tropical and subtropical regions. The relative distribution of C_3 and C_4 species depends on the temperature during the growing season of the plants [5].

Species with carbon metabolism by the cycle C_4 are usually more efficient in the use of water (higher WUE); as a consequence, they produce more biomass per unit of water consumed.

According to Patterson and Flint [39], *Amaranthus hybridus*, with C_4 metabolism, showed higher WUE compared to soybean plants. When soybean was compared to beans and with some weed species (*Euphorbia heterophylla*, *Bidens pilosa* and *Desmodium tortuosum*), bean was the plant which used the water more efficiently from the beginning of the cycle; soybean was the plant with the highest biomass accumulation rate and greater WUE along the cycle; *Desmodium tortuosum* was the most efficient in the capture and use of water during the vegetative stage and *Bidens pilosa* after flowering [40]. In another study, soybean and the weed *Xanthium strumarium* showed similar WUEs [41].

2.2.2. Competition for light

Light is the most disputed factor in competition, highlighting the importance of plant height in defining the competitive ability of crops [5]. The high ability of plants to intercept the incident light in the canopy is a desirable feature when crop is under competition with weeds [42]. Light interception by the canopy is dependent on plant density and arrangement, branching rate, plant height, leaf area, distribution of leaves, leaf angle, angle of leaf blades and dry mass accumulation [42]. Cultivars that concentrate photosynthates in leaves, i.e., high leaf area ratio (LAR), have greater potential for ground cover [43] and consequently the greater will be their competitive ability with weeds.

The initial growth rate is directly related to light interception and use in earlier stages of the plant cycle, allowing a greater leaf area development which provides to crop a higher competitive ability [32, 44].

The rate of biomass accumulation in shoots becomes a key factor for competitive success [40]. Earlier emergence of weeds in soybean, in relation to crop emergence, increased grain yield losses of soybean [13]. Evaluating the efficiency of capture and utilization of light by soybean and bean against the weeds *Bidens pilosa*, *Euphorbia heterophylla* and *Desmodium tortuosum*, Santos et al. [45] observed the highest accumulation rate of dry mass and the largest leaf area index for soybean, indicating its greater ability to capture light and shade the competing plants. Bean, especially after flowering, was more effective in draining its photosynthates for leaf formation.

2.2.3. Competition for nutrients

Nitrogen (N), phosphorus (P) and potassium (K), are of great importance for understanding yield losses by crops [46]. According to Anguinoni et al. [47] the capacity for absorption of nutrients in plants depends on the magnitude and the morphology of the root system and its efficiency in absorption of these elements. Crops with fast root growth maximizes the use of water and nutrients [48] so an accelerated growth of the root system constitutes a desirable feature for better nutrient use [49].

Under field conditions, in a study of competition for nutrients between soybean or bean with the weeds *Euphorbia heterophylla*, *Bidens pilosa* and *Desmodium tortuosum*, Bidens pilosa was the

plant species with higher leaf area increasing following N applications and soybean accumu-
lated the highest biomass in its root system, which tended to decrease with the addition of N.
B. pilosa and *E. heterophylla* increased its biomass as N was increased [46]. The total content of
N in soybean leaves decreased as the dose of N was increased; however, for all weed species,
N content in leaves increased according to the doses of N. The higher efficiency of roots in N
uptake was found for bean plants. *B. pilosa* and *E. heterophylla* were the most efficient species
in N use. The supply of N favored more the weed species not belonging to the legume family
than soybean and bean; therefore, an inadequate management of N in these crops may
exacerbate the problem of weed interference [46].

For the same species evaluated in the previous experiment, soybean was the species that
showed the largest increase of P in root biomass as the dose of this nutrient was increased.
Desmodium tortuosum, soybean and *B. pilosa* showed greater response to the addition of
increasing doses of P in relation to dry matter accumulation [50]. The efficiency of P uptake
by *D. tortuosum*, soybean and common bean decreased as the dose was increased. *E. heterophylla* and bean, performed worst on the efficient use of available P in soil.

3. Critical period of weed interference

The critical period of weed control (CPWC) has been defined by Silva et al. [51] as a window
in the crop growth cycle during which weeds must be controlled to prevent quantitative and
qualitative yield losses. In essence, the CPWC represents the time interval between two
separately measured crop-weed competition components: (1) the critical timing of weed
removal (CTWR) or the maximum amount of time early-season weed competition can be
tolerated by the crop before it suffers irrevocable yield reduction, and (2) the critical weed-free
period (CWFP) or the minimum weed-free period required from the moment of planting, to
prevent unacceptable yield reductions [52]. The former component is estimated to determine
the beginning of the CPWC, whereas the latter determines its end. Results from both components are combined to determine the CPWC. Theoretically, weed control before and after the
CPWC may not contribute to the conservation of the crop yield potential.

The beginning and end of the CPWC determined using the functional approach will depend
on the level of acceptable yield loss (AYL) used to predict its beginning and end. Many studies
report 5% as the maximum AYL. But it can be adjusted depending on the cost of weed control
and the anticipated financial gain [52].

Silva et al. [53], evaluated the CTWR in soybean, cv. BRS-244 RR in low, medium and high
weed density and observed that the CTWR was 17 days after emergence (DAE) in low
infestation area and 11 DAE in medium and high infestation area, considering 5% of tolerance
of crop yield decrease. According to the authors, weed interference during the full crop cycle
reduced soybean grain yield in 73%, 82% and 92%, for low, medium and high weed density,
respectively. Meschede et al. [54], evaluated the CPWC of *Euphorbia heterophylla* in soybean
crop, cv BRS 133, under low seeding rate, and observed that the presence of weeds caused
daily yield loss of 5.15 kg ha^{-1}, whereas their absence provided a daily yield gain of 7.27 kg

ha^{-1}. According to the authors, weed-crop coexistence for up to 17 DAE did not cause any negative effect on crop yield, and the maximum length of time in which weeds had to be controlled to prevent crop yield losses was 44 DAE; the CPWC was, therefore, from 17 to 44 DAE. Carvalho and Velini [55], observed that weeds germinated 20 days after the emergence of soybean, cv. IAC-11, did not affect crop yield.

Different results of CPWC showed that the degree of weed interference on crops depends on the infesting plant community (species, density and population), on the crop (cultivar, spacing and density) and environment (soil, climate and management). Thus, it is necessary a greater number of studies to create a data base and in the future create models to predict the adequate moment of weed control for each situation.

4. Weed control methods in soybeans

According to Hart [56], the population of weeds may be divided into three components: the active seed, the inactive/dormant seeds and plants.

The active seed (ready to germinate) can come from three sources: production by plants, seeds from outside the system and seeds that were dormant and that, for some reason, have become active. The dormant seed can also come from three sources: active seeds, plants and outside the system.

Weed management involves activities directed at the weeds (direct management) and, or, the system formed by soil and crop (indirect management). The direct management refers to the direct elimination of weeds using herbicides, manual or mechanical action and biological action. In soil management (indirect management) the relationship active and inactive seed can be worked. In this case, germination of the weeds should be increased before controlling them, using techniques such as the sequential application of desiccants.

According to Silva et al. [7], weed control possibilities include preventive, cultural, mechanical, biological and chemical methods. However, to maintain the sustainability of agricultural systems, it is important to integrate these control measures by observing the characteristics of soil, climate and socioeconomic aspects of the producer. The achievement of an environmentally and economically compatible integration requires deep knowledge of the available strategies, promoting balance with the management measures of soil and water, as well as the control of pests and diseases. To adopt any measure of control, the medium in which the weeds are should be treated as an ecosystem that can respond to any changes imposed, thus, not limited to the application of herbicides or using any other method alone. Furthermore, efforts will encourage the improvement of the quality of life, both of the farmer directly involved, as the whole population which will benefit from the supply chain.

4.1. Preventive control

It is harder to control weeds once they establish themselves, so preventing foreign weeds from entering a new area is usually easier and costs less than controlling after they have spread.

According to Silva et al. [57], the preventive control of weeds is the use of practices aimed at preventing the introduction, establishment and, or, spread of certain problematic species in areas not yet infested by them. These areas can be a country, a state, a municipality or a piece of land inside the farm.

In federal and state levels, there are laws regulating the entry of seeds into the country or state and its internal commercialization. Under these laws are the tolerable limits of seeds of each weed species and also the list of prohibited seeds per crop or crop group.

Locally, it is the responsibility of individual farmers or cooperatives, to prevent the entry and spread of one or more weed species that may become serious problems for the region. In summary, *the human element is the key to preventive control*. The efficient occupation of the agroecosystem space by the crop reduces the availability of appropriate factors for growth and development of weeds, and can be considered an integration between preventive and cultural method.

Choosing the right cultivars is actually the first step in successfully establishing a crop. In the soybean case, there is a large number of cultivars adapted to different regions of the world.

Some of the measures that can prevent the introduction of the species are: use of high purity seeds, clean thoroughly machines, harrows and harvesters; carefully inspect seedlings acquired with soil and also all the organic matter (manure and compost) from other areas; clean irrigation canals; quarantine of introduced animals, etc. [5].

Chauhan et al. [58] affirm that most crops have their seeds contaminated with weeds, especially when weed seeds resemble the size and shape of crop seeds. Contamination usually happens during the time of crop harvesting when weeds that have life cycles similar to those of crops set seeds. When even a small amount of weed seeds is present, it may be enough for a serious infestation in the next season. The idea should be to minimize the weed infestation area and decrease the dissemination of weed seeds from one area to another or from one crop to another. Control of weed species is achieved by reducing plants and propagules to the point at which their presence does not seriously interfere with an area of economic use. The planning of post-infested weed control programs should be done in such a way that the build-up of weed seeds is reduced drastically within a short period. Proper care should be taken to restrict the weed seed bank size in the area by using integrated methods of weed control. In undisturbed or no-till systems, seeds of weeds and volunteer crops are deposited in the topsoil [59, 60, 61]. Therefore, an appropriate strategy is needed to avoid high weed infestations and to prevent unacceptable competition with the emerging crop [60].

4.2. Cultural control

The competitive ability of weeds largely depends on the time of emergence in relation to the soybean, in such a way that, if the crop germinates faster, and also occurs a delay on the emergence of weeds, competition will be reduced [5].

According to Silva et al [57], cultural control is the use of common practices for the proper management of water and soil as crop rotation, variation of crop row spacing, living mulches, cover crops etc. Amending the soil, neutralizing the aluminum content and increasing the pH, favors the crop and not certain weed species adapted to acid soils conditions and high contents of Al. Fertilization applied at the planting furrow is a common practice, and also favors soybean, so the fertilizer do not stand so close to the weeds in the inter-rows. These practices help to reduce the seed bank of weeds. It consists, therefore, in using their own ecological traits, both from crops and weeds, in order to benefit the establishment and development of crops.

One of the main practices is crop rotation. Its benefits depend on the selection of crops and their sequence in the system. Continuous cultivation of a single crop or crops having similar management practices allows certain weed species to become dominant in the system and, over time, these weed species become hard to control [58]. According to Kelley et al. [62], soybean production is improved by using crop rotation as a management practice. Numerous studies have shown decreased yield when soybean was grown continuously in monoculture than when rotated with another crop [63, 64, 65]. In the short-term, benefit of crop rotation was increased soybean yield, which would likely increase soybean profitability. In the long-term, rotations with high residue-producing crops, such as wheat and grain sorghum, significantly increase total soil C and N concentrations over time, which may further improve soil productivity [62].

Variation of the spacing or plant density in the row is another practice that can contribute to the reduction of weed interference on the crop, depending on the architecture of the cultivated plants and weed species. The reduction of spacing between rows often provides competitive advantage for most crops over shading sensitive weeds. In this case, by reducing the spacing between rows, provided it does not exceed the minimum limit, there is increased light interception by the canopy of cultivated plants. This effect is dependent on factors like the type of species to be cultivated, morphophysiological traits of genotypes, weed species present in the area and season and weather conditions at the time of its emergence, as well as environmental conditions [66, 67, 68].

The main goal of using cover crops for weed control is replacing an unmanageable weed population with a manageable cover crop. This is accomplished by selecting the phenology of the cover crop to preempt the niche occupied by weed populations [69]. They have been used to manage weeds in soybean [70, 71, 72, 73]. According to Silva et al. [57], green covers are crops that usually are very competitive with weeds. Lupine, vetch, ryegrass, turnips, oats and rye are used in southern Brazil. In the subtropics, velvetbean, crotalarias, pigeon pea, jack-bean and lab-lab can be used. Its main effect is to reduce the seed bank and also improve soil physical-chemical conditions. However, these plants may also have inhibitory effects over others and can reduce infestations of some weed species after desiccation or incorporated in soil, and must be carefully chosen in each case. The presence of the mulch creates conditions for the installation of a dense and diverse microbiote in the soil, especially in the surface layer, with a high amount of microorganisms responsible for the elimination of dormant seeds by deterioration and loss of viability.

Both the composition and the population density of a weed community are influenced by the level of mulching in the production system [74]. The mulch has physical (interference on germination and seedling survival rate), chemical (allelopathic effect) and biological (installation of a dense and diverse microbiocenose in the topsoil) effects on weeds [75,76].

Thus, of the numerous known advantages of no-tillage - a practice that keeps the soil covered by crop residues - stands out the improvement in weed control. Trezzi & Vidal [77] found that the presence of residues of sorghum shoot (4 t ha⁻¹) was sufficient to reduce 91, 96 and 59% the population of *Sida rhombifolia*, *Brachiaria plantaginea* and *Bidens pilosa*, respectively.

According to Silva et al. [57], in no-tillage, using systemic herbicides as desiccants, together with not revolving the soil, whether to produce corn for grain or silage, excellent results were found in the management of purple nutsedge (*Cyperus rotundus*). In two years in this system, it is possible to reduce population levels of nutsedge in favor of no-tillage, compared to conventional tillage, for both corn and beans, to the order of 90 to 95%, being that in three years, the reduction on the bank of tubers in the soil can reach more than 90%. The greatest benefits of no-tillage system in the integrated management of purple nutsedge (*Cyperus rotundus*) are obtained due to the integration of chemical control provided by the use of the systemic herbicide for desiccation of the vegetation at pre-sowing, to the cultural control exercised by the lack of soil disturbance and consequent lack of fragmentation of the vegetative structures of the nutsedge, and to the adoption of highly competitive crops, mainly by light, such as corn and beans. Thus, the population levels of purple nutsedge can be reduced, especially during the crop development period that is sensitive to weed interference, or approximately 45 days after emergence, as not to cause reductions in infested crop yields. Furthermore, the ability of sprouting of the nutsedge tubers collected in the soil under integrated management is diminished over time, remaining dormant [78].

4.3. Mechanical control

According to Silva et al. [57], weed plucking, or weeding, is the oldest method of weed control. It is still used to control weeds in home gardens and in the removal of weeds between crop rows, when the main method of control is the use of a hoe.

The manual weeding made with a hoe is very effective and still widely used in our agriculture, especially in mountainous regions, where there is subsistence agriculture, and for many families, this is the only source of work. However, in a more intensive agriculture in larger areas, the high cost of manpower and the difficulty of finding workers when necessary and in the desired quantity, make this method only complementary to others, and should be done when the weeds are still young and the soil is not too humid. It can assume great importance in seed production fields, being a good alternative for using isolated or as a complement for other control methods [79].

According to Silva et al. [57], mechanized cultivation, made by cultivators pulled by animals or tractors, is widely accepted in Brazilian agriculture, being one of the main methods of weed control on properties with smaller areas planted. The main limitations of this method are the difficulty of controlling weeds in the crop rows, low efficiency when performed in wet conditions (wet soil), and it is also inefficient to control weeds that reproduce by vegetative parts. However, all the annual species, when young (2-4 pairs of leaves), are easily controlled in conditions of heat and dry soil. Cultivation breaks the intimate relationship between root and soil, suspending the absorption of water, and exposes the roots to unfavorable environmental conditions. Depending on the relative size of weeds and crops, the displacement of the soil on the row, using special hoe cultivators, can cause the burial of seedlings and thereby promote weed control even in the rows of the crop.

4.4. Biological control

Biological control is the use of natural enemies (fungi, bacteria, viruses, insects, birds, fish, etc.) capable of reducing weed populations, reducing their ability to compete. This is maintained by the population balance between the natural enemy and the host plant. It should also be considered as biological control the allelopathic inhibition of weeds [6].

According to Charudattan & Dinoor [80], bioherbicide is defined as a plant pathogen used as a weed-control agent through inundative and repeated applications of its inoculum. In the United States and many other countries, the prescriptive use of plant pathogens as weed control agents is regarded as a "pesticidal use" and therefore these pathogens must be registered or approved as biopesticides by appropriate governmental agencies. Currently, one fungus species is registered as bioherbicide in the United States for use in soybeans. Collego®, based on *Colletotrichum gloeosporioides* f.sp. *aeschynomene*, is used to control *Aeschynomene virginica* (northern jointvetch), a leguminous weed, in soybean and rice crops in Arkansas, Mississippi, and Louisiana [80].

Charudattan & Dinoor [80] also state that, among the limitations of biocontrol of weeds by plant pathogens, the most important are the limited commercial interest in this approach to weed control due to the fact that markets for biocontrol agents are typically small, fragmented, highly specialized, and consequently the financial returns from biocontrol agents are too small to be of interest to big industries; and the complexities in production and assurance of efficacy and shelf-life of inoculum can further stifle bioherbicide development. For instance, the inability to mass-produce inoculum needed for large-scale use is a serious limitation that has led to the abandonment of several promising agents. The authors conclude that plant pathogens hold enormous potential as weed biocontrol agents. In addition to the use of plant pathogens as biocontrol agents, it is likely that pathogen-derived genes, gene products, and genetic mechanisms (e.g., hypersensitive plant cell death and herbicidal biochemicals) will be exploited in the near future to provide novel weed management systems. On the other hand, the present over-reliance on chemical herbicides and the tendency to base weed-management decisions purely on economic considerations, at expense of the exclusion of ecological and societal benefits, is a serious limitation that could stifle biological control.

4.5. Chemical control

There are several advantages in using herbicides: pre-emergence control, eliminating the weeds precociously; hits targets that the hoe or cultivator does not reach, like the weeds in the crop row; reduces or eliminates the risk of damage to the roots and to young plants; do not alter soil structure and, therefore, reduces risk of erosion; controls more efficiently the perennial weeds; reduces the need for labor; increases the speed and efficiency of the control operation per unit area, reducing the cost per treated area; controls the weeds for a longer period, when the use of a cultivator is impossible in view of the crop growth; and can be used in rainy periods, when the mechanical control is not efficient and when labor is required for other activities. However, it has the disadvantage of requiring skilled labor, because, if done improperly, can poison the crop, the environment and, especially, the applicator himself. Although herbicides are very effective in controlling weeds, they may promote the development of resistant biotypes, a fact that would further exacerbate the problem within an area [81].

According to Oliveira Jr. et al. [82], the most common strategies used in the management of both cover crops and weed vegetation in areas of no-tillage are reduced to three: desiccation immediately before sowing, between seven and ten days before sowing or anticipated drying.

These authors undertook a study aimed to evaluate the interaction between tillage systems and weed control in post emergence in soybean with these three strategies. They concluded that, although desiccation in different management systems have been effective, the anticipation of desiccation in anticipated management favored the emergence and initial soybean development, providing greater productivity gains, given the infestation conditions. The management system also affected the flow of weed emergence after soybean emergence, with fewer reinfestations in the anticipated management system, due to the control of initial flows given by the second application of this management system. Management applied at planting and ten days before planting, hindered the development of soybean, resulting in lower productivity, while anticipated management provided the highest yield.

Procópio et al. [83] carried out a study in which they compared the effects of tillage systems on the control of the weeds *Digitaria insularis*, *Synedrellopsis grisebachii* and *Leptochloa filiformis* before soybean planted in no-till. The authors found satisfactory control and prevention of regrowth of *D. insularis* and *L. filiformis* when glyphosate was applied five days prior to soybean planting or when the sequential application of glyphosate and paraquat + diuron was done. Sequential applications of the mixture paraquat + diuron were not effective in controlling or preventing the regrowth of *D. insularis* and *L. filiformis* and the weed *S. grisebachii* proved to be tolerant to glyphosate. The use of a non-residual herbicide such as glyphosate fails by not controlling weeds emerged after application, and eventually produce seeds that can easily replenish the seed bank [84]. Adding a residual herbicide to glyphosate can be a consistent management to control the weeds as they germinate and promotes long-term activity which controls plants which emerge later [85].

According to Arregui et al. [86], there are several soil-applied broadleaf herbicides that effectively control weeds like *Ipomoea* spp., *Commelina* spp. and *Sida spinosa*. Chlorimuron and sulfentrazone reduce *Ipomoea* spp. density [87]; *S. spinosa* density decreased with imazaquin, metribuzin and sulfentrazone applications [87] and with cloransulam and diclosulam [88].

The same authors [86] affirm that soil-applied herbicides as metribuzin and imazaquin may be beneficial reducing early season competition of weeds, particularly those inherently more tolerant to glyphosate such as *Parietaria debilis* or *Commelina erecta*, which survive pre-planting glyphosate applications. Likewise, when dry conditions are observed during vegetative soybean growth, glyphosate applications could be less effective for weed control and the resulting competition could reduce soybean yields.

Hager et al. [89], in a study to examine the influence of herbicide application timing and dose on efficacy of six soil-applied herbicides for common waterhemp (*Amaranthus rudis*) control in soybean, found that sulfentrazone controlled this weed better and reduced its density more than other herbicides.

Nosworthy [90], evaluating broadleaved weed control and economics of conventional and glyphosate-containing herbicide programmes in glyphosate-resistant soybean planted in wide rows, found that pre-emergence herbicides followed by glyphosate, controlled *Ipomoea lacunosa* L. eight weeks after emergence (WAE). *I. hederacea* var. *integriuscula* Gray control with pre-emergence herbicides followed by glyphosate was 100% with similar control from chlorimuron plus sulfentrazone followed by lactofen, whereas control following the single glyphosate application was 84%. *Amaranthus palmeri* S. Wats. control nine WAE was 100% following single or sequential glyphosate applications, while control ranged from 76% to 96% with pre-emergence herbicides followed by lactofen. However, early season weed interference when a single application of glyphosate was delayed until four WAE reduced soybean yields an average of 389 kg ha^{-1} compared to pre-emergence herbicides followed by glyphosate.

5. Principles of integrated weed management

The concept of Integrated Weed Management (IWM), a component of Integrated Pest Management, has been proposed (i) to decrease the density of weeds emerging in crops, (ii) to reduce their relative competitive ability (in order both to preserve crop yields and to limit the replenishment of weed seed bank), and (iii) to control emerged weeds using non chemical techniques, with the overall aim of reducing the need for herbicide application at the cropping system level [91]. IWM advocates the use of all available weed control options such as: plant breeding, fertilization, crop rotation, tillage practices, planting pattern, cover crops and mechanical, biological and chemical controls. To define the correct weed management strategies, it is necessary to know the ability of the weed species, in relation to the crop, to compete for water, light and nutrients, which are factors responsible for decreasing crop yield [6].

Usually, it is not taken into consideration that a good program of weed management should allow for maximum production in the shortest time, the maximum sustainable production and minimal environmental and economic risk. Wilson et al. [92] in a study to compare the Ohio farmer model to a weed scientist decision model about management of weeds, concluded that farmers understand but do not practice IWM. The failure to adopt may be attributed to gaps in their understanding of the human role in weed dispersal, their focus on the risks associated with weeds without recognition of their ecological benefits, and the tendency to overlook risks associated with management.

Therefore, to accomplish the IWM, it is required knowledge in botany, plant physiology, molecular biology, climatology and application technology, among others.

The strategies for the integrated weed management in different weed species can be divided as short or long-term. Measures such as weeding or direct employment of herbicides (chemical control) can be considered as short-term, accounting for only temporary control, requiring new applications to each crop season. In the case of long-term measures, the use of cultural practices and control by other biological agents, has permanent character and take into account more pronounced changes in different agronomic practices. From this, results the integrated management, which should integrate prevention and other control methods that promote short (mechanical and chemical methods) and medium and long-term (cultural and biological methods) control.

According to Chauhan et al. [58], any single method of weed control cannot provide season-long and effective weed control. Therefore, a combination of different weed management strategies should be evaluated for widening the weed control spectrum and efficacy for sustainable crop production. The use of clean crop seeds and seeders and field sanitation (irrigation canals and bunds free from weeds) should be integrated for effective weed management. Combining good agronomic practices, timeliness of operations, fertilizer and water management, and retaining crop residues on the soil surface improve the weed control efficiency of applied herbicides and competitiveness against weeds. In Canada, for example, integrating superior cultivars with a high seeding rate and the earliest time of weed removal led to a 40% yield increase compared with the combination of a weaker cultivar, the lowest seeding rate, and the latest time of weed removal [93].

According to Bernards et al. [94], the development of an IWM program is based on a few general rules that can be used at any farm:

a.   use agronomic practices that limit the introduction and spread of weeds, preventing weed problems before they started;

b.   help the crop compete with weeds; and

c.   use practices that keep weeds off balance and do not allow weeds to adapt.

Combining agronomic practices based on these rules will allow the farmer to design an IWM program for his reality. There is not a single recipe for all conditions and years. The plan will need to be changed and adjusted to a particular farming operation and season. The goal is to manage, not eradicate weeds.

6. Herbicide resistant weeds in transgenic soybeans

Soybean is a crop characterized by the high consumption of herbicides. Chemical control is the most usual, given the characteristics of practicability, efficiency and speed on its execution.

Most of the farmers in Brazil and the world adopt the chemical method for weed control. This is because this technology is very efficient, has attractive cost compared to alternative methods, is easy to use and is professionally developed. However, most producers have only an immediatist and economical vision of weed control and this could lead to environmental problems in the medium and long-term. Although it is public domain that repeated applications of herbicides with the same mechanism of action on a genetically diverse population of weeds may cause strong selection pressure and evolution of resistance [95], it has been a common practice in many parts of the world. As a consequence, the population of herbicide-resistant weeds has expanded rapidly in several regions, making it a hard solution problem in many areas with intensive agriculture. Evidence suggests that the appearance of resistance to a herbicide, in a plant population, is due to the selection of pre-existent resistant biotypes, because of the selection pressure exerted by repeated applications of the same active ingredient, finding conditions for propagation and prevalence [96].

In 2005, transgenic soybean was officially released for planting in Brazil. From this moment on, several products and product combinations have been replaced by a single active ingredient, the glyphosate. Glyphosate is a systemic herbicide used for postemergence control of grasses and broadleaved weeds [97]. In transgenic soybean, it is used in single or sequential applications, at doses and times that will vary according to each scenario.

Currently, the technology of glyphosate-resistant soybean, readily accepted and adopted by the producers caused the use of this herbicide to expand, with average of three applications of glyphosate per cycle of soybean, at desiccation and two after crop emergence. Furthermore, the glyphosate is the primary herbicide for several crops such as fruits, coffee, eucalyptus and desiccation for no-tillage [96].

The technology of glyphosate-resistant soybean allows to reduce or eliminate the need to apply other herbicides for the management of different weed species, which contributes to increased selection pressure and emergence of resistant biotypes. Moreover, some aspects of population dynamics of weeds and the possibility of selecting glyphosate-tolerant species must be considered. The type of management and herbicides used in an area cause changes in the type and proportion of species which compose the local population. This is explained by the fact that herbicides do not control evenly the species in the area; so, some end up being benefited and multiply. In these situations, a low occurrence of plants in the area can become a serious problem for the producer. Thus, the repeated and continuous use of the same herbicide or herbicides with the same mechanism of action, makes the selection of species inevitable [98].

Conyza canadensis is an example of problematic weed in soybeans, in which were detected cases of resistance of biotypes from this species to glyphosate in various parts of the world in transgenic soybeans fields. Experiments conducted by Vargas et al. [99], Moreira et al. [100] and Lamego & Vidal [101] demonstrated that application of 360 g a.e. ha^{-1} of glyphosate is enough, under greenhouse studies, to distinguish between resistant or susceptible biotypes of Conyza bonariensis and C. canadensis.

The resistance factors (GR$_{50}$) ranged between 7 and 11 for *C. canadensis* [100] and between 10 and 15 [100] and 2.4 [101] for *C. bonariensis*. It is noteworthy that determining the resistance level of suspect populations supports the decisions on strategies to control these biotypes.

Up to date, 23 cases of glyphosate resistant weeds were found in weed species worldwide, described in Table 1.

Weed species	Occurrence/observation	Type of resistance
1. Amaranthus palmeri	USA/2005*	Multiple resistance to ALS and EPSPs inhibitors
2. Amaranthus tuberculatus	USA/2005	Multiple resistance, triple (ALS, Protox and EPSPs inhibitors) and double (ALS and EPSPs inhibitors)
3. Ambrosia artemisiifolia	USA/2004	Multiple resistance to ALS and EPSPs inhibitors
4. Ambrosia trifida	USA/2004	Multiple resistance to ALS and EPSPs inhibitors
5. Bromus diandrus	Australia/2011	
6. Chloris truncata	Australia/2010	
7. Conyza bonariensis	South Africa, Spain, Brazil, Israel, Colombia, USA/2003, Australia, Greece, Portugal	Multiple resistance to Photosystem I and EPSPs inhibitors
8. Conyza canadensis	USA/2000, Brazil, China, Spain, Czech Republic, Poland and Italy	Multiple resistance to ALS and EPSPs inhibitors and to Photosystem I and EPSPs inhibitors
9. Conyza sumatrensis	Spain and Brazil/2009	
10. Cynodon hirsutus	Argentina/2008	
11. Digitaria insularis	Paraguay and Brazil/2005	
12. Echinochloa colona	Australia/2007, USA and Argentina	
13. Eleusine indica	Malaysia/1997, Colombia and USA	Multiple resistance to ALS and EPSPs inhibitors
14. Kochia scoparia	USA/2007 and Canada	Multiple resistance to ALS and EPSPs inhibitors
15. Leptochloa virgata	Mexico/2010	
16. Lolium multiflorum	Chile/2001, Brazil, USA, Spain and Argentina	Multiple resistance to ALS and EPSPs inhibitors, ACCase and EPSPs inhibitors, triple resistance to ALS, ACCase and EPSPs inhibitors
17. Lolium perenne	Argentina/2008	
18. Lolium rigidum	Australia/1996, USA, South Africa, Spain, Israel and Italy	Multiple Resistance, double (Photosystem II and EPSPs inhibitors), triple (ACCase, Photosystem I and EPSPs inhibitors), quadruple (ALS , ACCase, EPSPs and dinitroanilines inhibitors)
19. Parthenium hysterophorus	Colombia/2004	
20. Plantago lanceolata	South Africa/2003	
21. Poa annua	USA/2010	
22. Sorghum halepense	Argentina/2005 and USA	
23. Urochloa panicoides	Australia/2008	

* Observation year of the first resistance case. Source: Weed Science [103]

Table 1. Glyphosate (EPSPs inhibitor) resistant weed species, countries of occurrence and type of resistance.

6.1. Management of herbicide resistant weeds in soybean

The rational management of herbicides with different mechanisms of action is a very important practice. Furthermore, the use of herbicides with little soil residual activity and optimization of doses and number of applications reduces the selection pressure, decreasing the risks of selection of plant resistance to herbicides. Another very efficient technique for the management of weeds consists in using mixtures of herbicides with different mechanisms of action. In this case, the prevention of resistance is based on the fact that the active ingredients efficiently control both biotypes of the same species, i.e., the biotype resistant to a herbicide is controlled by another active ingredient of the mixture [98]. It is noteworthy that the herbicide mixture of different mechanisms of action as a means of management and prevention of resistance is more efficient when the reproductive system of the weed is self pollination, since the genetic recombination of different alleles which confer resistance is less likely to occur in relation to allogamous plants.

Due to the numerous cases of herbicide-resistant weed biotypes in Brazil, several studies were performed looking for alternatives to control these plants, finding that the use of herbicides with different mechanisms of action is a viable alternative for managing resistance [103]. Table 02 shows alternative herbicides suitable for soybean according to the resistant species in the area.

Weed	Alternative herbicides
Lolium multiflorum/ post-emergence	Fluazifop-p, Haloxyfop-r, Clethodim, Sethoxydim
Lolim multiflorum/ desiccation	Paraquat and Ammonium-Glufosinate
Conyza bonariensis and *Conyza canadensis*/ post-emergence	Clorimuron-ethyl
Conyza bonariensis and *Conyza canadensis* / desiccation	Paraquat + Diuron, Ammonium-Glufosinate, Clorimuron-ethyl and 2,4-D

Table 2. Alternative herbicides to control glyphosate resistant weeds in soybean crop used in Brazil.

Weed resistance is an evolving phenomenon in world and, in certain cases, may restrain the use of some herbicides. Therefore, weed resistance to herbicides should be managed through the use of alternative strategies associated to the application of herbicides. Crop rotation is a good strategy to break the life cycle of weed, preventing its dominance in the area. When the same cultural techniques are applied, year after year, in the same soil, the interference of these weeds is greatly increased. When the main goal is the weed control, the choice of the rotating crop should fall on plants with very contrasting growth habits and cultural characteristics [98]. Thus, when using crops with different physiological needs, a change occurs in weed species from one crop to another and, if it becomes necessary to use herbicides, there is a greater chance they will have different mechanisms of action. The rotation is an effective method both in preventing the appearance of resistant biotypes as in managing installed resistance.

Only with a rational management and using several control methods will the resistance be mitigated and the likelihood of the emergence of new cases minimized [98].

7. Final comments

The challenge of agriculture sustainability requires solving the trade-off between producing satisfying levels of agricultural products, both in terms of quantity and quality, and reducing the environmental impacts and preserving non renewable resources. Weed management is a key issue, because herbicides are the most sprayed pesticides around the world and they are some of the mostly found contaminating substances in the surface and below-ground waters. Therefore, it is necessary to adopt correct strategies for weed management, but for that it is necessary to know the ability of weed species, present in a given area, in relation to the crop, to compete for water, light and nutrients, factors responsible for decreasing crop yield. Simple measures like choosing the correct cultivar, adopting correct tillage practices, using cover crops and crop rotation are responsible for decreasing the use of herbicides and, consequently, contribute for environmental sustainability.

Author details

Alexandre Ferreira da Silva[1], Leandro Galon[2], Ignacio Aspiazú[3], Evander Alves Ferreira[4], Germani Concenço[5], Edison Ulisses Ramos Júnior[6] and Paulo Roberto Ribeiro Rocha[7]

1 Embrapa Milho e Sorgo, Sete Lagoas-MG, Brazil

2 Universidade Federal da Fronteira Sul, Erechim-RS, Brazil

3 Universidade Estadual de Montes Claros-MG, Brazil

4 Universidade Federal dos Vales do Jequitinhonha e Mucuri, Diamantina-MG, Brazil

5 Embrapa Agropecuária Oeste, Dourados-MS, Brazil

6 Embrapa Soja, Londrina-PR, Brazil

7 Universidade Federal Rural do Semi-Árido, Mossoró-RN, Brazil

References

[1] Vargas, L, & Roman, E. S. Controle de plantas daninhas na cultura da soja. Unaí: (2000).

[2] Ghersa, C. M, Benech-arnold, R. L, Satorre, E. H, & Martínez-ghersa, M. A. Advances in weed management strategies. Field Crops. (2000). 0378-4290, 67, 95-104.

[3] Wilson, R. S, Hooker, N, Tucker, M, Lejeune, J, & Doohan, D. Targeting the farmer decision making process: A pathway to increased adoption of integrated weed management. Crop Protection. (2009). 0261-2194, 28, 756-764.

[4] Swanton, C. J, & Weise, S. F. Integrated weed management: the rationale and approach. Weed Techonol. (1991). 1550-2740, 5, 657-663.

[5] Radosevich, S, Holt, J, & Ghersa, C. Ecology of weeds and invasive plants: relationship to agriculture and natural resource management. New York: Wiley. (2007). p. 978-0-47016-894-3

[6] Silva, A. A, Ferreira, F. A, Ferreira, L. R, & Santos, J. B. Biologia de plantas daninhas. In: Silva AA, Silva JF. (Eds.). Tópicos em manejo de plantas daninhas. Viçosa: Universidade Federal de Viçosa; (2007). 978-8-57269-275-5, 18-61.

[7] Bennett, A. C, & Shaw, D. R. Effect of Glycine max cultivars and weed control weed seed characteristic. Weed Science. (2000). 1550-2759, 48(4), 431-435.

[8] Bussan, A. J, Burnside, O. C, Orf, J. H, Ristau, E. A, & Puettmann, K. J. Field evaluation of soybean (Glycine max) genotypes for weed competitiveness. Weed Science. (1997). 1550-2759, 45(1), 31-37.

[9] Jannink, J. L, Orf, J. H, Jordan, N. R, & Shaw, R. G. Index selection for weed suppressive ability in soybean. Crop Science. (2000). 1435-0653, 40(4), 1087-1094.

[10] Jordan, N. Prospects for weed control through crop interference. Ecological Applications. (1993). 1051-0761, 3(1), 84-91.

[11] Olofsdotter, M. My view. Weed Science (2000). 1550-2759

[12] Bussan, A. J, Burnside, O. C, Orf, J. H, Ristau, E. A, & Puettmann, K. J. Field evaluation of soybean (Glycine max) genotypes for weed competitiveness. Weed Science (1997). 1550-2759, 45(1), 31-37.

[13] Lamego, F. P, Fleck, N. G, Bianchi, M. A, & Vidal, R. A. Tolerância à interferência de plantas competidoras e habilidade de supressão por cultivares de soja: I. Resposta de variáveis de crescimento. Planta Daninha. (2005). 0100-8358

[14] Bianchi, M. A, Fleck, N. G, Lamego, F. P, & Agostinetto, D. Papéis do arranjo de plantas e do cultivar de soja no resultado da interferência com plantas competidoras. Planta Daninha. (2010). n.spe): 0100-8358, 979-991.

[15] Fleck, NG, Bianch, I MA, Rizzardi, MA, & Agostinetto, D. . Interferência de Raphanus sativus na produtividade de cultivares de soja. Planta Daninha. 2011; 29 (4): 783-792. ISSN 0100-8358.

[16] Costa, J. A. Pires JLF, Thomas AL, Alberton M. Comprimento e índice de expansão radial do hipocótilo de cultivares de soja. Ciência Rural. (1999). 0103-8478, 29(4), 609-612.

[17] Passini, T, & Christoffoleti, P. J. Dourado Neto D. Modelos empíricos de predição de perdas de rendimento da cultura de feijão em convivência com Brachiaria plantaginea. Planta Daninha. (2002). 0100-8358, 20(2), 181-187.

[18] Benjamin, L. R. Variation in time of seedling emergence within populations: a feature that determines individual growth and development. Advances in Agronomy. (1990). 978-0-12000-795-0

[19] Fischer, R. A, & Miles, R. E. The role of spatial pattern in the competition between crop plants and weeds. A theoretical analysis. Mathematical Biosciences. (1973). 0025-5564

[20] Gurevitch, J, Scheiner, S. M, & Fox, G. A. Ecologia vegetal. Porto Alegre: Artmed; (2009). p. 8-53631-918-6

[21] Dieleman, A, Hamill, A. S, Weise, S. F, & Swanton, C. J. Empirical models of pigweed (Amaranthus spp.) interference in soybean (Glycine max). Weed Science. (1995). 1550-2759, 43(4), 612-618.

[22] Carranza, P, Saavedra, M, & Garcia-torres, L. Competition between Ridolfia segetum and sunflower. Weed Research. (1995). 1365-3180, 35(5), 369-375.

[23] Pires JLFCosta JA, Thomas AL. Rendimento de grãos de soja influenciado pelo arranjo de plantas e níveis de adubação. Pesquisa Agropecuária Gaúcha. (1998). 0104-9070, 4(2), 183-188.

[24] Thomas, A. L, & Costa, J. A. Pires JLF. Rendimento de grãos de soja afetado pelo espaçamento entre linhas e fertilidade do solo. Ciência Rural. (1998). 0103-8478, 28(4), 543-546.

[25] Board, J. E, Harville, B. G, & Saxton, A. M. Branch dry weight in relation to yield increases in narrow-row soybean. Agronomy Journal. (1990). 2090-7656, 82(3), 540-544.

[26] Board, J. E, & Harville, B. G. A criterion for acceptance of narrow-row culture in soybean. Agronomy Journal. (1990). 2090-7656, 86(6), 1103-1106.

[27] Legere, A, & Schreiber, M. M. Competition and canopy architecture as affected by soybean (Glycine max) row width and density of redroot pigweed (Amaranthus retroflexus). Weed Science. (1989). 1550-2759, 37(1), 84-92.

[28] Burnside, O. C. Soybean (Glycine max) growth as affected by weed removal, cultivar, and row spacing. Weed Science. (1979). 1550-2759, 27(5), 562-564.

[29] Nice GRWBuehring NW, Shaw DR. Sicklepod (Senna obtusifolia) response to shading, soybean (Glycine max) row spacing and population in three management systems. Weed Technology. (2001). ISNN 1550-2740., 15(1), 155-162.

[30] Young, B. G, Young, J. M, Gonzini, L. C, Hart, S. E, Wax, L. M, & Kapusta, G. Weed management in narrow- and wide-row glyphosate-resistant soybean (Glycine max). Weed Technology. (2001). 1550-2740, 15(1), 112-121.

[31] Silva, A. F, Ferreira, E. A, Concenço, G, Ferreira, F. A, Aspiazu, I, Galon, L, et al. Densidades de plantas daninhas e épocas de controle sobre os componentes de produção da soja. Planta Daninha. (2008). 0100-8358, 26(1), 65-71.

[32] Ni, H, Moody, K, & Robles, R. P. Oryza sativa plant traits conferring ability against weeds. Weed Science. (2000). 1550-2759, 48(2), 200-204.

[33] Galon, L, & Agostinetto, D. Comparison of empirical models for predicting yield loss of irrigated rice (Oryza sativa) mixed with Echinochloa spp. Crop Protection. (2009). 0261-2194

[34] Kissmann, K. G, & Groth, D. Plantas infestantes e nocivas. Tomo II, 2.ed. São Paulo: BASF, (1999). p. 858829902

[35] Rizzardi, M. A, Fleck, N. G, & Ribas, A. V. Merotto Jr A, Agostinetto D. Competição por recursos do solo entre ervas daninhas e culturas. Ciência Rural. (2001). 0103-8478, 31(4), 707-714.

[36] Rizzardi, M. A, Roman, E. S, Borowski, D. Z, & Marcon, R. Interferência de populações de Euphorbia heterophylla e Ipomoea ramosissima isoladas ou em misturas sobre a cultura de soja. Planta Daninha. (2004). 0100-8358, 22(1), 29-34.

[37] Griffin, B. S, Shilling, D. G, Bennett, J. M, & Currey, W. L. The influence of water stress on the physiology and competition of soybean (Glycine max) and Florida Beggarweed (Desmodium tortuosum). Weed Science. (1989). 1550-2759, 37(4), 544-551.

[38] Holm, L. Weeds and water in world food production. Weed Science. (1997). 1550-2759

[39] Patterson, D. T, & Flint, E. P. Comparative water relations, photosynthesis, and growth of soybean (Glycine max) and seven associated weeds. Weed Science. (1983). 1550-2759, 31(3), 318-323.

[40] Procópio, S. O, Santos, J. B, Silva, A. A, & Costa, L. C. Análise do crescimento e eficiência no uso da água pelas culturas de soja e do feijão e por plantas daninhas. Acta Scientiarum. (2002). 1679-9275, 24(5), 1345-1351.

[41] Scott, H. D, & Geddes, R. D. Plant water stress of soybean (Glycine max) and common cocklebur (Xanthium pensylvanicum): A comparison under field conditions. Weed Science. (1979). 1550-2759, 27(3), 285-289.

[42] Seavers, G. P, & Wright, K. J. Crop canopy development and structure influence weed suppression. Weed Research. (2002). 1365-3180, 39(4), 319-328.

[43] Fleck, N. G. Balbinot Jr AA, Agostinetto D, Vidal RA. Características de plantas de cultivares de arroz irrigado relacionadas á habilidade competitiva com plantas concorrentes. Planta Daninha. (2003). 0100-8358, 21(1), 97-104.

[44] Merotto Jr AFischer AJ, Vidal RA. Perspectives for using light quality knowledge as an advanced ecophysiological weed management tool. Planta Daninha. (2009). 0100-8358, 27(2), 407-419.

[45] Santos, J. B, Procópio, S. O, Silva, A. A, & Costa, L. C. Captação e aproveitamento da radiação solar pelas culturas da soja e do feijão e por plantas daninhas. Bragantia. (2003). 0006-8705, 62(1), 147-153.

[46] Procópio, S. O, Santos, J. B, Pires, F. R, Silva, A. A, & Mendonça, E. S. Absorção e utilização do nitrogênio pelas culturas da soja e do feijão e por plantas daninhas. Planta Daninha. (2004). 0100-8358

[47] Anghinoni, I, Volkart, K, Fattore, C, & Ernani, P. R. Morfologia de raízes e cinética da absorção de nutrientes em diversas espécies e genótipos de plantas.Revista Brasileira de Ciência do Solo (1989). 0100-0683

[48] Seibert, A. C, & Pearce, R. B. Growth analysis of weed and crop species with reference to seed weight. Weed Science. (1993). 1550-2759, 41(1), 52-56.

[49] Balbinot Jr AAFleck NG, Agostinetto D, Rizzardi MA, Merotto Jr A, Vidal RA. Velocidade de emergência e crescimento inicial de cultivares de arroz irrigado influenciado a competitividade com as plantas daninhas. Planta Daninha. (2001). 0100-8358, 19(3), 305-316.

[50] Procópio, S. O, Santos, J. B, Pires, F. R, Silva, A. A, & Mendonça, E. S. Absorção e utilização do fósforo pelas culturas da soja e do feijão e por plantas daninhas. Revista Brasileira de Ciência do Solo. (2005). 0100-0683, 29, 911-921.

[51] Silva, A. F, Concenço, G, Aspiazú, I, & Ferreira, E. A. Freitas MA Silva, AA, et al. Período anterior a interferência na cultura da soja-RR em condições de baixa, média e alta infestação. Planta Daninha. (2009). 0100-8358, 27(1), 57-66.

[52] Knezevic, S. Z, Evans, S, & Blankenship, E. E. Acker RCV, Lindquist JL. Critical period for weed control: the concept and data analysis. Weed Science. (2002). 1550-2759, 50, 773-786.

[53] Silva, A. F, Concenço, G, Aspiazú, I, Ferreira, E. A, Galon, L, et al. Interferência de plantas daninhas em diferentes densidades no crescimento da soja. Planta Daninha. (2009). 0100-8358, 27(1), 75-84.

[54] Meschede, D. K. Oliveira Jr RS, Constantin J, Scapim CA. Período Crítico de Interferência de Euphorbia heterophylla na cultura da soja sobre baixa densidade de semeadura. Planta Daninha. (2002). 0100-8358, 20(3), 382-387.

[55] Carvalho, F. T, & Velini, E. D. Período de interferência de plantas daninhas na cultura da soja. I- Cultivar IAC-11. Planta Daninha. (2001). 0100-8358, 19(3), 317-322.

[56] Hart, R. D. El subsistema malezas. In: Hart RD. ed. Conceptos básicos sobre agroecossistemas. Turrialba: CATIE, (1985). , 103-110.

[57] Silva, A. A, Ferreira, F. A, Ferreira, L. R, & Santos, J. B. Métodos de controle de plantas daninhas. In: Silva AA, Silva JF. (Eds.). Tópicos em manejo de plantas daninhas. Viçosa: Universidade Federal de Viçosa; (2007). 978-8-57269-275-5, 64-82.

[58] Chauhan, B. S, Singh, R. G, & Mahajan, G. Ecology and management of weeds under conservation agriculture: A review. Crop Protection. (2012). 0261-2194, 38, 57-65.

[59] Locke, M. A, Reddy, K. N, & Zablotowicz, R. M. Weed management in conservation crop production systems. Weed Biology and Management. (2002). 1445-6664, 2, 123-132.

[60] Lyon, D. J, Miller, S. D, & Wicks, G. A. The future of herbicides in weed control systems of great plains. Journal of Production Agriculture. (1996). 0890-8524, 9, 209-215.

[61] Swanton, C. J, Shrestha, A, Roy, R. C, Ball-coelho, B. R, & Knezevic, S. Z. Effect of tillage systems, N, and cover crop on the composition of weed flora. Weed Science. (1999). 1550-2759, 47, 454-461.

[62] Kelley, K. W. Long Jr JH, Todd TC. Long-term crop rotations affect soybean yield, seed weight, and soil chemical properties. Field Crops Research. (2003). 0378-4290, 83(1), 41-50.

[63] Crookston, R. K, Kurle, J. E, Copeland, P. J, Ford, J. H, & Lueschen, W. E. Rotational cropping sequence affects yield of corn and soybean. Agronomy Journal. (1991). 1435-0645, 83, 108-113.

[64] Meese, B. G, Carter, P. R, Oplinger, E. S, & Pendleton, J. W. Corn/soybean rotation effect as influenced by tillage, nitrogen, and hybrid/cultivar. Journal of Production Agriculture. (1991). 0890-8524, 4, 74-80.

[65] West, T. D, Griffith, D. R, Steinhardt, G. C, Kladivko, E. J, & Parsons, S. D. Effect of tillage and rotation on agronomic performance of corn and soybean: twenty-year study on dark silty clay loam soil. Journal of Production Agriculture. (1996). 0890-8524, 9, 241-248.

[66] Herbert, S. J, & Litchfield, G. V. Growth response of short-season soybean to variations in row spacing and density. Field Crops Research. (1984). 0378-4290, 9, 163-171.

[67] Anaele, A. O, & Bishnoi, U. R. Effects of tillage, weed control method and row spacing on soybean yield and certain soil properties. Soil and Tillage Research. (1992). 0167-1987, 23(4), 333-340.

[68] Knezevic, S. Z, Evans, S. P, & Mainz, M. Row spacing influences the critical timing for weed removal in soybean (Glycine max). Weed Technology. (2003). 1550-2740, 17(4), 666-673.

[69] Teasdale, J. R. Contribution of cover crops to weed management in sustainable agricultural systems. Journal of Production Agriculture. (1996). 0890-8524, 475-479.

[70] Ateh, C. M, & Doll, J. D. Spring-planted winter rye (Secale cereale) as a living mulch to control weeds in soybean (Glycine max). Weed Technology. (1996). 1550-2740, 10, 347-353.

[71] Liebl, R, Simmons, F. W, Wax, L. M, & Stoller, E. W. Effect of rye (Secale cereale) mulch on weed control and soil moisture in soybean (Glycine max). Weed Technology. (1992). 1550-2740, 6, 838-846.

[72] Moore, M. J, Gillespie, T. J, & Swanton, C. J. Effect of cover crop mulches on weed emergence, weed biomass, and soybean (Glycine max) development. Weed Technology. (1994). 1550-2740, 8, 512-518.

[73] Samarajeewa KBDPHoriuchi T, Oba S. Finger millet (Eleucine corocana L. Gaertn.) as a cover crop on weed control, growth and yield of soybean under different tillage systems. Soil and Tillage Research. (2006). 0167-1987, 0167-1987.

[74] Correia, N. M, Durigan, J. C, & Klink, U. P. Influence of type and amount of crop residues on weed emergence. Planta Daninha. (2006). 0100-8358, 24(2), 245-253.

[75] Barnes, J. P, & Putnam, A. R. Rye residues contribute to weed suppression in no-tillage cropping systems. Journal of Chemical Ecology. (1983). 0098-0331, 9, 1045-1057.

[76] Bhowmika, P. C. Inderjit. Challenges and opportunities in implementing allelopathy for natural weed management. Crop Protection. (2003). 0261-2194, 22(4), 661-671.

[77] Trezzi, M. M, & Vidal, R. A. Potential of sorghum and pearl millet cover crops in weed suppression in the field: II- Mulching effect. Planta Daninha. (2004). 0100-8358, 22(1), 1-10.

[78] Jakelaitis, A, Ferreira, L. R, Silva, A. A, Agnes, E. L, & Miranda, G. V. Machado AFL. Weed population dynamics under different corn and bean production systems. Planta Daninha. (2003). 0100-8358, 21(1), 71-79.

[79] Gazziero DLPPrete CEC, Sumiya M. Manejo de Bidens subalternan aos herbicidas inibidores da acetolactato sintase. Planta Danihna. (2003). 0100-8358, 21(2), 283-291.

[80] Charudattan, R, & Dinoor, A. Biological control of weeds using plant pathogens: accomplishments and limitations. Crop Protection. (2000). 0261-2194, 0261-2194.

[81] Zimdahl, R. L. WEEDS/Weed Technology and Control. IN: Murphy DJ, Thomas B, Murray BG. Encyclopedia of Applied Plant Sciences. 978-0-12227-050-5Academic Press, (2000). p.

[82] Oliveira Jr RSConstantin JI, Costa JM, Cavalieri SD, Arantes JGZ, Alonso DG, et al. Interaction between burndown systems and post-emergence weed control affecting soybean development and yield. Planta daninha. (2006). 0100-8358, 24(4), 721-732.

[83] Procópio, S. O, & Pires, F. R. Menezes CCE, Barroso ALL, Moraes RV, Silva MVV et al. Efeitos de dessecantes no controle de plantas daninhas na cultura da soja. Planta Daninha. (2006). 0100-8358, 24(1), 193-197.

[84] Puricelli, E, & Tuesca, D. Weed density and diversity under glyphosate-resistant crop sequences. Crop Protection, (2005). 0261-2194, 2, 533-542.

[85] Tuesca, D, & Puricelli, E. Effect of tillage systems and herbicide treatments on weed abundance and diversity in a glyphosate resistant crop rotation. Crop Protection. (2007). 0261-2194, 26(12), 1765-1770.

[86] Arregui, M. C, Scotta, R, & Sánchez, D. Improved weed control with broadleaved herbicides in glyphosate-tolerant soybean (Glycine max). Crop Protection. (2006). 0261-2194, 25(7), 653-656.

[87] Ellis, J. M, & Griffin, J. L. Benefits of soil-applied herbicides in glyphosate-resistant soybean (Glycine max). Weed Technology. (2002). 1550-2740, 16, 541-547.

[88] Reddy, K. N. Weed control in soybean (Glycine max) with cloransulam and diclosulam. Weed Technology. (2000). 1550-2740, 14, 293-297.

[89] Harger, A. G, Wax, L. M, Bollero, G. A, & Simmons, F. W. Common waterhemp (Amaranthus rudis Sauer) management with soil-applied herbicides in soybean (Glycine max (L.) Merr.) Crop Protection. (2002). 0261-2194, 21(4), 277-283.

[90] Norsworthy, J. K. Broadleavedweedcontrol in wide-row soybean (Glycine max) using conventional and glyphosate herbicide programmes. Crop Protection. (2004). 0261-2194, 23(12), 1229-1235.

[91] Deytieux, V, Nemecek, T, Knuchel, R. F, Gaillard, G, & Munier-jolain, N. M. Is the weed management efficient for reducing environmental impacts of crop systems? A case study based on life cycle assessment. Europ. J. Agronomy. (2012). 1161-0301, 36, 55-65.

[92] Wilson, R. S, Hooker, N, Tucker, M, Lejeune, J, & Doohan, D. Targeting the famer decision making process: A pathway to increased adoption of integrated weed management. Crop Protection. (2009). 0261-2194, 28, 756-764.

[93] Harker, K. N, Clayton, G. W, Blackshaw, R. E, Donovan, O, & Stevenson, J. T. FC. Seeding rate, herbicide timing and competitive hybrids contribute to integrated weed management in canola (Brassica napus). Canadian Journal of Plant Science. (2003). 0008-4220, 83, 433-440.

[94] Bernads, M. L, Gaussoin, R. E, Klein, R. N, Knezevic, S. Z, Lyon, D, Sandell, L. D, et al. Guide for weed management in Nebraska. EC-130. Lincoln, NE. Extension, University of Nebraska- Lincoln; (2009).

[95] Powles, S. B, & Shaner, D. L. Hebicide resistance and world grains. ((2001). CRC-Press, Printed in the USA, 328p. ISBN/84932-2197

[96] Ferreira, E. A, Germani, C, Vargas, L, Silva, A. A, & Galon, L. Resistência de Lolium multiflorum ao Glyphosate. In: Agostinetto D, Vargas L. Resistência de plantas daninhas no Brasil. Passo Fundo: Gráfica Berthier; , 271-289.

[97] Rodrigues, B. N, & Almeida, F. S. Guia de Herbicidas, 4 ed., Londrina: (1998). p. 859053211

[98] Silva, A. A, Ferreira, F. A, Ferreira, L. R, & Santos, J. B. Herbicidas: Resistência de plantas daninhas. In: Silva AA, Silva JF. (Eds.). Tópicos em manejo de plantas daninhas. Viçosa: Universidade Federal de Viçosa; (2007). 978-8-57269-275-5, 279-324.

[99] Vargas, L, Bianchi, M. A, Rizzardi, M. A, & Agostinetto, D. Dal Magro T. Buva (Conyza bonariensis) resistente ao glyphosate na região Sul do Brasil. (2007). Planta Daninha. (2007). 0100-8358, 25(3), 573-578.

[100] Moreira, M. S, & Nicolai, M. Carvalho SJP, Christoffoleti PJ. Resistência de Conyza canadensis e C. bonariensis ao herbicida ghyphosate. Planta Daninha. (2007). 0100-8358, 25(1), 157-164.

[101] Lamego, F. P, & Vidal, R. A. Resistência ao glyphosate em biótipos de Conyza bonariensis e Conyza canadensis no Estado do Rio Grande do Sul, Brasil. Planta Daninha. (2008). 0100-8358, 26(2), 467-471.

[102] Weed Science- International Survey Of Herbicide Resistant WeedsDisponível em: <http://www.weedscience.org/in.asp>.acessed 17 march (2012).

[103] Guaratini, M. T. Toledo REP, Christoffoleti PJ. Alternativas de manejo de populações de Bidens pilosa e Bidens subalternans resistentes aos herbicidas inibidores da ALS. In: Congresso Brasileiro da Ciencia das Plantas Daninhas, 25, 2006, Resumos expandidos.... Brasília: SBCPD, (2006). p. (CD-ROM).

Chapter 3

Nutritional Requirements of Soybean Cyst Nematodes

Steven C. Goheen, James A. Campbell and
Patricia Donald

Additional information is available at the end of the chapter

1. Introduction

Soybeans [*Glycine max*] are the second largest cash crop in US Agriculture, but the soybean yield is compromised by infections from *Heterodera glycines*, also known as Soybean Cyst Nematodes [SCN]. SCN are the most devastating pathogen or plant disease soybean farmers confront. This obligate pathogen requires nutrients from the plant to complete its life cycle. To date, SCN nutritional requirements are not clearly defined. Growth media supporting SCN still contain soy products. Understanding the SCN nutritional requirements and how host plants meet those requirements should lead to the control of SCN infestations. The nutritional requirements of SCN are reviewed in this chapter and those requirements are compared to those of other nematodes. Carbohydrates, vitamins, amino acids, lipids, and other nutritional requirements are discussed.

The survival of parasitic nematodes requires adequate nutrition. These essential nutrients are at least partially supplied by the host. But, availability of nutrients may not alone be sufficient for survival and reproduction. The parasite must also be able to establish a feeding site. Both the establishment of the feeding site and the presence of adequate nutrients for the soybean cyst nematode [SCN] are discussed below.

1.1. Feeding site establishment

Nematodes have differing mouth part structures which are adapted to their food source [1]. In the case of plant-parasitic nematodes, a stylet [analogous to a hypodermic needle], is used to puncture plant cells and a pump mechanism located in the nematode esophagus allows for exchange of fluids between the nematode and plant [1]. Most studies of the economically important root-knot and cyst-forming plant-parasitic nematodes have focused on what fluids are secreted by the nematode and how this facilitates establishment of a feeding site [2-4].

Specific information on the essential nutrients provided by the plant is lacking. In this chapter we focus on what is known about nutrient requirements for soybean cyst nematode, SCN.

The SCN is an obligate parasite requiring a host plant to complete its life cycle (see Figure 1). The cysts are found in the soil and contain eggs and first stage juveniles. The second stage juvenile hatches from the egg and penetrates plant roots. If the roots are a plant that is a host for SCN, the third and fourth stage juveniles molt into an enlarged shape called a sausage once a feeding site is successfully established where the primary goal is removing nutrients from the plant for use by the nematode. After enough nutrients have been obtained by the nematodes, those destined to become males molt into a worm-shape again and migrate out of the roots in search of a female. As the females mature, their size increases breaking root epidermal cells and the nematode is exposed to the soil where she emits pheromones to attract the males already in the soil. Once fertilization of the eggs has occurred, the female dies and her hardened body becomes the cyst which protects the eggs from environmental extremes and organisms which can kill the eggs. Some eggs are extruded into the soil in a gelatinous matrix and these eggs are thought to hatch once conditions favor hatch. The eggs within the cyst go through diapause and can survive within the cyst for more than a dozen years under the right conditions. Juveniles which enter nonhost plant roots may molt into a third stage juvenile but a successful feeding site will not be established and the plant will recognize the nematode as an invader and form necrotic cells surrounding the nematode effectively killing the nematode. Alternatively, some plants are slower to recognize the nematode as an invader and a molt to the third stage may occur but no further development of the nematode will occur. Once the nematode reaches the sausage stage, it lacks the muscles to leave the root and it dies.

As an important crop in the United States [5], there are over 120 soybean lines which have some level of resistance to SCN [6]. Commercial soybean varieties primarily contain one or more different sources of resistance but 95% of all resistance is found from one source, PI 88788. Peking [PI 548402] and Hartwig [PI 437654] are also found in a few commercial varieties. Genetics of resistance is complex with multiple genes involved and interaction of minor genes or nongenetic sources complicates understanding of the process. In a resistant reaction, cytological changes occur and these have been documented [7-19]. Initial reaction to the nematode during the formation of the syncytium in both susceptible and certain resistant lines is identical for the first 4 days after infection [7. 9. 11]. Resistant reactions can be seen about day 4-5 [7, 9-11].

Cyst nematode juveniles hatch from eggs within the cyst or in the soil and enter plant roots typically in the zone of root elongation. They migrate to the pericycle and establish a feeding site [20]. Cellulases break polysaccharide chains and associated proteins in the plant cell walls. Other enzymes have been shown to be secreted by the nematodes as they move through plant tissue [21]. Rapid response by the plant to the nematode inhibits formation of a successful feeding site. A successful feeding site initiation results when the plant fails to respond or responds slowly to the presence of the nematode. One of the ways plant-parasitic nematodes protect themselves from plant responses to the nematodes is through secretion of peroxiredoxin, glutathione periosidase, and secreted lipid binding proteins within the surface coat of the nematode [22]. Although considerable knowledge is now available on the morphological

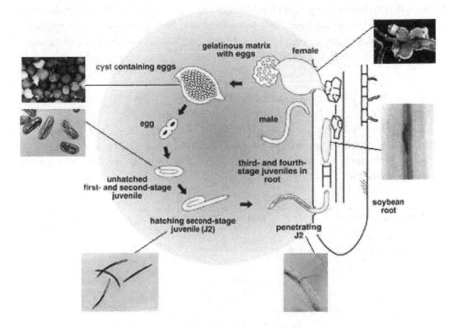

Figure 1. The life cycle of the soybean cyst nematode (SCN) is shown. Soil contains cysts with eggs as well as first stage juveniles. Second stage juveniles hatch from eggs and can then penetrate plant roots. The third and fourth stage juveniles feed off the plant. Males migrate out of the roots in search of a female. Maturing females rupture the root, releasing pheromones to attract males from soil. Females die after egg fertilization and her body becomes the cyst. This figure was obtained with permission from www.extension.umn.edu.

changes in the plant cells due to the presence of the nematode feeding site and molecular studies have advanced our understanding of the interactions on a molecular level, the details of host specificity are unknown [23].

Information is available on the changes that occur within soybean plants when a compatible interaction between SCN and the plant occur. Information is also present on incompatible reactions when plant resistance inhibits SCN reproduction through either a hypersensitive response or formation of small syncytia which limit SCN reproduction. Infection of plant-parasitic nematodes is thought to alter plant products from the shikimic pathway. Infection by SCN increases the concentration of glucose, K, Ca and Mg in the roots but information is not available on whether these increases are products SCN then extracts from plant cells or whether these are responses by the plant to the presence of the nematode.

1.2. Nutritional requirements

Heterodera glycines is considered to have a wide host range. Riggs and Hamblen tested 1152 entries from the Leguminosae family and found that 399 of these entries from 23 genera were susceptible. Poor hosts included 270 entries in 12 other genera [24]. Additional host studies have been conducted by Riggs and Hamblen [25-26], Miller and Gray [27-28], Venkatesh et al, [29], and Venkatesh et al [30]. Variability in host status within a plant species potentially makes identification of necessary nutrients required for establishment of the obligate feeding site easier but to date the specifics have eluded scientists.

A summary of the plants invaded by SCN are shown in Table 1. Most hosts of SCN are legumes and are limited to three subfamilies of the Leguminosae; however, approximately 50 genera in 22 families including nonlegumes are also hosts [31-32]. Some plants allow SCN to penetrate plant roots but limit reproduction of SCN [33]. The reason for this could be nutritional, or it could be due to other barriers within the plant. To determine which of those two possibilities are controlling virulence of SCN, nutritional requirements should be investigated more fully.

Host Common Name	Host Scientific Name	Use
azuki bean	*Vigna angularis*	edible
. bean tree	*Laburnum sp*	ornamental
beans, green, dry	*Phaseolus vulgaris*	edible
beard tongue	*Penstemon digitalis*	ornamental
begger tick	*Desmodium ovalifolium*	weed
bells of Ireland	*Mollucella laevis*	ornamental
bitter cress	*Barbarea vulgaris*	spice
bladder senne	*Colutea arborescens*	shrub -ornamental
bush clover	*Lespeza capitata*	prairie plant
California burclover	*Medicago hispida*	weed
common chickweed	*Stellaria media*	weed
common lespedeza	*Lespedeza striata*	weed
coral bells	*Heuchera sangiunea*	ornamental
cranesbill	*Geranium maculatum*	weed
largeflowered beardtongue	*Penstemon gradiflorus*	wildflower
field pea tuberous vetch	*Lathyrus tuberosus*	edible/weed
fennugreek	*Trigonella goenum-gracum*	spice
foxglove	*Digitalis sp.*	weed
. geranium	*Pelargonium sp*	ornamental
gold apple	*Lycopersicon esulentum*	weed
golden chain	*Laburnum anagyroides*	ornamental

Host Common Name	Host Scientific Name	Use
grass pea vine	*Lathyrus sativa*	edible/ornamental
green pea	*Pisum sativum*	edible
hairy vetch	*Vicia villosavillosa*	forage /cover crop
hemp sesbania	*Sesbania exaltata*	weed
henbit	*Lamium amplexicaule*	weed
hog peanut	*Amphicarpa bracteata*	weed
Indian joint vetch	*Aeschynomene virginica*	weed
indigo	*Indigofera parodiana*	shrub/herbaceous/small tree
clover Kenyan clover	*Trifolium*	ornamental
Korean lespedeza	*Lespedeza stiulacea*	forage
lance leaf rattlebox	*Crotalaria lanceolata*	weed
large flowered beard tongue	*Penstemon grandiflorus*	wild flower
large leaf lupine	*Lupinus polyphyllus*	wild flower
licorice milk vetch	*Astragalus glaucophyllus*	forage
little bur clover	*Medicago minima*	weed
milk vetch	*Astragalus canadensis*	forage
milky purslane	*Euphorbia supino*	weed
mouse ear chickweed	*Cerastium vulgatum*	weed
Common mullein	*Verbascum thapsus*	weed
nasturtium	*Tropaelum pergrinum*	ornamental
old field toadflax	*Linaria canadensis*	weed
pigeon pea	*Cajanus cajan*	edible
Americana pokeweed	*Phytolacca*	weed
purple deadnettle	*Lamium purpureum*	weed
purslane	*Portulaca oleracea*	weed
rainbow pink	*Dianthus chinensis*	ornamental
river bank lupine	*Lupinus rivularis*	edible
Rusian sickle milk vetch	*Astragalus falcate*	weed
service lespedeza	*Lespedeza cuneata*	weed
shrub lespedeza	*Lespedeza bicolor*	ornamental

Host Common Name	Host Scientific Name	Use
Siberian pea tree	*Caragana arborescens*	ornamental
sicklepod	*Cassia tora*	weed
small flowered buttercress	*Cardamine parviflora*	weed
soybean	*Glycine max*	edible
Spanish broom	*Spartium junceum*	ornamental
speedwell	*Veronica peregrine*	weed
spider flower	*Cleome spinosa*	ornamental
spotted burclover	*Medicago arabica*	forage
stinking clover	*Cleome serrulata*	weed
sweet clover	*Melilotus taurica*	weed
sweet pearl lupine	*Lupinus mutabilis*	edible
tiny vetch	*Vicia hirsute*	ornamental vine
white horsehound	*Marrubium vulgare*	medicinal plant
white lupine	*Lupinus albus*	livestock feed
white pea	*Lathyrus ochrus*	wild flower
Wilcox penstemon	*Penstemon wilcoxi*	wilflower
winged pigweed	*Cycloloma atriplicifolia*	weed
yellow lupine	*Lupinus lateus*	wild flower

Table 1. Common names for plants that have been identified as good hosts for soybean cyst nematode [24-31].

In many ways, it is inappropriate to compare humans to nematodes. But, from a nutritional perspective, much more is known about human nutrition than what is known about nutritional requirements of nematodes. For humans, numerous biochemical and mineral components are essential nutrients. But, for nematodes, only a few are known. Yet, nematodes have a comparatively simple digestive system. So, it would be reasonable to predict that nutritional requirements for these organisms are more extensive than what is currently known.

It is also inappropriate to generalize nutritional needs from studies on one nematode to all the nematodes within the various trophic categories. Certainly there should be similarities, but it is clear from the literature that animal parasitic nematodes have different needs from the plant parasites. And, it may also be that those plant parasites infecting specific organisms, such as SCN might have nutritional needs that synergize with the contents of the host soybean plant.

Survival is best understood when chemically defined culture media can be shown to not only sustain life, but also to promote reproduction. Chemically defined media have been identified for the survival of some nematodes and this work has recently been reviewed [34]. The successful media originally included all the amino acids in *Escherichia coli*, and in the amino acid ratios found in *E. coli*. Nematode growth media has been since modified to include a greater number of constituents including glucose, minerals, growth factors, nucleic acid precursors, vitamins, a sterol and heme source. However, SCN has not yet been shown to survive or reproduce on these media. Currently, the only growth media known to sustain SCN includes soy products [35].

Articles published on the nutritional requirements of a wide range of nematodes, generally do not specify SCN [1. 36-37]. While a few nutritional requirements for individual nematode species have been studied, these requirements are limited and their applicability to SCN is unknown. It is assumed that plant- and animal-parasitic nematodes may have different nutritional requirements from entomopathogenic, and microbivorous nematodes.

2. Lipids

Lipids consist of many non-water soluble components including free fatty acids, phospholipids, triglycerides, sterols, and other species. Many of these classes have been studied at least in one host-nematode relationship and are the most studied with the exception of nucleic acids due to their great structural variety and importance as food reserves. For example, Krusberg [38] reported the total lipids and fatty acids from 5 species of plant parasitic nematodes, and their common hosts. They found that the nematodes had the same fatty acids as the hosts, with the exception of the polyunsaturated fatty acids. These appeared to be synthesized by the nematodes. There was also some speculation that nematode fatty acid synthesis resembled that of bacterial pathways rather than that of higher animals. It was not clear from the study whether intestinal flora of the nematode could have been at least partially responsible for this difference, or whether the nematode itself synthesized the fatty acids. Some nematodes are clearly capable of synthesizing longer chain fatty acids from shorter chain precursors. They are also capable of desaturating the fatty acids [39].

Entomopathogenic nematodes infecting locusts consume host fat and protein [40]. A decrease in lipid reserves has been seen in starved nematodes which can be related to decreased infectivity [41]. Lipid content is also known to decrease when nematodes come out of anhydrobiosis [42]. Lipids associated with the nematode surface [cuticle] are triacylglycerols, sterols, specific phospholipids, and other glycolipids [43-45].

The most widely known class of essential nutrients for nematodes is sterol [36,46]. This nutritional requirement was first discovered by Dutky et al. [47] and thought to be potentially a means for control of plant parasitic nematodes. A recent review further confirms this nutritional sterol requirement for the nematode *C. elegans* [48]. Nematode parasites of animals also require sterol for larval development [49]. The biochemical mechanism which converts sitosterol to cholesterol appears to be lacking in nematodes [50]. Nematodes are capable of modifying sterols obtained from their diet [46] but degradation of sterols to CO_2 by nematodes is not clear [51]. More than 63 sterols have been identified from free-living and plant-parasitic nematodes. Characteristics of sterols which can be used by nematodes include those which have a hydroxyl group at C-3, a trans-A/B ring system and an intact nonhydroxylated side chain but lack methyl groups at C-4 [52]. Plant sterols are different than animal sterols with plants being unique in methyl, ethyl or related alkyl groups at the C-24 position of the sterol side chain [52]. There are also differences between plant sterols and plant-parasitic nematode sterols. These findings suggest that nematodes ingest plant sterols and remove the C-24 side chain. In addition, the nematode saturates the double bonds in the four-membered ring system to produce stannols [52]. Steroid hormones are important in development processes and in transition to different life stages [53]. Most likely genetic and biochemical methods will be needed to determine the function of hormones found in nematodes [54]. Novel genes involved in the production of 17β-hydroxysteroid dehydrogenase in the soybean cyst nematode have been reported [55].

Sterols were first reported in soy oil by Kraybill et al. [56]. Formononetin is an o-methyl-isoflavone mainly produced in legumes, including soybean plants [57]. It helps stimulate the production of steroids in mammals, and possibly also in nematodes. Research in this area by the USDA was reviewed by Chitwood [58].

3. Amino acids and proteins

There are no clearly defined requirements for proteins, amino acids, or peptides for SCN. However, it is unlikely that nematodes synthesize all the amino acids. For humans, there are 9 essential amino acids [phenylalanine, valine, threonine, tryptophan, isoleucine, methionine, leucine, lysine, and histidine]. Some others are required under special circumstances [arginine, cysteine, glutamine, proline, serine, tyrosine, and asparagiene]. Cysteine, tyrosine, and arginine are required during rapid growth, such as in infancy. And, arginine, cysteine, glycine, glutamine, histidine, proline, serine and tyrosine are required by some individuals because these amino acids are not adequately synthesized by these individuals. These are essential components for the synthesis of many essential enzymes and structural proteins ; it is anticipated there are similar needs in the nematode diet.

Protein consumed by parasitic nematodes can severely damage the host. Juveniles have high protein requirements and consuming the host protein can severely weaken the plant [46].

There have been efforts to identify the essential amino acids of nematodes [59-61], but so far common requirements have not been identified. However, protein synthesis in cotton roots is modified when the root-knot nematode [RKN] infects susceptible plants. These plant-parasitic nematodes influence the distribution of amino acids in cotton root galls [61]. Also, there is one genetic modification of the cotton plant which makes them less susceptible to infection by the RKN. This modification is responsible for the synthesis of a 14 kDa protein [60].

For the snail parasitic nematode, *Rhabditis maupasi*, five essential amino acids have been identified. These include lysine, methionine, phenylalanine, tryptophane, and valine [62]. In the entomophilic locust parasite, *M. migrescens*, essential nutrients include protein nitrogen [63]. Essential amino acids have also been identified for the nematode *C. briggsae* [64].

4. Vitamins

There are 13 essential vitamins required by humans. These include Vitamin A [Retinol] Vitamin B_1 [Thiamine] Vitamin C [Ascorbic acid] Vitamin D [Calciferol] Vitamin B_2 [Riboflavin] Vitamin E [Tocopherol] Vitamin B_{12} [Cobalamins] Vitamin K_1 [Phylloquinone] Vitamin B_5 [Pantothenic acid] Vitamin B_7 [Biotin] Vitamin B_6 [Pyridoxine] Vitamin B_3 [Niacin] Vitamin B_9 [Folic acid]. Of these, vitamin E is known to be a nutritional requirement for the gastrointestinal parasite, *Heligmosomoides bakeri* [65], and several of the B vitamins are known to be essential nutrients of *C. elegans* [66-68].

For SCN, DNA sequences responsible for the biosynthesis of enzymes that can produce some of the B vitamins *de novo* have been discovered [69]. Therefore, SCN may not need the same B vitamins as *H. bakeri*, for example. And, it is likely that there are other differences in vitamin and supplement requirements across all nematodes.

5. Minerals

Considerable research on mineral requirements for nematodes has been reported in mammalian parasites. For example, the gastrointestinal nematode, *H. bakeri*, requires boron [70], zinc [71], and selenium [65] for survival. And, other nematodes have similar mineral requirements [72-74]. For example, magnesium, sodium, potassium, manganese, calcium and copper are required nutrients of *C. elegans* [5]. However, SCN mineral requirements remain unclear.

Whether minerals, influence nematode survival may not help in their control if necessary minerals are readily available in soil, and essential to the host organisms. But, elements not essential to survival of the host could be controlled in soils to help control SCN survival.

6. Carbohydrates

Nematodes require carbohydrates for energy, usually in the form of glycogen. One study showed that several different carbohydrates were sufficient to provide a carbon, or energy source for *C. elegans*, and that glucose was more effective than fructose or sucrose [76]. For *C. elegans*, glucose along with cytochrome c and β-sitosterol were sufficient to sustain a healthy population.

One of the most striking features of soybean chemistry is the abundance of pinitol [77-79]. Pinitol is a carbohydrate with unusual nutritional properties [77]. Figure 2 shows a total ion chromatogram of a derivatized extract of soybean roots. It is unusual for a plant to have so much pinitol. The levels shown in this study indicate pinitol is present at a concentration of 26 mg/g (dry weight) compared to peanuts with only 4.7 mg/g or clover with 14 mg/g [79]. However, there is no evidence that pinitol, or any of the related inositols are needed for SCN survival [79].

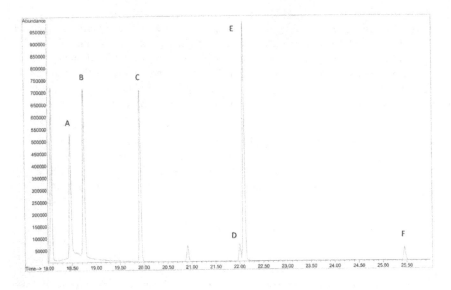

Figure 2. A total ion chromatogram of derivatized soybean root extract is shown. A = D-(-)-Fructose, B = D-Pinitol, C = D-(+)-Glucose, D = D-*chiro*-inositol, E = β-D-(+)-Glucose, F = *Myo*-inositol. Reproduced with permission from [79].

7. Other nutrients or feeding requirements

The nematode *Rhabditis maupasi* requires hemin or another iron porphyrin for survival [62]. Similarly, *C elegans* also requires a heme source for survival [34]. It is likely that many other nematodes require heme, or a closely related hemin. There is also good evidence that SCN requires a heme source [80].

8. Discussion

In comparison to our knowledge of human nutrition, our understanding of nutritional requirements of SCN is in its infancy. Limited information is available for members of the Nematoda Phyllum, but such a small amount of information is available that extrapolation across trophic groups and even within genera may be misleading. Finding a successful artificial diet would be a reasonable first step in defining the nutritional needs of SCN. But, this data needs to be coupled with a good understanding of feeding site establishment and plant responses to SCN infections.

Studying biochemical pathways would be a valuable approach, and could also help identify pathways that could be blocked to help minimize SCN survival. Our laboratory began by examining the chemistry of the plant to identify unique nutrients necessary for SCN survival, but that approach was not immediately successful. Another approach is to continue to use DNA mapping to better understand potential plant and parasite pathways. While this approach is less direct, it is currently a very active area of investigation, and can reveal more information than simply nutritional requirements.

Details of the SCN host-parasite responses during infection and feeding site establishment have been more extensively investigated than nutritional requirements. Relationships between the available nutrients from host plants compared to non-hosts could provide valuable clues on these requirements. And, once an adequate media for SCN survival has been well defined, methods to control this pest should follow.

Acknowledgements

The authors acknowledge support from the USDA and Battelle.

Author details

Steven C. Goheen[1], James A. Campbell[1] and Patricia Donald[2]

1 Pacific Northwest National Laboratory, Richland, Washington, USA

2 U. S. Department of Agriculture/ARS, Jacksonville, Tennessee, USA

References

[1] Munn, E.A. Munn, P. D. Feeding and digestion in Lee, D. L. (ed). The Biology of Nematodes. Taylor and Frances. New York; 2002. p211-232.

[2] Davis, E. L., Hussey, R. S., Baum, T. J. Getting to the roots of parasitism by nematodes. *Trends in Parasitology*. 2004; 20:134-141.

[3] Hussey, R. S., Mims, C. W. Ultrastructure of esophageal glands and their secretory granules in the root-knot nematode, *Meloidogyne incognita*. *Protoplasma*. 1990; 156: 9-18.

[4] Wyss, U. Feeding behavior of plant-parasitic nematodes.. In: Lee, D. L. (ed.). The Biology of Nematodes. Taylor and Francis: New York; 2002. p233-259.

[5] Wrather, A., Koenning, S. Effects of Diseases on Soybean Yields in the United States 1996 to 2007. *Plant Management Network*. 2009; (http://www.plantmanagementnet-work.org/pub/php/research/2009/yields/).

[6] Shannon, J.G., Arelli, P. R., Young, L. D. *Breeding for resistance and tolerance. In Schmitt, D. P., Wrather, J. A., Riggs, R. D. eds, Biology and Management of Soybean Cyst Nematode, Ed 2. Schmitt & Associates of Marceline, Marceline, MO*, 2004: p155–180.

[7] Endo, B.Y. Histological responses of resistant and susceptible soybean varieties, and backcross progeny to entry development of *Heterodera glycines*. *Phytopathology*. 1965; 55:3.75-381.

[8] Endo, B.Y. Ultrastructure of initial responses of resistant and susceptible soybean roots to infection by *Heterodera glycines*. *Revue of Nematology*. 1991; 14:73-94.

[9] Riggs, R.D., Kim, K.S., Gipson, I. Ultrastructural changes in Peking soybeans infected with *Heterodera glycines*. *Phytopathology*. 1973; 63:76-84.

[10] Acido, J. R., Dropkin, V. H., Luedders, V. D.. Nematode population attrition and his-topathology of *Heterodera glycines*-soybean associations. *Journal of Nematology*. 1984; 16:48-57.

[11] Kim, Y. H., Riggs, R. D., Kim, K. S. Structural changes associated with resistance of soybean to *Heterodera glycines*. *Journal of Nematology*. 1987; 19:177-187.

[12] Halbrendt, J.M., Lewis, S.A., Shipe, E. R.. A technique for evaluating *Heterodera gly-cines* development in susceptible and resistant soybean. *Journal of Nematology*. 1992; 24:84-91.

[13] Kim, Y. H., Riggs, R. D. Cyclopathological reactions of resistant soybean plants to nematode invasion. in Wrather, J.A., Riggs, R. D. (eds). Biology and management of the soybean cyst nematode. APS Press. St. Paul. 1992; p157-168.

[14] Mahalingam, R., Skorupska, H.T.. Cytological expression of early response to infec-tion by *Heterodera glycines* Ichinohe in resistant PI 437654 soybean. *Genome*. 1996; 39:986-998.

[15] Klink, V.P. Overall, C. C., Alkharouf, N., MacDonald, M. H., Matthews, B. F. Laser capture microdissection (LCM) and comparative microarray expression analysis of synctial cells isolated from incompatible and compatible soybean roots infected by soybean cyst nematode (*Heterodera glycines*). *Planta*. 2007; 226:1389-1409.

[16] Klink, V.P., Overall, C. C., Alkharouf, N., MacDonald, M. H., Matthews, B. F. A com-parative microarray analysis of an incompatible and compatible disease response by soybean (*Glycine max*) to soybean cyst nematode (*Heterodera glycines*) infection. *Plan-ta*. 2007; 226:1423-1447.

[17] Klink, V.P., Hosseini, P., Matsye, P., Alkharouf, N., Matthews, B. F. A gene expres-sion analysis of syncytia laser microdissected from the roots of the *Glycine max* (soy-

bean) genotype PI 548402 (Peking) undergoing a resistant reaction after infection by *Heterodera glycines* (soybean cyst nematode). *Plant Molecular Biology.* 2009; 71:525-567.

[18] Klink, V.P., Hosseini, P., Matsye, P., Alkharouf, N. W. Matthews, B. F. Syncytium gene expression in *Glycine max* (PI 88788) roots undergoing a resistant reaction to the parasitic nematode *Heterodera glycines. Plant Physiological and Biochemistry.* 2010; 48:176-193.

[19] Klink, V.P., Overall, C. C., Alkharouf, N., MacDonald, M. H., Matthews, B. F. Micro-array detection calls as a means to compare transcripts expressed within synctial cells isolated from incompatible and compatible soybean (*Glycine max*) roots infected by the soybean cyst nematode (*Heterodera glycines*). *Journal of Biomedicine and Biotechnology.* 2010:491217 (1-30).

[20] Wyss, U., Stender, C., Lehmann, H.. Ultrastructure of feeding sites of the cyst nematode *Heterodera schachtii* Schmidt in roots of susceptible and resistant *Raphanus sativus* L. var. *oleiformis*. Pers. cultivars. *Physiological Plant Pathology.* 1984; 25: 21-37.

[21] Davis, E. L., Hussey, R. S. Baum T. J., Bakker, J., Schots, A. Nematode parasitism genes. *Annual Review of Phytopathology.* 2000; 38:365-396.

[22] Gheysen, G., Jones, J. T. Molecular aspects of plant-nematode interactions. In Perry, R. N., Moens, M. (eds). *Plant Nematology.* CABI. 2006; p234-254.

[23] Vanholme, B., De Meutter, J., Tytgat, T., Van Montagu, M., Coomans, A. Gheysen, G. Secretions of plant-parasitic nematodes: a molecular update. *Gene.* 2004; 332: 13-27.

[24] Riggs, R.D., Hamblen, M.L. Soybean cyst nematode host studies in the family Leguminosae. Arkansas Agricultural Experiment Station Report Series 110. Fayetteville, AR 1962; 17p.

[25] Riggs, R.D., Hamblen, M.L. Additional weed hosts of *Heterodera glycines. Plant Disease Reporter.* 1966; 50:15-16.

[26] Riggs, R.D., Hamblen, M.L. Further studies on the host range of the soybean cyst nematode. Arkansas Agricultural Experiment Station Bulletin 718. Fayetteville, AR 1966; 19p.

[27] Miller, L.I., Gray, B. J. Reaction of lambsquarter, Swiss chard, and spinach to eleven isolates of the soybean cyst nematode. *Virginia Journal of Science.* 1966; 17: 246.

[28] Miller, L.I., Gray, B. J. Development of eleven isolates of *Heterodera glycines* on six legumes. *Phytopathology.* 1967; 57:647.

[29] Venkatesh, R., Harrison, S. K., Riedel, R. M. Weed hosts of soybean cyst nematode (*Heterodera glycines*) in Ohio. *Weed Technology.* 2000; 14:156-160.

[30] Venkatesh, R., Harrison, S. K., Regnier, E. E., Riedel, R. M. Purple deadnettle effects on soybean cyst nematode populations in no-till soybean. North Central Weed Science. Society. 2004; 59:56.

[31] Riggs, R.D., Host Range. In Riggs, R. D., Wrather, J. A. (eds). Biology and management of the soybean cyst nematode. APS Press, St Paul, MN. 1992; p107-114.

[32] Mock, V.A., Creech, J.E., Davis, V.M., Johnson. W. G. Plant growth and soybean cyst nematode response to purple deadnettle (*Lamium pupureum*), annual ryegrass, and soybean combinations. *Weed Science*. 2009; 57:489-493.

[33] Riggs, R.D., Nonhost root penetrations by soybean cyst nematode. *Journal of Nematology*. 1987; 19:251-254.

[34] Braeckman, B. P., Houthoofd, K., Vanfleteren, J. R. Intermediary metabolism, WormBook, (ed. The C. elegans Research Community), WormBook, doi/10.1895/wormbook. 2009; 1.146.1, http://www.wormbook.org.

[35] Krusberg, L.R. Studies on the culturing and parasitism of plant-parasitic nematodes, in particular *Ditylenchus dipsaci* and *Aphelenchoides rizemabosi* on alfalfa tissues. *Nematologica*. 1961; 6:181-200.

[36] Chitwood, D. J. Biochemistry and function of nematode steroids. *Critical Reviews in Biochemistry and Molecular Biology*. 1999; 34(4) 273-284.

[37] DeLey, P., Mundo-Ocampo, M. Cultivation of nematodes.. In Nematology: Advances and Perspectives Vol 1. Nematode morphology, physiology and ecology. CABI Publishing, Cambridge, Mass. 2004; p541-608.

[38] Krusberg, L. R. Analyses of total lipids and fatty acids of plant-parasitic nematodes and host tissues. *Comparative Biochemistry and Physiology*. 1967; 21: 83-90.

[39] Ruess, L., Haggblom, M. H., Zapta, E. J. G., Dighton, J. Fatty acids of fungi and nematodes – possible biomarkers in the soil food chain?" *Soil Biology & Biochemistry*. 2002; 34: 745-756.

[40] Gordon, R, Webster, J. M., Hislop, T. G. Mermithid parasitism, protein turnover and vetillogenesis in the desert locust, *Schistocerca gregaria* Forskal. *Comparative Biochemistry and Physiology*. 1973; 46B: 575-593.

[41] Barrett, J., Wright, D. J. Intermediary metabolism. In Perry, R. N., Wright, D. J. (eds). The physiology and biochemistry of free-living and plant-parasitic nematodes. CABI. London. 1998; p331-353.

[42] Crowe, J. H., Madin, K. A. C., Loomis, S. H. Anhydrobiosis in nematodes: metabolism during resumption of activity. *Journal of Experimental Zoology*. 1977; 201:57-63.

[43] Blaxter, M. L. Cuticle surface proteins of wild type and mutant *Caenorhabditis elegans*. *Journal of Biological Chemistry*. 1993; 2638: 6600-6609.

[44] Bird, A.F., Bird, J. The Structure of Nematodes. 2nd edition. Academic Press, San Diego. 1991.

[45] Spiegel, Y. and M.A. McClure.. The surface coat of plant-parasitic nematodes: chemical composition, origin, and biological role- a review. *Journal of Nematology*. 1995; 27:127-134.

[46] Chitwood, D. J., Lusby, W. R. Metabolism of plant sterols by nematodes. *Lipids*. 1991; 26 (8): 619-627.

[47] Dutky, S. R., Robbins, W. E., Thompson, J. V. The demonstration of sterols as requirements for the growth, development, and reproduction of the DD-136 nematode. *Nematologica*. 1967; 13: 140.

[48] Entchev, E., Kurzchalia, T. V. Requirement of sterols in the life cycle of the nematode *Caenorhabditis elegans"* Seminars in Cell & Developmental Biology. 2005; 16: 175-182.

[49] Bolla, R.I, Weinstein, P.P., Cain, G.D. Fine structure of the coelomocyte of adult *Ascaris suum*. *Journal of Parasitology*. 1972; 58: 1025-1036.

[50] Cheong, C. M., Na, K., Kim, H., Jeong, S., Joo, H., Chitwood, D. J., Paik, Y. A potential biochemical mechanism underlying the influence of sterol deprivation stress on *Caenorhabditis elegans* longevity. *Journal of Biological Chemistry*. 2011; 286: 7248-7256.

[51] Rothstein, M. Nematode biochemistry. IX. Lack of sterol biosynthesis in free-living nematodes. *Comparative Biochemistry and Physiology*. 1968; 27: 309-317.

[52] Chitwood, D. J. Biosynthesis in Perry, R. N., Wright, D. J. (eds). The physiology and biochemistry of free-living and plant-parasitic nematodes. CABI. London. 1998; p303-330.

[53] Riddle, D.L. and P.S. Albert. Genetic and environmental regulation of dauer larva development. In *C. elegans* II. Riddle, D., Meyer, B., Priess, J., and Blumenthal, T. (ed). Cold Spring Harbor: Cold Spring Harbor Press. 1997; p739-768.

[54] Chervitz, S.A., Aravind, L., Herloc, G. S., Ball, C. A., Koonin, E. V., Dwight, S. S.. Comparison of the complete protein sets of worm and yeast: orthology and divergence. *Science*. 1998; 282: 2022-2028.

[55] Skantar, A. M., Guimond, N. A., Chitwood, D. J. Molecular characterization of two novel 17β-hydroxysteroid dehydrogenase genes from the soybean cyst nematode *Heterodera glycines*. *Nematology*. 2006; 8(3) 321-333.

[56] Kraybill, H. R., Thornton, M. H., Eldridge, K. E. Sterols from crude soybean oil. *Industrial and Engineering Chemistry*. August, 1940:1138-1139.

[57] Medjakovic, S. and Jungbauer, A. "Red Clover Isoflavones Biochanin A and Formononetin are Potent Ligands of the Human Aryl Hydrocarbon Receptor". *The Journal of Steroid Biochemistry and Molecular Biology*. 2008; 108 (1–2) 171–177.

[58] Chitwood, D. J. Research on plant-parasitic nematode biology conducted by the United States Department of Agriculture-Agricultural Research Service." *Pest Management Science*. 2003; 59: 748-753.

[59] Young, J. R., Riggs, R. D. Identification of the free amino acids of nematode resistant and susceptible soybeans. *Arkansas Academy of Science Proceedings*. 1964; 18: 46-49.

[60] Hedin, P. A., Creech, R. G. Altered amino acid metabolism in root-knot nematode inoculated cotton plants. *Journal of Agriculture and Food Chemistry*. 1998; 46: 4413-4415.

[61] Hedin, P. A., Creech, R. G. Effects of root-knot nematodes on distribution of amino acids in cotton root galls. Mississippi State University Experiment Station Bulletin 1103 Office of Agricultural Communications, Mississippi. 2001; p1-6.

[62] Brockelman, C. R., Jackson, G. J. Amino acid, heme, and sterol requirements of the nematode, *Rhabditis maupasi Journal of Parasitology*. 1978; 64(5): 803-809.

[63] Baylis, H. A. Observations on the nematode *Mermis migrescens* and related species. *Parasitology*. 1944; 36: 122-132.

[64] Vanfleteren, J. R. Amino acid requirements of the free-living nematode *Caenorhabditis briggsae*. *Nematologica*. 1973; 19: 93-99.

[65] Smith, A., Madden, K. B., Yeung, K. J. A., Zhao A., Elfrey, Finkelman, , F. J., Lavender, O., Shea-Donohue, T., Uban, J. F. Deficiencies in selenium and/or vitamin E lower the resistance of mice to *Heligmosomides polygyrus* infections. *Journal of Nutrition*. 2005; 135: 830-836.

[66] Liu , C., Lu, N. Biotin requirement and its biosynthesis blockage in the free-living nematode, *Caenorhabditis elegans*. *Federation of American Societies for Experimental Biology*. 2008; 22: 1102-1104.

[67] Nicholas, W. L. Hansen, E., Dougherty, E. C. The B-vitamins required by *Caenorhabditis briggsae (Rhabditidae)*. *Nematologicia*. 1962; 8: 129-135.

[68] Szewczyk, N. J. , Kozak, E., Conley, C. A. Chemically defined medium and *Caenorhabditis elegans*. *BMC Biotechnology*. 2003; 3: 19.

[69] Craig, J. P., Bekal, S. , Niblack, T. , Domier, L., Lambert, K. N. Evidence for horizontally transferred genes involved in the biosynthesis of vitamin b1, b5, and b7 in *Heterodera glycines*. *Journal of Nematology*. 2009; 41(4) 281–290.

[70] Bourgeois, A. C., Scott, M. E., Sabally, K., Koski, K. G. Low dietary boron reduces parasite (nematoda) survival and alters cytokine profiles but the infection modifies liver minerals in mice. *Journal of Nutrition*. 2007; 137(9): 2080-2086.

[71] Shi, H. N., Scott, M. E., Koski, K. G., Boulay, M., Stevenson, M. Energy restriction and severe zinc deficiency influence growth, survival and reproduction of *Heligmosomoides polygyrus* (Nematoda) during primary and challenge infections in mice. *Parasitology*. 1995; 110: 599-609.

[72] Coop, R. L., Kyriazakis, I. Influence of host nutrition on the development and consequences of nematode parasitism in ruminants. *Trends in Parasitology*. 2001; 17(7) 325-330.

[73] Koski, K. G., Scott, M. E. Gastrointestinal nematodes, trace elements, and immunity. *Journal of Trace Elements in Experimental Medicine*. 2003; 16: 237-51.

[74] McClure, S. J., McClure, T. J., Emery, D. I. Effects of molybdenum intake on primary infection and subsequent challenge by the nematode parasite *Trichostromgylus colubriformis* in weaned Merino lambs. *Research in Veterinary Science*. 1999; 67: 17-22.

[75] Lu, N. C., Cheng A. C., Briggs, G. M. A Study of Mineral Requirements in *Caenorhabditis elegans*. *Nematologica*. 1983; 29: 425-434.

[76] Lu, N. C., Goetsch, K. M. Carbohydrate Requirement of *Caenorhabditis elegans* and the Final Development of a Chemically Defined Medium. *Nematologica*. 1993; 39: 303-311.

[77] Campbell, J.A., Goheen, S.C., Donald, P. Extraction and Analysis of Inositols and Other Carbohydrates from Soybean Plant Tissues. in Recent Trends for Enhancing the Diversity and Quality of Soybean Products, Krezhova, D. ed. InTech publisher http://www.intechopen.com/articles/show/title/extraction-and-analysis-of-inositols-and-other-carbohydrates-from-soybean-plant-tissues. 2011; p421 – 446.

[78] Garland, S., Goheen, S., Donald, P.A., Campbell, J. Application of derivatization gas chromatography/mass spectrometry for the identification and quantitation of pinitol in plant roots. *Analytical Letters*. 2009; 42: 2096-2105.

[79] McDonald IV, L. W., Goheen, S. C., Donald, P. A., Campbell J. A. Identification and quantitation of various inositols and o-methylinositols present in plant roots related to soybean cyst nematode host status. *Nematropica*. 2012; 42: 1-8.

[80] Ko, M. P., Huang, P., Huang, J., Barker, K. R. The occurrence of phytoferritin and its relationship to effectiveness of soybean nodules. *Plant Physiology*. 1987; 83: 299-305.

Chapter 4

Weed Management in Soybean — Issues and Practices

Rafael Vivian, André Reis, Pablo A. Kálnay,
Leandro Vargas, Ana Carolina Camara Ferreira and
Franciele Mariani

Additional information is available at the end of the chapter

1. Introduction

Weed management is essential for any current system of agricultural production, especially for large monoculture areas, which exert high pressure on the environment. Soybean is among the largest monoculture registered worldwide, with 102 million hectares harvested only in 2010. The leading countries of production are Argentina, Brazil and the United States, with more than 70% of the total cultivated area. Along with China and India, these five countries represent 90% of all produced soybean. The production incentive is related to growing global demand for oil and protein for food and feed, as well as the feasibility of crops for biodiesel production, extremely important for the global economy.

Meanwhile, weeds are considered the number one problem in all major soybean producing countries. Even with advanced technologies, producers note high losses due to interference by weeds. According to estimates, weeds, alone, cause an average reduction of 37% on soybean yield, while other fungal diseases and agricultural pests account for 22% of losses [1]. In the United States, it is considered that weeds cause losses of several millions of US dollars annually. In Brazil, with an average production of 75 million tons, it is estimated that expenses on weed control represent between 3% and 5% of total production cost, which means more than US$ 1.2 billion used in that country, only for weed chemical control in soybeans.

Disregarding the high cost, weed might be controlled in soybean crop using good management practices of all available methods, combining them in an integrated weed management (IWM). Crop rotation is a rather efficient method, since it allows an easy control of the most troublesome weeds. In order to achieve success on crop rotation, weeds must be managed throughout the growing soybean season. Using full capacity of crop competition is another alternative, yet this tool is often overlooked.

Despite differences between soybean cultivars used worldwide and the main weed species which attack these cultivars, there are many resemblances in management practices and control. The species hairy fleabane, *Conyza bonariensis* (L.) Cronq., horseweed, *Conyza canadensis* (L.) Cronq., goosegrass, *Eleusine indica (L.)* Gaertn., barnyardgrass, *Echinochloa crusgalli (L.)* Beauv., johnsongrass, *Sorghum halepense* (L.) Pers., beggarticks, *Bidens pilosa* L. and common ragweed, *Ambrosia artemisiifolia* L., are common weeds in Argentine, Brazilian and American soybean crops. The burndown and subsequent post-emergence (POST) spraying of crop with glyphosate usually occur from south to north in the American continent, with some distinctions among products used in mixture with glyphosate for managing resistant weeds. All these factors increase the selection pressure even more.

The introduction of GR (glyphosate-resistant) soybean, genetically modified (GM), contributed to standardization of weed management. With a large adoption of this technology, there are many concerns regarding the control and the high selection pressure on common weed species in soybean. In the US, more than 93% of soybean has the GR technology. In Brazil and Argentina, these values represent 80% and 99%, respectively.

The use of very similar technologies as well as the facility of proliferation of weeds has intensified reported herbicide resistance. Since the first report of *E. indica* resistance, in Malasia (1997), 22 species (biotypes) are already not controlled by glyphosate and 10 show multiple resistance. The number of reports increases every year and, in 2011, 7 weed resistance cases were recorded. The evolution of weed resistance to glyphosate also worries members of the Weed Science Society of America, mainly by the spread rate and by the impact on ecosystems.

New technologies derived from genetic alteration of cultivars resistant to herbicides are part of management alternatives to glyphosate. Many of them still under test should be available on short notice. In Brazil, both soybean resistant to ALS (acetolactate synthase) inhibitors and those resistant to 2,4-D should take up areas with a history of weed glyphosate resistance. In the US, besides soybean resistant to dicamba and that resistant to glyphosate + ALS, mixtures are used on crop pre-emergence (PRE), for example, dimethenamid and saflufenacil (new active ingredient). Spraying of encapsulated ingredients (acetochlor) at soybean POST and at weed PRE also come up as management alternatives.

Despite efforts on weed control in soybeans, the benefits of IWM based on preventive and cultural controls will always be fundamental to the maintenance of monocultures. However, it appears that much of what is discussed about IWM is slightly practical, with corrective measures mostly. This chapter aims to present some focal issues related to weed management in soybean growing areas, which include weed potential to cause severe damages and yield losses by weeds, the evolution of resistant weeds in GR soybean monoculture, the soybean management characterization in the main producing countries and discussions about the benefits of IWM use as an accurate control measure. It presents a set of information for researchers and experts on weed management service area, reporting clear and objectively the major impacts of the current management used and the outlook for soybean farming.

2. Implication of weed management in soybean

Weed control is a practice of great importance for obtaining high soybean yields. Weed species is a serious problem for the soybean crops and its control is needed especially in infested sides. Therefore, weed management is an integral part of soybean production. Recently, research has reported that the density and distribution of weed species in the soybean plantations are significant parameters on yield losses. This happens because the weed species competes with the sunlight, water and nutrients, and may, depending on the level of infestation and species, hamper harvesting operations and compromise the quality of soybean grains [2]. Current studies on weed biology are changing, largely due to the effects of agricultural practices on weeds, cropping systems, and the environment. Research emphasis has been altered based on the need to understand basic weed biology [3]. It is our job to predict how weed species, populations, and biotypes evolve in response to selection pressure primarily due to agricultural practices. This knowledge helps developing weed management practices in the soybean crops. Other important biological factors in weed management decisions include weed and crop density, seedbank processes, demographic variation, weed-crop competition, and reproductive biology [4]. Development of economic thresholds for weed species made significant progress in the last decade. Integrated weed management has focused on the effects of crop planting dates, row spacing, cultivators, use of cover crops and reduced herbicide rates.

Selection and adaptation of weed populations occur at the level of the individual. Weeds interfere with crop production, and the yield losses incurred are the aggregate consequence of competition between heterogeneous weed phenotypes and homogeneous crop phenotype [5]. Because weed selection results in diversity, a population of weeds on a field consists of a heterogeneous collection of genotypes and phenotypes that allows exploitation of many niches left available by crops. Weed species respond to these opportunities with an impressive array of adaptions: phenotypes plasticity in response to microsite resource availability, somatic polymorphism of plant and seed form and function, density-dependent mortality (population size adjustment), density-independent mortality (disease, predator, stress resistances), and chemical inhibition of neighbors by allelopathic interference [6]. When all else fails, many weed seeds can remain dormant and extend their life for several years in the soil seedbank, waiting for the right opportunity to grow [7].

Weed populations possess considerable heterogeneity at many levels, consequence of adaptation for colonization and survival. In order to select the most appropriate herbicides or devise the optimum weed control system, one must be able to properly identify the weeds present within a field. Weed identification immediately following emergence is essential since the effectiveness of most herbicides depends on weed size. Maps of weeds by species in fields prior to harvest will aid in the choice of herbicide program for the following year.

2.1. Issues on weed management

All the characteristics cited are essential for soybean weed management. However, starting from the identification of species, three leading questions must be answered in order to suitably handle weeds: i) What are the available tools for weed management? ii) How should one use them for reducing weed interference? and iii) When should one use them?

The available tools are those that enable the reduction of weed-crop competition. It integrates all traditional control — cultural, physical, chemical, among others — and it should be evaluated in accordance with locally grown system. Currently, due to countless resistance cases, preferences are for those that integrate cultural and physical controls together with chemical ones, and the following ones can be cited: no tillage system, crop rotation, using of cover crops, autumnal herbicide management directed to key-weeds, and new GM soybean resistant to herbicide from different modes of action.

All tools should be adapted to use availability, particularly considering the ratio income/investment. Many of these tools are easy to be used and have high impact. The no tillage system, for example, changes weed management completely, so that the mulch formed reduces weed survival [8] and also encourages the germination of negative photoblastic species [9], in addition to all other benefits found in the tropical regions of soybean production [10]. The advantages of no tillage over conventional tillage systems in improving soil quality are generally accepted, resulting in benefits for physical, chemical and biological properties of the soil [11]. Nowadays, no tillage is practiced on over 100 million ha worldwide, mostly in North and South America, but also in Australia and in Europe, Asia and Africa [12,13]. Among the advantages, one can cite the control of soil erosion, moisture conservation, favorable soil temperatures, increased efficiency in nutrient cycling, improvement on soil structure, machinery conservation and time saving in terms of human and animal labor [12,14]. The system also ensured changing among the population of arthropods, which are usually favored by the system because they find greater protection to natural enemies or use many of weed seeds as a feed source.

The crop rotation system constitutes another important management tool, often overlooked by producers. It allows the variation primarily at chemical control. Corn rotating, despite inconvenient profitability decreases, compared with soybean, allows an important POST emergent apply against glyphosate-tolerant weeds, in areas where GR corn is not used. Several studies carried out from 1970 to 1990, associated with cultivation of soybeans in crop rotation systems with diverse grasses (rice, maize, sorghum, wheat, sugar cane) and cotton, have shown that nitrogen residual effect, fixed by soybean crop and its residues, replaces partial the nitrogen on following crop, resulting in field optimization and alleviating part of the production costs [15]. In China, for example, soybean is commonly grown continuously in monoculture rather than rotated with other crops, like maize or wheat. The soybean monoculture results in yield decline, as well as its quality. The yield reduction on soybean in 2, 3 and 4-year monoculture was 15%, 20%, and 30%, respectively [16,17], highlighting the significance of rotational system in the preservation of crop production. Furthermore, several experiments suggest that carbon and nitrogen from microbial biomass (particularly nitrogen) are sensitively affected by soil- and crop-management regimens, being directly influenced by crop rotation [18].

Using cover crops between the main crops (fallow period) is also part of conservation practices and it represents a breakthrough in weed management. Besides, competing against weeds, many cover crops allow using selective herbicide in the fallow period, reducing hard-to-control species. Despite the high costs, it saves on using herbicides along cultivation years for the primary crop, as the infestation plant is reduced by ongoing practice of this system. Nutrient cycling is also favored by means of cover crops, especially for those who exhibit high mobility on the ground, such as nitrogen [19]. For other nutrients, arbuscural mycorhizal development is favored in areas in which cover plants are used. This arbuscural mycorhizal promotes phosphorus absorption [20]. Nitrate loss in annual row crops could also be significantly mitigated by the adoption of no tillage and cover crops or greater reliance on biologically based inputs, according to [21]. In general, cover crops increase the primary productivity of the system and diversify basal resources for higher trophic levels.

However, the selection of proper cover crop is essential for the success of the system. Plant-feeding nematodes, for example, were less abundant in plots with Poaceae cover crops, while bacterivorous, omnivorous and root-hair-feeding nematodes were more abundant with Fabaceae cover crops than with bare soil, indicating that cover crop identity or quality greatly affects soil food web structure [22]. Other species, such as those from genus Desmodium, may be used suppressing *Striga hermonthica* (Del.) Benth. by means of an allelopathic mechanism. Their root exudates contain novel flavonoid compounds, which stimulate suicidal germination of *S. hermonthica* seeds and dramatically inhibit its attachment to host roots [23].

Herbicides, in the broad action spectrum, are and will be essential tools in weed management, even for those with a great number of resistant weeds. But the trend is that using different herbicide is increasingly related to GM crops which show resistance to more than one active ingredient. For new GM soybean, 2,4-D and dicamba resistance traits will always be used in stacks with at least one other herbicide-resistant trait. Glyphosate and ALS trait stack, recently deregulated in the US, possibly will allow the use of ALS-inhibiting herbicides with soil residual that are too phytotoxic to use on conventional crop cultivars [24]. In reference [25], diversification may make weed management more complex, but growers must not use new GM crop resistant to herbicides in the same way that some used initial GM crops, in order to rely only on one herbicide until it is no longer effective and then switch herbicides. Research alerts that "if growers use the new GM crops and the herbicides that they enable properly, GM crops will expand the utility of currently available herbicides and provide long-term solutions to manage resistant weeds".

Answering the question related to the period when control tools should be used, different opinions arise. Many specialists recommend to use tools, especially chemical control, only when economic loss level is reached, ie, when population density finds a minimum threshold at which costs of controlling are lower than economic damage coming from losses by weed interference. Nevertheless, by following the concept of integrated management, it is recom-mended the use of many available tools, even at fallow periods or at low weed densities. In reference [26], as opposed to pest and pathogens which attack crops in epidemic cycles, weeds are endemic, regenerating from the seed and/or vegetative propagules that are introduced into the soil; thus, the continuous management allows the best result. Besides, confining weed management to a narrow temporal window increases the risk of unsatisfying weed manage-ment outcomes due to unfavorable weather [27]. Coupled with this agreement, good management models for weed control may join forces to the definition of weed control periods according to their competitive ability and the local crop conditions set out during the growing (climate, cultivar, sowing density, etc).

So far, absence of management or misuse of control tools may undermine the productivity, the sustainability of system production and the agricultural activity, also interfering in the preservation and balance between species. Thus, interactions among weeds and further organisms (fungi, viruses, bacteria, mites, insects, nematodes, etc.) as well as their handling may have a direct or indirect impact into the production system.

2.2. Impact of weed management on nontarget organisms

Many studies have attempted to relate the intensification of certain pathogenic diseases of shoot plants in areas annually treated with herbicides, being placed on proof the intensive use of those mainly in no tillage system. Glyphosate, for example, is a highly effective broad-spectrum herbicide that is phytotoxically active on a large number of weeds and crop species across a wide range of taxa [28]. Glyphosate inhibits the biosynthesis of aromatic aminoacids, thereby reducing biosynthesis of proteins, auxins, pathogen defense compounds, phytoalex-ins, folic acid, precursors of lignins, flavonoids, plastoquinone, and hundreds of other phenolic and alkaloid compounds [29]. These effects could increase the susceptibility of glyphosate-sensitive plants to pathogens or other stress agents [30]. Engineered to express enzymes that are insensitive to or are able to metabolize glyphosate, GR crops have enabled farmers to easily apply this herbicide in soybean, corn, cotton, canola, sugar beet and alfalfa, besides controlling problematic weeds without harming the crop [28].

For glyphosate and its interspecific transfer from weeds to nontarget organisms, in [31] it was related the increasing remark number of plant diseases growing in long term [32]. But the herbicide influence on disease incidence at glyphosate-resistant crops has varied. While in [33,34] it was observed an increase of *Fusarium solani* (Mart.) Sacc. in soybean, others showed a reduction of *Phakopsora pachyrhizi* Sidow at this crop [35]. For nitrogen-fixing microorganisms in soybean, negative interference of glyphosate has been proven by different authors [36-39], usually in laboratory experiments, with clear differences among rhizobial strains, as well as among glyphosate formulations, having roughly deleterious effects according to combinations of these.

Disease caused by *Sclerotinia sclerotiorum* (Lib.) de Bary, for example, occurs in numerous weeds considered plant hosts. Crop rotation is essential in this case, specially when it uses non host crops and some herbicides with effects over the weed hosts and, consequently, the disease. In [40], the use of chemical weed management with sethoxydim, an important herbicide on soybean system, had the biggest toxicity rate together with cycloxydim. Other herbicides tested, such as cycloxidim and haloxyfop-ethoxy-ethyl, had less impact on *S. sclerotiorum*, but negative action on *Trichoderma* sp..

In other cases, not only herbicides, but also weeds, can supply the decrease of several crop dis-eases, so that their management is extremely important. In [41] it was investigate the efficacy of three common weeds, i.e., *Amaranthus viridis* (L.), *Lantana camara* (L.) and *Malvastrum coroman-*

delianum (L.) Garcke against four bacterial species, *Xanthomonas axonopodis*, *Pseudomonas syringae*, *Corynebacterium minutissium*, *Clostridium difficile* and major seed-born fungi *Aspergillus niger*, *Alternaria alternata*, *Drechslera biseptata*, *Fusarium solani* in vitro. Leaf extracts of these weeds exhibit antimicrobial effects and all were moderately active against seed-born fungi.

Some experiments found preliminary details, which suggest that the presence of weeds that serve as hosts of both tobacco rattle virus (Corky ringspot disease) and *Paratrichodorus allius* (root nematode) may nullify the positive effects of growing alfalfa or Scotch spearmint for Corky ringspot control conducted [43]. For all species researched, *Solanum sarrachoides* Sendtn presented positive correlation with Corky ringspot disease.

Weed management is also associated with most pests on crop cultivation; ecological relationships set out among organisms (weeds, insects, mites, etc.) allow their maintenance and proliferation. Examples of pest and weed interactions established in soybean has been reported by [44], who found anticipation of 14 days at critical period of weed control when crop was 60% defoliated by insects. Increasing of *Anticarsia gemmatalis* Hübner oviposition was also logged in [45] when soybean presented a high infestation of *Sesbania exaltata* (Raf.) Rydb. ex A.W. Hill. Thus, S. exaltata management reduces *A. Gemmatalis* population. Overall, monoculture areas tend to present higher mites and pest infestation and reduced biological diversity when maintained free of weeds. At the same time, weeds help insect diversity and natural biological control [46]. Mites, important arthropods in agricultural systems, currently constitute themselves key pests for soybeans in regions of hot and dry weather. Some predatory mites can be used against them into the biological management scope. Therefore, a fundamental aspect is the alternative feed sources for predatory mites during periods in which mite pests are at low populations. Among feed sources, there are many weeds, especially *Ageratum conyzoides* L., commonly encountered in citrus orchards and further agricultural areas. Overall, dicotyledonous weeds that produce a lot of pollen are preferred by predatory mites, in particular the genus Euseius [47]. Phytophagous mites, especially the web mite family Tetranychidae, were traditionally considered secondary pests in soybean. However, in recent years, it has been recorded severe and frequent attacks of these in different producing regions in Brazil [48]. Into surveys about GR soybean carried out in the state of Rio Grande do Sul, six phytophagous mite species were identified, five tetranychid — *Mononychellus planki* (McGregor), *Tetranychus desertorum* (Banks), *T. gigas* (Pritchard & Baker), *T. ludeni* (Zacher) and *T. urticae* (Koch) — and white mite tarsonemid *Polyphagotarsonemus latus* (Banks) [49,50]. In most of the sampled sites, more than one tetranychid were reported, being directly influenced by weed management.

The integrated management of weeds and pests, despite essential, is not easy to be performed on extensive production systems, especially because there are interactions of many species having various relations, either symbiosis, predation or parasitism. Knowing the interactions and the organisms that comprise production system is the great challenge and it can bring good results. Examples can be viewed in [51], with the reporting of lepidopterous in corn, in cotton [52], in *Heliothis zea* [53] and *H. virescens* [54] with *Bemisia tabaci* (Genn), among others. Maintaining biodiversity and sustainable production are some of the main advantages of using these systems [55].

3. Evolution of weed resistance in soybean

Herbicide-resistant weeds represent the evolution of plants as a consequence of environmental changes, which are usually caused by human action. This process is aligned with the theory of evolution. The process of natural selection, according to Darwin's theory of evolution, may be summarized by three guiding principles: i) principle of variation – there are variations in physiology, morphology and between behavior of individuals of any population, ii) principle of heredity – descendents are more similar to their parents than unrelated individuals, and iii) principle of selection – some individuals are more successful at survival and reproduction than others in a particular environment [56].

Therefore, a whole species keeps changing its composition because the individuals evolve in the same direction. The next generation will have a higher frequency of individuals that have been most successful in surviving and multiplying on environmental conditions. Frequencies of individuals within a population will change over time and those better adapted to the environment become predominant [56]. The biotype selection in a population by the same repeated herbicide application and its multiplying are shown bellow (Figure 1).

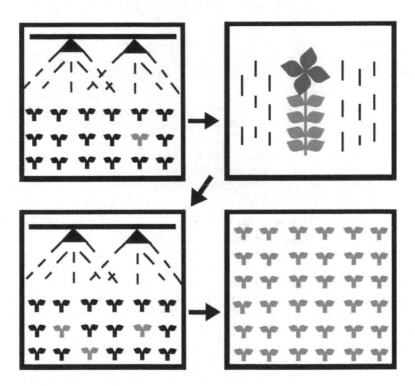

Figure 1. Illustration of a resistant biotype selection of a sensitive species [57].

Considerable evidence suggests that the appearance of herbicide resistance in a plant population comes with the selection of a resistant biotype, which is pre-existing. According to the selection pressure, this individual finds favorable conditions to reproduce [58]. The perception of resistance is only possible when the number of resistant plants or failure in control are clearly identified (Table 1). Unfortunately, for most cases, the seedbank already has seedlings of the resistant biotype in this time and eradication becomes arduous and expensive. The resistant biotypes may exhibit less ecological adaptation in these environments and become predominant due to elimination of sensitive plants. In terms of natural selection, biotypes with greater ecological adaptation reveal greater production than less adapted biotypes [59].

Years	N° resistant plants	N° sensitive plants	Control (%)	Progress
0	1	1,000,000	99.9999	unnoticeable
1	1	100,000	99.999	unnoticeable
2	1	10,000	99.99	unnoticeable
3	1	1,000	99.9	unnoticeable
4	1	100	99.0	unnoticeable
5	1	10	90.0	barely noticeable
6	1	5	80.0	noticeable
7	1	2	50.0	apparent

Table 1. Evolution of resistance in a population of resistant weed biotypes [60].

Most of the ecological issues associated with evolution of herbicide resistance involve the understanding of relationship between adaptation, gene frequency, inheritance and gene flow [61] because the interactions among these factors shall determine the time required for resistant biotypes to become predominant.

The time for resistant plants' appearance and resistant and non-resistant weed proportion change frequently with herbicide use and its biological effects, which may be fairly short (two years from commercial use — ALS inhibitors) or take more than 20 years, as happened with glyphosate (EPSP – 5-enolpyruvylshikimate-3 phosphate synthase inhibitors) [62] (Table 2). Weeds resistant to sulfonylureas were identified after four or five years of a continuous use of this herbicide group [63]. In Australia, *Lolium rigidum* Gaudin biotypes resistant to diclofop-p-methyl have been selected into three generations, starting from a sensitive population and by using a normal herbicide dose.

Herbicides with a high level of safety, i.e., high efficiency and specificity play a huge selective pressure. Examples include inhibitors of the enzymes ALS and ACCase (acetyl coA carboxylase), which have great chances to select resistant weed biotypes, since any change in its action point (enzyme) may result on activity losses and resistant weed increase.

Herbicide	Introduction year	Confirmation year	Place
2,4-D	1948	1957	EUA and Canada
Triazines	1959	1970	EUA
Propanil	1962	1991	EUA
Paraquat	1966	1980	Japan
EPSP syntase inhibitor	1974	1996	Australia
ACCase inhibitor	1977	1982	Australia
ALS inhibitor	1982	1984	Australia

Table 2. Year of introduction and its first confirmation of weed resistance to different herbicide action mode [64].

There are six factors related to plant population, which interact and determine the probability as well as the time of resistance evolution. They are the following: *number of alleles* involved in strength expression, *resistant allele frequency* in an initially sensitive population, mode of *resistance inheritance* (cytoplasmatic or nuclear), *reproductive traits* of species, *rate crosses between resistant and sensitive biotypes* and *selection pressure* [65].

The **number of genes** that confer resistance is important because, when inheritance is polygenic, the likelihood of the resistance to appear is low. However, when a single gene is responsible for resistance (monogenic), there is a high probability of occurrence. Most cases of resistance are conferred on a single gene. It is due to two factors. First of all, modern herbicides are specific, acting upon specific enzymes in metabolic pathways. Incidence of gene mutations responsible for coding that enzyme may change the plant sensitivity to the product, resulting in resistance. The second factor refers to the high selection pressure exerted by high efficiency of these herbicides. In order to occur polygenic resistance, the recombination between individuals for several generations would be necessary to obtain adequate number of alleles and to confer high plant resistance level [66].

Frequency of resistant allele(s) in sensitive population is usually between 10^{-16} and 10^{-6} [65]. So, the higher is the frequency of these alleles, the greater is the probability of selecting a resistant biotype. The frequency of resistant allele in the population becomes more significant in the evolutionary process when herbicide requires low selection pressure. However, if allele frequency is high, evolution of resistance may be faster, regardless of selection pressure.

Inheritance resistance type is fundamental for the establishment of resistance in a plant population. There are two basic types of inheritance: cytoplasmatic (maternal) and nuclear. Cytoplasmatic inheritance happens when hereditary traits are transmitted by cytoplasm, so only the mother plant can pass the trait to the descendants, as an example, resistance to triazines. On the other hand, if the inheritance is nuclear, transmission is by chromosomes, and both father and mother might forward its resistance, such as resistant plants to ALS inhibitors. In case of maternally inherited resistance, allelic migration between adjacent populations does not occur [66], so the development of this type of resistance is slower than nuclear, where migration of alleles occurs via pollen.

The **reproductive traits**, such as pollen scattering and number of propagules generated, influence directly the spread of resistant plants. Dispersal of resistance by pollen is affected by scattering efficiency and pollen longevity [67].

The **cross rate between resistant and sensitive biotypes** determines the spread of resistant alleles in a population. Pollen exchange between resistant and sensitive plants allows dispersion of the resistance, mainly in plants with high cross-fertilization rate, since the contribution of seed displacement is relatively small [59]. Gene flow is correlated with pollen flow distribution and varies on the species, with pollination mechanism and climatic conditions during flowering [68]. Species which presents more resistant biotypes and effective propagule dispersion may spread itself quickly, even though the inheritance of this resistance is maternal.

Repeated use of herbicides to plant control exerts high **selection pressure**, causing changes on flora of some regions, especially those with predominance of monoculture, such as soybean in major producing countries. Usually, better biotypes of a species adapted to a particular practice are selected and then they multiply rapidly [69]. Species exhibit different features and several responses to herbicide treatment. Therefore, an association of species characteristics with those of herbicides creates different periods needed for selection of resistant biotypes (Table 3).

Weed	Herbicide sprayed	Years
Alopecurus myosuroides Huds.	Chlortoluron	10
Avena fatua L.	Diclofop methyl	4-6
Avena fatua L.	Triallate	18-20
Carduus nutans L.	2,4-D or MCPA	20
Hordeum leporinum Link	Paraquat or Diquat	25
Kochia scoparia (L.) Schrad	Sulfonylurea	3-5
Lolium multiflorum Lam.	Diclofop methyl	7
Lolium rigidum Gaudin	Diclofop methyl	4
Lolium rigidum Gaudin	Amitrole + Atrazine	10
Lolium rigidum Gaudin	Sethoxydim	3
Senecio vulgaris L.	Simazine	10
Setaria viridis (L.) Beauv.	Trifluralin	15

Table 3. Number of years required for natural selection of resistant biotypes of a weed population according to the herbicide used [61].

In summary, the evolution process of herbicide resistance goes through three stages: removal of biotypes highly sensitive, remaining only the most tolerant and resistant; elimination of all biotypes except those resistant and selecting them in a population with high tolerance; intercrossing among survivors biotypes, generating new individuals with higher level of resistance, which may be selected later [65]. This process resulted in 383 resistant biotypes, 208 species (122 dicotyledonous and 86 monocotyledonous) and over 570,000 fields [62].

There are no doubts that selection pressure by use of herbicides at cultivated soybean areas contributed to the increasing of resistant weeds. Among the main representative countries, Argentina, Brazil and the USA, there is a positive correlation between soybean expansion areas and intensive use of herbicides, as well as between the increasing of resistance incidence and massive adoption by the same technology in these countries, i.e., one or few herbicide action modes.

In the USA, country with the largest number of resistance cases, 139 occurrences have been recorded, approximately 119 resistant species to different states and mechanism actions. From the 139 cases, around 25.9% are resistant species to two or more herbicide mechanism actions [62]. The first resistance case in the US, to auxin herbicides, was *Commelina diffusa* Burm. f., in 1957. Then, in 1964, it was reported a *Convolvus arvensis* L. case, resistant to 2,4-D. In the years 1970 and 1972, resistance cases of *Senecio vulgaris* L. and *Amaranthus hybridus* L. to PSII inhibitors were reported. In 1973, it was recorded *Eleusine indica* (L.) Gaertn resistant to dinitroanilines and, in 1979, *Chenopodium album* L. resistant to PSII inhibitors. With soybean advance in the 80s, the resistance cases increased to 28 reports with PSII inhibitors and 10 cases with ALS inhibitors. In the 90s, the intensive use of ALS and ACCase inhibitors in soybean contributed to 68 ALS resistance events and 26 to ACCase. From 2000, resistance cases to glyphosate became more common. Between 2000 and 2011, it was registered more than 70 resistance events to glycine group as the result of larger glyphosate use at GR soybean, genetically modified to glyphosate resistance (RR1). Among reports so far, the largest number of species is related to ALS inhibitors (44), triazines (25), ACCase inhibitors (15) and glycines (13).

In Brazil, selection of tolerant or resistant species started in the 70s, with repeated metribuzin use. This herbicide was introduced to control *Bidens pilosa* L., but it had low efficiency against *Euphorbia heterophylla* L.. *E. heterophylla* showed tolerance to metribuzin and so was selected and became a major weed to be fought in crops. Concerns with *E. heterophylla* control were solved by imazaquin herbicide (ALS inhibitor) in the 80s, which had been used widely, becoming the main herbicide used in soybean fields. But at the end of the 90s, *E. heterophylla* and *B. pilosa* became resistant to imazaquin, including the selection of *Cardiospermum halicacabum* L..

The control of resistant species to ALS inhibitors was solved with GR soybean. History repeated itself with glyphosate and this has become practically the only herbicide hold on soybeans, imposing great selection pressure of tolerant and resistant species. Thus, the continuous glyphosate spraying has selected tolerant weeds such as *Ipomoea* sp., *E. heterophylla*, *Richardia brasiliensis* (Moq.) Gomez and *Commelina* sp., as those resistant species, such as *Lolium multiflorum* Lam., *Conyza bonariensis (L.)* Cronq., *C. canadensis*, *C. sumatrensis* (Figure 2) and *Digitaria insularis* (L.) Mez ex Ekman.

Resistance of *L. multiflorum* to glyphosate was identified in Brazil in 2003, and this forced ALS inhibitors and ACCase to become the main control options for this species. The continuous use of ALS inhibitors (iodosulfuron-methyl) and ACCase to *L. multiflorum* control resulted in biotypes resistant to ALS and ACCase in 2010 and in 2011, respectively. These biotypes have multiple resistance to glyphosate, glyphosate + ALS and glyphosate + ACCase. Certainly, the resistance to the three mechanisms in the same biotype will not take long to happen.

For soybeans and wheat, ACCase inhibitors are the main alternative to *L. multiflorum* control. Thus, impact selection of resistant and tolerant species in Brazil is mainly focused on cost production, since the farmer will have to use alternative herbicides in the area, usually more expensive than glyphosate and less efficient.

(a)Photograph: Leandro Vargas, Embrapa Wheat,   (b)Photograph: Marlene Lazzaretti, Unnoba,

Figure 2. (a) Illustration of *Conyza* sp. resistant to glyphosate in Brazilian soybean field; b) rossettes of *Conyza bonariensis* (L.) Cronq. (smaller, smooth lobes) and *C. sumatrensis* (wider, serrated lobes), germinating in the fall, in Argentina.

In general, weed resistance to herbicides in Argentina became important after 2005 and is also related to the intensive use of glyphosate in GR soybean crop. The introduction of the RR technology in 1996 quickly masked the incipient problem of herbicide resistance in the country, marked by the appearance, in the northern part of Argentina, of an *Amaranthus* sp. resistant to ALS inhibitors herbicide (sulfonylureas and imidazolinones), officially confirmed as resistant in 1996. For many years, the problem faded into obscurity and farmers enjoyed the efficacy of an herbicide that seemed to elude the perils of resistance selection, again ignoring the advice of the few experts that protested against the practice of monoculture and lack of herbicide rotation. Reports of *Sorghum halepense* (L.) Pers. escapes in the province of Salta (NW Argentina), even after repeated applications of glyphosate, started in 2003, and the resistance was confirmed in 2006. This was the first case of resistance to glyphosate in Argentina, followed by *Lolium rigidum* Gaudin (and *L. multiflorum*) in 2007. In 2010, it was reported the case of *Avena fatua* L. resistant to ACCase inhibitors and two cases of multiple resistance *L. multiflorum*, resistant to ALS inhibitors plus glycine and ACCase inhibitors plus glycine as well [62], followed by *Echinochloa colonum* (L.) Moench (2011) and *Cynodon hirsutum* (2012).

The outlook is that the main crops (soybean, corn, cotton) from Brazil, the USA and Argentina will be resistant to glyphosate. In this context, succession and crop rotation with conventional seeds is a strong chance in the field. There is the necessity to convince farmers that repeated and continuous use of glyphosate-resistant crops in few years could cripple the weed control with the use of glyphosate-based products.

Evolution of glyphosate-resistant populations is an imminent threat in areas where there is dominance of glyphosate-resistant crops, intense selection pressure and no diversity [70]. Certainly other glyphosate-resistant weeds will be identified in the coming years. But when and how it is related to use of glyphosate-resistant crops? The use of practices to reduce selection pressure and switch mechanisms is important to protect and prolong the use of important molecules such as triazines, ALS inhibitors, ACCase, and glycines.

4. Management of weeds in soybean areas: Argentina, Brazil and the USA

4.1. Weed management in Argentina

The first recorded experience with soybeans in Argentina was in 1862, just a few years after their introduction to the US, but back then the country was a stronghold of cattle production, and there was little interest in agriculture. The first variety trials and commercial harvests occurred during the 60s. At the turn of the century, soybeans in Argentina were reaching the 10,000,000 hectares mark, coinciding with the adoption of transgenic GR soybeans. Soybean production increased over 1000-fold to a record of 52 million metric tons in 2010. The most productive area for soybeans is comprised by the northern portion of the province of Buenos Aires, the central and southern part of the province of Santa Fe, and the southeastern part of Córdoba (humid pampas), but in recent years the expansion has been more noticeable in other provinces, like Entre Ríos, Santiago del Estero, Tucumán, Salta and Chaco, in the northern part of Argentina. Another factor influenced by the adoption of GR soybeans was the oversimpli-fication of the weed control programs, which eventually led to the selection of resistant biotypes and hard-to-control weeds.

The development and early expansion of the crop in Argentina was accompanied by the constant introduction of new herbicide molecules. During the 70s, as farmers in Argentina were learning how to grow this crop, the most common weed control methods in soybeans were a combination of tillage and pre-emergent (PRE) herbicides such as trifluralin, dinitra-mine (dinitroanilines), cloramben (benzoic acid), naptalam (amide), flucloralin (chloroanilin), vernolate (thiocarbamate), metribuzin, prometrin (triazines), alaclor (chloracetamide), and linuron (phenylurea). Bentazon, one of the first post-emergent options, did not become available until the end of that decade. The dinitroanlinies, flucloralin, and vernolate were used on pre-planting incorporated (PPI) for annual grasses and broadleaves control, clormaben was one of the few burndown options for broadleaves, naptalam was applied PRE for annual grasses and broadleaves, the triazines also PRE, for small seeded broadleaves, often in combination with alachlor to improve annual grass control, and linuron offered broad spectrum control also applied PRE.

As a result of the limited choices in herbicides in soybean, there were several weed problems, such as the perennial grasses *Sorghum halepense* (L.) Pers. and *Cynodon dactylon* (L.) Pers., several annual grasses, such as *Digitaria sanguinalis* (L.) Scop., *Echinochloa crus-galli* (L.) Beauv., *E. colonum* (L.) Moench, *Eleusine indica* (L.) Gaertn., and the typical broadleaf weeds of summer crops — *Amaranthus* sp., *Chenopodium album* L., *C. cordobense* Aellen, *C. pumilio* R. Br., *Datura ferox* auct. non L., *Tagetes minuta* L., *Ipomoea* spp.. It was mention at least 6 species of Ipomoea, *Xanthium strumarium* L., *X. cavanillesii* Shouw, *Anoda cristata* (L.) Schlecht. and *Portulaca oleracea* L. [71] — among the broadleaf weeds in the humid pampas (Table 4). These plants represented a challenge and slowed the initial expansion of the crop. Most of the weeds described here are the same or very similar to the weeds commonly found in conventional-tillage systems around the world. A very interesting point is that none of the broadleaf weeds that are posing a challenge today to glyphosate in the temperate region is in this list, and most of the emerging weeds are local weeds, not common in other regions.

Economically important weeds	Secondary weeds	Emerging weeds
Echinochloa crus-galli (L.) Beauv; *E. colonum* (L.) Moench	*Xanthium* spp.	*Physallis angulata* L.
Cyperus rotundus L.	*Sida rhombifolia* L., *S. spinosa* L.	*Solanum sisymbriifolium* Lam.
Datura ferox auct. *non* L.	*Galinsoga parviflora* Cav.	*Aeschynomene virginica (L.) B.S.P.*
Tagetes minuta L.	*Bidens* spp.	*Nicandra physalodes (L.) Gaertn.*
Cynodon dactylon (L.) Pers	*Ipomoea nil* (L.) Roth., *I. purpurea* (L.) Roth.	*Abutilon theophrasti* Medik
Anoda cristata (L.) Schlecht	*Setaria viridis* (L.) Beauv., *S. verticillata* (L.) Beauv.	*Solanum chacoense* Bitter, *S. nigrum* L.
Digitaria sanguinalis (L.) Scop.	*Helianthus annuus* L. (volunteer)*	*Acanthospermum hispidum* DC.
Chenopodium album L.	*Eleusine indica* (L.) Gaertn.	*Flaveria bidentis (L.)* Kuntze
Portulaca oleracea L.	*Alternanthera philoxeroides (Mart.)* Griseb.	
Amaranthus quitensis Kunth	*Euphorbia heterophylla* L.	
Sorghum halepense (L.) Pers.	*Wedelia glauca* (Ort.) Hoffm. ex Hicken	

*Sunflower was a common component of the rotation systems.

Table 4. Most important weeds in the humid pampas in 1997, before the adoption of GR soybeans [72].

Usually a moldboard plow was used in the fall to incorporate the previous crop residue and destroy existing vegetation. Herbicides were part of the control methods from the beginning, given the timing of the introduction of soybeans in Argentina, so a mechanical-only technology was never developed for the region, except for specific purposes, like organic soybeans. In the spring, residual herbicides were applied after the preparation of the seedbed, incorporating them if needed. There were several escape problems given the limitation of POST options, especially with large seeded broadleaf weeds like *D. ferox*, *A. cristata*, and *Ipomoea* spp. The problem was so common that in many areas a special device called "Chamiquera" (Figure 3) was used to separate the harvested soybeans from "Chamico" (*D. ferox*) and sometimes "Bejucos" (*Ipomoea* spp.) seeds before the beans could be delivered at the grain elevators.

Figure 3. Special device "Chamiquera", Rojas, Buenos Aires, circa 1980.

During the 80s and 90s, until the introduction of GR soybeans, the development of several new molecules improved the control of many weeds, but still in combination with mechanical methods, leading to a steady expansion of both the area planted with soybeans and the average yields (Figure 4). Gradually, new herbicides allowed technology developments that replaced, at least in part, mechanical control methods with chemical ones. The need of field cultivators was reduced or replaced by the application of pre-emergent combinations of alachlor and metribuzin that offered a wide spectrum of control and proven residuality, replacing other herbicides — like trifluralin — that required mechanical incorporation.

Acifluorfen became a common tool for rescuing treatment, even though it caused severe crop injury. This herbicide allowed the control of large seeded broadleaf weeds — *Xanthium* spp., *D. ferox*, late flushes of *Ipomoea* spp. —, all common problems in most of the soybean area, but the injury it caused to the crop was something the farmer was not used to dealing with. It was replaced in part by another diphenylether, fomesafen, although it did not have the same efficacy or control spectrum. The registration of ALS inhibiting herbicides (sulfonylureas, imidazolinones and triazolopyrimidines) ushered a new era of weed control in soybeans in Argentina, allowing for PRE/POST combinations that offered effective and lasting control of the most important weeds with less crop injury than the previous options. Imazaquin, imazethapyr, chlorimuron, diclosulam and flumetsulam were launched in Argentina during the second half of the 80s (the first registration of imazaquin was actually in Argentina, in 1984) and allowed the first approach to no tillage in soybeans. The resistance problems associated with this group were not noticeable in Argentina, although the first resistant weed in the country is resistant to this herbicide group (ALS inhibitors), because it coincided with the introduction of the GR technology, and the quick adoption of the new varieties masked the problem.

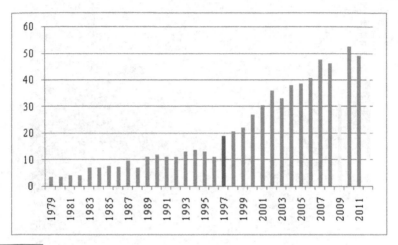

Sources: 1979-2004, Secretaría de Agricultura, Ganadería, Pesca y Alimentos de la República Argentina. 2005-2012, Diario La Nación, May 24, 2012.

Figure 4. Soybean production, in million metric tons, from 1979 to 2012. In red: first year with commercial GR soybeans. In yellow: droughts of the 08-09 and 11-12 seasons.

New inhibitors of the protoporphyrinogen oxidase enzime herbicides were introduced during the late 90s. Carfentrazone, sulfentrazone (aryl triazinones) and flumioxazin (N-phenylphtalimides derivative) offered new options for burndown (carfentrazone) and residual control (sulfentrazone, flumioxazin), but the introduction of the GR soybean varieties prevented its adoption, thus the most dramatic expansion of soybean production in Argentina was the introduction of the glyphosate resistant varieties in 1996.

Nearly all the soybeans in Argentina are transgenic (GR1). Argentina had the fastest adoption of glyphosate-resistant soybeans in the world. This fast adoption coincided with the expansion of no tillage technology in the region, fueling a synergism between GR soybeans and no tillage. AAPRESID, the national association of no tillage farmers, had held its first national symposium a few years prior to the launching of this technology, and its members welcomed and quickly embraced a new biotech development that allowed them to fully implement their preferred technology.

Until the adoption of GR soybeans, tillage was an important weed control method, complementing chemical control options, but it had a negative impact on erosion, soil structure and organic matter mineralization. The introduction of herbicide-resistant varieties increased not only the use of glyphosate, but also the practice of no tillage as well, replacing mechanical control almost completely in soybean production. The high efficacy of this herbicide combined with the simplicity of the system resulted in a quick replacement of other herbicides used in soybeans, both over the top applications and during the chemical fallow period. In 2005, over 92% of the herbicide volume used in chemical fallow was glyphosate, while some hormonal herbicides were commonly tank-mixed with glyphosate to improve the control of thistles and other "new" weeds. Overall costs of weed control in soybeans decreased dramatically as new generic glyphosate brands entered the Argentine market. Another aspect that contributed to the simplification of the system, including soybean monoculture, was the general economic situation of the country. Corn required a higher investment, while soybeans, especially GR soybeans, as described above, allowed farmers to plan their soybean season with less financial requirements (in Argentina, the law allows farmers to save seeds for their own use) in times when the prices of commodities were uncertain and financial means were limited, or expensive.

Glyphosate effectively controlled not only the most problematic weeds in soybean fields; it replaced herbicide combinations that required a deep knowledge of the weed spectrum, careful planning to avoid escapes, tank mix problems, timing concerns and crop injury, and still did not offer the satisfaction of a field completely clean of weeds. RR technology simplified the business of growing soybeans like no other technology ever developed. Today, soybean system is characterized by over-reliance on glyphosate, low crop rotation, absence of mechanical control methods and limited monitoring (of both weeds present at the time of application and results). The lack of monitoring practices is a direct result of the high efficacy of glyphosate control in the early years of the biotech age. As a result, the weed spectrum has shifted and there are several glyphosate-resistant weeds, combined with hard-to-control ones, while the presence of weeds with resistance to other modes of action is still limited.

Glyphosate is still a very valuable weed control tool, in spite of the weed shift that Argentina has experienced due to its over-use. In [73] it was studied the effectiveness of glyphosate applications at two stages (vegetative and reproductive) on 31 weeds that represented the typical weed spectrum of the region. The herbicide had complete control on 58% of the species at both stages, complete control at the vegetative stage but deficient control at the reproductive stage on 32% of the species and poor control on only 10% of the species at either stage. Disregarding the poor control at the reproductive stage-only, which is not recommended, it is clear that glyphosate satisfactorily controlled 90% of the weeds. The remaining 10% can be managed easily combining glyphosate with the proper herbicides, providing a cost-effective complement. The control of some of these difficult weeds is improved when glyphosate is combined with atrazine or metsulfuron applied during fall [74]. For example, *Bowlesia incana* Ruiz & Pavón and *Parietaria debilis* G. Forst., increased when glyphosate was applied as a tank mix to these herbicides, compared to glyphosate by itself. These herbicides are readily available and are cost-effective alternatives to combine with glyphosate.

The selection of herbicide-resistant biotypes was a consequence of lack of crop and herbicide rotations. Daniel Tuesca, a weed scientist in the University of Rosario, states that, in the years preceding the introduction of the GR soybean, farmers mentioned, on surveys, the use of 16 different herbicides in the fallow process and 13 on the crop (either PRE or POST), but a few years after the introduction of the technology, there were only 3 herbicides applied in fallow, and only glyphosate over the crop. From the three herbicides mentioned in the surveys, there were no specific graminicides (Table 5), so it is not a surprise that all the weeds that have been confirmed as resistant to glyphosate are grasses.

Before 1997 (1995-1997)		After 1997	
	Picloram		Atrazine
	Flumetsulam		2,4-D
	Metribuzin		Metsulfuron
	MCPA		
Fallow applications	Dicamba	Fallow applications	
	Atrazine		
	2,4-D		
	Metsulfuron		
	Other herbicides		
	No applications		
	Flumioxazim		
	Clorimuron		
	2,4-DB		
	Imazaquin		
	Acetochlor		
PRE/Over the top	Graminicides (FOP's)	PRE/Over the top	
	Flumetsulam		
	Diclosulam		
	Imazetapyr		
	Other herbicides		
	No applications		

Based on individual responses to surveys, % for each answer is omitted. FOPs (aryloxyphenoxypropionate herbicide group)

Source: courtesy of Professor Daniel Tuesca, Universidad Nacional de Rosario, AR.

Table 5. Herbicides, other than glyphosate, used in soybean production before and after the introduction of GR soybeans in Argentina, according to surveys with farmers.

Apart from the confirmed cases of glyphosate-resistant weeds, there are several problems caused by the excessive use of glyphosate. To better understand the problem, it is important to state that in Argentina about 70% of the farming is done in rented land, and during the last decade the rental price has increased constantly. In many cases, this situation prevented the traditional early fallow procedures and resorted to burndown practices with weeds that had grown beyond their optimal control stage. One particular case is *C. bonariensis*, which has been confirmed to be resistant to glyphosate in Brazil, although the resistant biotype is not present in Argentina yet. When treated at the rosette stage, the plant is susceptible to be controlled with glyphosate, but when it has elongated (early in the spring), it becomes resilient, even when using 2 and 3 times the dose of glyphosate. The situation changes when residual herbicides are applied in the fall (flumioxazin, metsulfuron, atrazine, and diclosulam have proved to be effective). There was a lot of confusion when this weed began to emerge as a problem because it co-exists with another species, *C. sumatrensis*, more susceptible to be controlled by glyphosate applications at later stages, leading to a general belief that there are resistant biotypes that escape control. Again, the lack of monitoring practices is evident here. These weeds are strongly associated with no-tillage practices, since they do not progress at all in tilled soil.

Today, there are many efforts to revert the reliance on glyphosate and the selection of resistant biotypes and hard-to-control weeds. Universities, professional associations and the industry are advocating the rational use of herbicides with different sites of action, in a crop rotation program, to prevent the selection of new resistant biotypes, not only to glyphosate but to others as well, especially biotypes with multiple resistance. It is only fair to mention that academics from different institutions such as INTA, Universidad Nacional de Buenos Aires, Universidad Nacional de Rosario, Universidad Católica de Córdoba, Universidad Nacional de Tucumán, Estación Obispo Colombres, just to mention a few, have been working hard on this matter in the previous years, when glyphosate was still the undisputed weed control method of choice. Argentina is shifting from a simple and effective system to a more complex one that requires a stronger commitment from farmers, advisors, the academic sector and the industry. The soybean sector is facing a turning point, and this new reality will have to include more crop rotations, more herbicides and also mechanical and cultural weed control methods.

4.2. Weed management in Brazil

According to professor Gustavo Dutra, from Cruz das Almas, Bahia, it may be inferred that, since its introduction in the country in 1882, soybean crop has transformed the Brazilian agriculture. Initially planted in the state of Rio Grande do Sul, first recorded in 1914, in Santa Rosa, the soybean "tropicalization" has found space coming out from southern pampas to the midwestern region of the country. While only 2% of national soybean production had been recorded in this region in the 70s, more than 47% of national production was reported in midwestern region in 2010/2011 harvest. Hence, Brazil represents one of the most important regions with a growing potential in soybean production. Probable areas to produce soybean ponder between 20° S and 20° N. However, the largest portion of this production belt is concentrated in the Brazilian lands, with estimated increases of 2.3% up to the year 2020.

Weed control has bothered growers from the beginning of soybean cultivation, especially since 1950, with the expansion of southern region. Adaptation of production system allowed the satisfactory management, even when using only mechanical tools to control. Cost constraints and limitations set by this control led to its quick replacement by the chemical control, which became a primary tool of weed management. Due to its importance, Brazilian pesticide market has expanded from 1977 to 2006, on average, 10% per year. Even after many decades, the use of soybean herbicide has been restricted to spraying in incorporated pre-plant (eg trifluralin) and pre-emergence (eg metribuzin, alachlor and linuron) along with plowing and harrowing, to prepare conventional soybean field.

Few herbicides used previously restricted the implementation period, affecting more specific actions for managing the weeds emerged in advanced stages of the crop. The launch of bentazon POST herbicide revolutionized the market, allowing the control of major dicotyled-onous weeds on soybean. Introduction of new molecules from the 80s and 90s afforded efficiency on the control of several species, in particular those belonging to genders Amaran-thus, Digitaria, Brachiaria, Euphorbia and Bidens. The main herbicides applied belonged to the chemical groups ALS and ACCase inhibitors, with monocotyledonous and dicotyledonous actions.

Since the introduction of no tillage system, weed management has changed and, as a conse-quence, moved to consider factors other than chemical control on the production system. The main benefit of no tillage system is the reduction of weed germination over time [75] and greater use of crop control. Furthermore, species not commonly observed in the conventional system demand better preparation and expertise of producers. Such modifications are related to the absence of soil disturbance, favoring perennial cycle weeds, as well as changes in patterns of temperature and light incidence, influencing seeds' mechanisms of dormancy. Cover crops result in greater amount of organic residue, with higher C/N ratios, and are more efficient in weed management, by composing a thicker layer of mulch on surface soil [76]. The weed density decreases linearly with organic residues increasing on surface soil, mainly by reduction on weed germination.

Originally, no tillage system in Brazil used 2,4-D and paraquat herbicides as burndown to prepare cultivation areas. At the time there was no product like glyphosate, with non-selective and desiccant action. Despite the effective action, there were limited control with paraquat and some residual effects of 2,4-D on soybeans, hindering the sowing immediately after spraying. With glyphosate releasing in Brazil in 1982, the technology suited local and producers' needs, gaining the market by its control efficiency. But the POST application was still limited to the same herbicides (bentazon, imazethapyr, setoxydin, tepraloxydym, etyl-chlorimuron, diclo-sulam, clorasulan-methyl, etc.). Doses were necessarily higher and the number of resistance cases to ALS inhibitors started to increase, since the first record of Bidens pilosa L., which is resistant to imazaquin and chlorimuron-ethyl, appeared in 1993.

With the introduction of GR soybean, most of the herbicides were replaced in 2003/2004 harvest in Brazil. The system that provides a single application of glyphosate at early stages of the crop gained market for its easy adoption, undeniable efficiency in weed control and guarantee of profitability. According to data, nearly 81% of all soybeans cultivated area in Brazil is GR and its contribution to farmers is unquestionable (Figure 5). The impact of using GR soybeans has been similar to that identified in the US and Argentina, although the net savings on herbicide costs are larger in Brazil, due to higher average costs of weed control [77]. The average cost savings originated from a combination of reduced herbicide use, fewer spray runs, labor and machinery savings, were between US$30/ha and US$81/ha in the period 2003-2010, which means that the net cost saving after deduction of the technology fee (assumed to be about US $19/ha in 2010) has been between US$9/ha and US$61/ha in recent years, with increased farm income levels of US$694 million in 2010 by the GR soybean adoption.

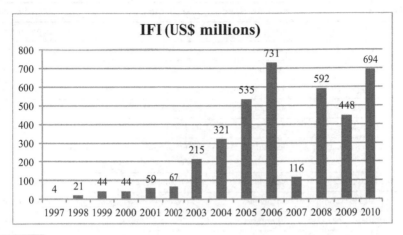

Source: adapted [77].

Figure 5. Impact of using GR soybean on farm income (IFI), at a national level. Brazil, 1997-2010.

Unfortunately, the overuse of the technology (GR soybean + glyphosate) in tillage and no tillage system led to strong selection pressure. Apart from the variation of biotypes selectivity, the level of herbicide application also contributes to the tolerance of species. It was checked the Brazilian herbicide usage data for the periods 2001-2003 and 2007-2009, as well as information from industry and extension advisers and was conclud-ed that the annual average use of herbicide active ingredient per ha in the early years of GR soybean was lesser than 2007-2009, an estimated difference of 0.22 kg/ha [77]. From 2007-2009 data, it was observed an average active ingredient use of 2.37 kg/ha for GR soybean compared to 1.96 kg/ha for conventional soybeans.

These data clearly illustrate the current weed management in soybean in the country. Nowa-days, Brazilian producers are using sequential spraying of glyphosate in order to control species which are difficult to manage in crops, such as *Bidens* spp., *Chamaesyce hirta* (L.), *Spermacoce latifolia* Aubl., *Chloris polydactyla* (L.) Sw., *Ipomoea grandifolia* (Dammer) O'Donell, *Commelina benghalensis* L., etc. along with glyphosate herbicides. They also associate herbicides of other chemical control groups and, especially on southern and southeastern regions, producers are using the autumn management, in areas where these species are present [2]. Other herbicides — such as imazethapyr and imazapic — are frequently applied to reduce the emergence of weeds during the fallow period and/or associated with the herbicide 2,4-D on burndown, about 15-20 days before the sowing, for dicotyledonous management of complex control by glyphosate. A relevant number of not highlighted weed species are worrying Brazilian soybean producers. *Borreria* spp., *Tridax procumbens* L. and *Alternathera tenella* Colla, among others (Table 6), are species with high adaptability to different ecological niches throughout the national territory and they are on the list of species likely to be capable of developing resistance to herbicides used in cultivation, being it a GMO or not.

Scientific name	Common name	Scientific name	Common name
Acanthospermum hispidum DC.	Starbur	Eleusine indica (L.) Gaertn	Goosegrass
Amaranthus retroflexus L.	Pigweed	Euphorbia heterophylla (L.)	Wild poinsettia
Bidens pilosa L.	Hairy beggarticks	Galinsoga parviflora Cav.	Smallflower
Brachiaria plantaginea L.	Alexandergrass	Ipomoea purpurea (L.) Roth	Morningglory
Cenchrus echinatus L.	Sandbur	Panicum maximum Jacq	Urochloa maxima
Commelina benghalensis L.	Dayflower	Pennisetum setosum Rich	Bufflegrass
Cynodon dactylon (L.) Pers	Bermudagrass	Setaria geniculata auct. non (Willd.) Beauv.	Foxtail
Conyza bonariensis (L.) Cronq.	Hairy fleabane	Sida rhombifolia L.	Sida
Conyza canadensis (L.) Cronq.	Horseweed	Sorghum halepense (L.) Pers.	Johsongrass
Digitaria insularis (L.) Mez ex Ekman	Sourgrass	Spermacoce latifolia Aubl.	Buttonweed
Digitaria horizontalis Willd.	Jamaica crabgrass		

Table 6. Some weed species on soybean Brazilian crop [2].

For the management of weeds resistant to glyphosate, the alternative control, besides herbicide mixtures, includes crop rotation, autumnal management or even return of non transgenic soybeans, as well as herbicides spray recommended in 80s and 90s. To reduce *Conyza* spp. competition, which can cause yield losses above 70% for soybean [78] it is recommended winter management by mixing residual herbicides and glyphosate + 2,4-D [79] ever sprayed on initial growth stage and on plant less than 10 cm-height. For *L. multiflorum* control in the south region, clethodin or haloxyfop-p-methyl herbicides in a glyphosate mix can be used. *S. halepense* is another glyphosate-resistant species which has a reasonable control with haloxyfop-p-methyl application. Nevertheless, this last management must be made with young plants.

In the US, saflufenacil is being used as a major product mixed to glyphosate for controlling resistant weeds. This PPO inhibitor empowers the action of glyphosate as desiccant and it is applied on off-season management or before crop sowing. Though, its release in Brazil has not occurred yet and it should be soon on the market to assist the producers. One of its advantages is the low residual rate in the soil at recommended doses, which allows its implementation and subsequent planting without requiring longer intervals before sowing.

The steady cost increase in weed control by intensive herbicides use and their mixtures emphasizes the need of changing. Since introduction of GR soybean technology in 2003, until 2006, there has been a reduction in herbicide application in soybeans in the country, deriving mainly from efficiency control and range of action of the glyphosate (Table 7). However, amount of active ingredients utilized on crop has risen since 2006, as a result of the intense use of glyphosate and other herbicides. New generic glyphosate brands entered the Brazilian market and it contributed to glyphosate use indefinitely.

Year	ai saving (kg; negative sign denotes increase in ai use)*	% decrease in ai (- = increase)
1997	22,333	0.1
1998	111,667	0.3
1999	263,533	0.7
2000	290,333	0.7
2001	292,790	0.7
2002	389,145	0.8
2003	670,000	1.2
2004	1,116,667	1.7
2005	2,010,000	2.9
2006	2,546,000	4.0
2007	-5,808,563	-8.8
2008	-5,704,705	-17.6
2009	-6,642,000	-18.7
2010	-7,529,650	-20.0

Sources: Kleffmann & AMIS Global;

* Including herbicides (mostly glyphosate) used in no/low tillage production systems for burndown.

Table 7. National level changes in herbicide use (active ingredient – ai) by GR soybean. Brazil, 1997-2010 [77].

In spite of weed shift in Brazil, glyphosate is still a helpful weed control tool. To extend its use as a major tool in chemical control strategies on tillage and no tillage sowing, GR and no-GR soybean, current management in soybean aims to integrate methods that minimize the effects to the environment and offer adequate security control. Therefore, in addition to new technologies afforded by the chemical industry, producers should also cooperate in the process, even though this implies the return of already used tools, as the conventional soybean (no GM). Among the alternatives, there is the rotation area with conventional soybeans, the use of offseason management (autumn), the spraying of non-selective herbicides that reduce shifts on further glyphosate applications, the advanced management in spraying installment — being the first 30 days before sowing and the second between five and seven days prior of planting —, the sowing of cover crops in fallow period and the spraying of recommended herbicide doses in order to avoid progressive biotypes selection [30,80].

4.3. Weed management in the USA

Soybean production in the US is undoubtedly part of the greatest productions worldwide and it has an expressive occupation of agriculture area in the country. According to the USDA projections, last average yield was around 3.7 tons/ha crop; in 2012, there will be about 29.9 million hectares crop in the country. Most of soybean cultivated area in the US (93%) uses GR soybeans. The first scientific record of soybean cultivation in the US took place in 1879 at the Rutgers Agricultural College, in New Jersey [81]. Initially, the crop was mainly used as animal fodder. However, the growing interest in culture, sponsored by the demand for oil and meat, forced soybean to expand rapidly and occupy many areas previously cultivated with corn, in the extensive Corn Belt.

Despite high yields, the country also passed through difficulties at the beginning of crop establishment. Even with great advances in farmland during the 50s, farming tools were limited, especially the ones related to weed management. There was no PRE or POST herbicides. Usual control practices were restricted to the use of mechanical weeding, fundamental on conventional crop system. Wide-row spacings were used in order to provide effective mechanical weeding and post-sowing. The 2,4-D was used over-the-top at the end of crop growing, prior to the harvest. This allowed reduction on dicotyledonous weeds and on subsequent crops, but did not control the monocotyledonous ones. These have become the main weeds and *Sorghum halepense* (L.) Pers was a major problem weed in many fields.

Until glyphosate and, mainly, GR soybean advents, weed management in the US was restricted to mechanical control and some PRE and POST herbicides to monocotyledonous and dicotyledonous control. Trifuralin was a major narrowleaf herbicide used for years, which was applied in autumn or in spring before sowing. Its use requires tillage system but did not aid weed management in early-season, especially *S. halepense* and *Amaranthus* sp. control. Between the 70s and 80s, glyphosate and paraquat came into use as preplant burndown, being helpful on no tillage system. These herbicides replaced preplant tillage and fostered the currently used stale seedbed planting system. Not so far, PRE and POST selective herbicides became available to most monocotyledonous and dicotyledonous weed control. Narrow-row and no tillage system challenged soybean farmers to introduce a new management concept. In the US, the first POST herbicide formulations were available years before their release in Argentina and Brazil and some chemical alternatives on weed management in the country had always been more accessible. Nevertheless, the order of release was followed, initially by bentazon, with a broad spectrum of action, and after, ACCase inhibitors, diphenylethers, imidazolinones and sulfonylureas (ALS inhibitors).

Traditionally, soybean is the rotational crop with rice in most farming areas, particularly in midsouth region. Prior to rapid rice expansion area in the 70s, the common rotation involved 2-year soybean and 1-year rice. Today, rice is often grown for 2 or 3 years before another crop, especially where the land is unsuited for other crops, and soybean is predominant. Major conventional herbicides that have been used in soybean include trifluralin, pendimethalin, metolachlor, alachlor, dimethenamid, clomazone, imazethapyr, sethoxydim, fluazifop, quizalofop, and clethodim [82]; many of them are useful against *S. halepense*.

The main herbicides such as trifluralin, pendimethalin, imazethapyr and imazaquin were widespread until the mid 90s, but with glyphosate effectiveness, mainly linked to GR soybean, there was a massive replacement of the "out-of-fashion" herbicides. During the period considered, 1995-2006, the treated areas with pendimethalin decreased from 26% to 3%; areas treated with imazethapyr suffered a reduction from 44% to 3% [83]. Especially for imazethapyr, whose decrease was greater than pendimethalin, many resistant weeds had been selected, even in the first using years, encouraging technology exchangings.

Many advantages provided by glyphosate on GR soybean weed control overlapped other management tools, leading to a replacement of herbicides and conventional soybean for the new technology. No tillage systems became widely used and weed control costs were lowered. Total applied herbicides and labor inputs declined initially and narrow-row on soybean became the standard. In 1995 the GR soybean areas treated with glyphosate were only 20%, but they took over 96% in 2006 [84]. Currently, the GR soybean represents over 94% of the soybeans grown in the US, and more than 90% of soybeans produced worldwide are considered GR.

The initial advice for GR soybean system was only one spray and its late application would not undermine crop yield. In extremely wet sites with late sowing — Iowa, for example —, weeds emerged early and single POST glyphosate spray was enough for effective control till the end of the cycle [85]. But for the midwest region, the sowing scheduled occurred earlier, thus only one application was unsuitable for weed control, usually requiring additional sprays.

Concerns about the definition of better periods of spraying, along with the appearance of the first glyphosate resistance case, registered for *Lolium rigidum* in 1998, have collaborated with gradual increase in herbicide use. For the period 2003-2009, herbicides applied to GR soybean increased 30%, whereas consumption remained stable for conventional soybeans [83]. Among changes observed in the global agricultural production, there is the search for socioeconomic and environmental efficiency. Farmers want new tools for weed management. New GM crops have allowed simple and effective solutions, but if producers keep outdated manners when using new tools with GR soybean and glyphosate, these tools will soon become obsolete [25].

As a result, a second generation of GR soybean was launched recently in the US in 2009. Although this technology offers the same soybean resistance to glyphosate as the first gener-ation (RR1), it has a higher yield potential, between 7% and 11%. Some farmers reported no increasing yield in relation to first GR soybean generation; perhaps others found positive yield effect. In 2010, soybean farmers pointed that second GR soybean generation has, on average, about 5% of yield improvement.

Many soybean farmers currently use glyphosate mixed with residual herbicides employed previously. The increase of these mixtures permits earlier glyphosate sprays promoting weed management for a larger period. Using conventional herbicides into new GM soybeans are also essential to ensure its resilience, since new traits will be released to use with former herbicides. New technologies include GM soybeans resistant to glufosinate "Liberty Link", to 2,4-D "Optimum GAT", to dicamba and also to glyphosate plus ALS inhibitors. Despite the creation of technologies for landing efficiency and easy management on weed control, good practices at all soybean crop system are rather necessary. Also, weaknesses and difficulties on weed management in many regions of the US have attracted the interest for non-GM soybeans. Differentiated prices in the international market have also stimulated this substitution, yet it is constrained to small and middle producers.

5. Benefits of integrated weed management

Effective weed management is very important to maintain agricultural productivity. By competing for light, water and nutrients, weeds can reduce crop yield and quality and can lead to billions of dollars in global crop losses annually. Because of their ability to persist and spread through the production and dispersal of dormant seeds or vegetative propagules, weeds are virtually impossible to eliminate from any given field. The importance of weed management to successful farming systems is demonstrated by the fact that herbicides account for the large majority of pesticides used in agriculture, eclipsing inputs for all other major pest groups. To no small extent, the success and sustainability of our weed management systems shapes the success and sustainability of agriculture as a whole [86].

Integrated pest management (IPM) concept was introduced in the 60s comprising many definitions from then. The primary goals of IPM programs are to reduce pesticide use and the subsequent environmental impact and to rely more on alternative strategies to control pests [87]. Integrated weed management (IWM) comes as a secondary effect of IPM, but it has similar proposal of using multiple management tactics and incorporating the knowledge of weed biology and crop physiology into the weed management system. The goals of IWM range from maximizing profit margins to safeguarding natural resources and minimizing the negative impact of weed control practices on the environment [88].

Integrated Weed Management combines multiple management tools (biological, chemical, mechanical and others) to reduce a pest population to an acceptable level while preserving the quality of existing habitat, water, and other natural resources. The integrated management provides connection of all the involved organisms, whether weeds, pests or diseases, and should focus on decision-making with case studies. There are many practices set out in the integrated management systems, whose benefits have been extensively studied by several authors (Table 8). These studies demonstrate many benefits and the efficiency of integrated tools in crop management systems.

Practices evaluated in IWM	Study
Monitoring weeds in crop fields	[90,91]
Use economic thresholds to determine when to apply herbicides	[91-93]
Crop rotation	[80,91]
Using the biological and chemical control	[94,95]
Using cultural and chemical control	[96]
Using mechanical and chemical control	[97]
Using rotation of herbicides	[90,91]
Plant cover crops	[90,98]
Using the tillage, no-tillage or reduced tillage system	[90,92]

Table 8. Practices evaluated in previous studies as part of an Integrated Weed Management (IWM).

However, there are no more ready-made and generalized solutions without risk of errors. IWM is characterized by reliance on multiple weed management approaches that are firmly underpinned by ecological principles [89]. As its name implies, IWM integrates tactics, such as crop rotation, cover crops, competitive crop cultivars, the judicious use of tillage, and targeted herbicide application, to reduce weed populations and selection pressures that drive the evolution of resistant weeds. Under an IWM approach, a grain farmer, instead of relying exclusively on glyphosate year after year, might use mechanical practices such as rotary hoeing and interrow cultivation, along with banded PRE and POST herbicide applications in a soybean crop one year, which would then be rotated to a different crop, integrating different weed management approaches.

Earlier studies have also demonstrated that IWM strategies are effective in managing herbicide-resistant weeds. For example, glyphosate-resistant horseweed in no tillage soybean can be controlled by integrating cover crops and soil-applied residual herbicides [100]. In a recent experiment in which the integration of tillage and cover crops was evaluated for controlling glyphosate-resistant *Amaranthus palmeri* in Georgia, the combination of tillage and rye cover crops reduced *A. palmeri* emergence by 75% [101]. In addition to cultivation and cover crops, other practices can be used to manage resistant-weed populations.

In another experiment, it was experienced biological and chemical control to *Sesbania exaltata* [Raf.] Rydb. ex A.W. Hill in soybean field. Different concentrations of *Colletotrichum truncatum* (Schwein.) Andrus & Moore were tested alone and in combination with glyphosate. Positive results suggest that it might be possible to utilize additive or synergistic herbicide and pathogen interactions to enhance *S. exaltata* control [94]

Despite many results, researchers suggest that implementation has been slow, and that farmers rarely move beyond incorporating cost-effective, targeted pesticides application [102]. Many growers are not adopting integrated management because current assessment methods are inadequate [99]. In their study, evaluating data from eastern North Carolina, US, they considered four components of the integrated management: weed, pest, environmental and general management of the properties. The component weed had the highest percentage (79%), indicating that growers were undertaking its management.

In [97] it was evaluated a cropping system, including various combinations of seeding rate and date, herbicide timing and rate, and tillage operations, by measuring weed response to six IWM systems, in a wheat–oilseed rape–barley–pea rotation. Changes in weed communities assessed over 4 years indicated a gradual increase of *Thlaspi arvense*, *Chenopodium album*, *Amaranthus retroflexus* and *Fallopia convolvulus* in the no herbicide/high tillage system. Winter and early spring annuals and perennials increased in most systems, but particularly in the low herbicide /zero tillage and medium herbicide/zero tillage systems. This study confirms the potential of contrasting IWM systems under the challenging environmental conditions.

Some mathematical models are also used into IWM. It allows to model scenarios and to compare long-term economic and weed population outcomes of various integrated management tools. In southern Australia, species like *Lolium rigidum* and *Raphanus raphanistrum* were managed for many years with selective herbicides. But these species became resistant and are widespread now. In [93] it was tested an integrated model to compare the management over a 20-year period and found that differences between scenarios are not due to weed densities but differences in total cost on weed control.

In fact, despite all the benefits, the implementation of IWM is extremely challenging for researchers and especially for farmers. In a recent paper — *True integrated weed management* — was highlighted in glowing way the need for a single platform development, including sensors and decision-support software, that has multiple application technologies for weed management [103]. According to the actor, "*Ideally, a self-guided machine is needed that could comb the field in a systematic way to identify weeds and then apply the necessary control tool (eg spray, mow, cultivate) at the individual plant or patch scale*". The illustration of a machine model (Figure 6), which allows the required operations case by case is utopian, although it is believed that efforts to achieve this goal are unlimited.

Figure 6. Illustration of a robotic weed control using multiple tools designed [103].

6. Conclusions

Weed management has always been inserted into the soybean crop system, contributing decisively to the success of this crop in major producing countries nowadays. The evolution of weed management practices in Argentina, Brazil and the US has been developed similarly, by means of mechanical growers and massive use of GM soybean. However, weeds also have evolved and as new tools were used, new species or new biotypes appeared.

Despite the persistent search for weed control in the soybean areas, it is observed that management of those has increased considerably in the last 10 years. There are numerous cases of weed resistance to various chemical herbicide groups used in the crop and some weed species are resistant to more than two chemical groups.

Even with the biotechnology advances and other GM soybean introduction, history must repeat itself, since the tendency to standardize production systems favors the weeds, allowing better adaptation response as it increases the selection pressure. The application of glyphosate to GM crops like soybeans, corn, cotton, canola, wheat, among others — all resistant to this herbicide — is not the best alternative to properly manage weeds. In regions where RR technology is predominant, shifts on weed control are increasing, as well as new weed problems, including weeds resistant to glyphosate which are infesting other crops. In this case, soybean producers must use all available technologies, considering both socioeconomic and environmental efficiency.

The use of IWM is the most suitable alternative to maintain weed populations below damage threshold on the soybean crop. Besides difficulties on IWM implementation, there are concerns about farmers' awareness and variations into each farm. The use of IWM without considering the integration of control methods of other organisms (pests and diseases) does not allow the sustainability of used practices.

Even with prediction models to IWM implementation, weed control is not indefinitely assured if it is not continuously adapted to new changes in soybean production system. In this context, there is no single solution, ready and with indeterminate validity on weed management. Choosing intelligent systems, which integrate the basic concepts of ecology and biology of species to the available tools (GM crops, herbicides, biological control, etc.), should assist weed management.

Author details

Rafael Vivian[1*], André Reis[2], Pablo A. Kálnay[3], Leandro Vargas[1], Ana Carolina Camara Ferreira[4] and Franciele Mariani[5]

*Address all correspondence to: rafael.vivian@cpamn.embrapa.br

1 Brazilian Department of Agriculture, Agriculture Research Service – Embrapa Mid-North, Teresina, PI, Brazil

2 Department of Civil and Environmental Engineering, Waseda University, Shinjuku-ku, Okubo, Tokyo, Japan

3 National University of Buenos Aires/Northwest, Pergamino, Buenos Aires, Argentina

4 Federal University of Piauí, M.Sc. student – Soil conservation – Teresina – PI, Brazil

5 Federal University of Pelotas, Ph.D. student – Weed management – Passo Fundo – RS, Brazil

References

[1] Oerke, E. C, & Dehne, H. W. Safeguarding production losses in major crops and the role of crop protection. Crop Protection (2004). , 3, 275-285.

[2] Reis, A. R, & Vivian, R. Weed competition in the soybean crop management in Brazil. In: Soybean- Applications and Technology, Eds. (2011). , 185-210.

[3] Bhowmik, P. C. Weed biology: importance to weed management. Weed Science (1997). , 45(3), 349-356.

[4] Buhler, D. D, Hartzler, R. G, & Forcella, F. Implications of weed seedbank dynamics to weed management. Weed Science (1997). , 45(3), 329-336.

[5] Thill, D. C, & Mallory-smith, C. A. The nature and consequence of weed spread in cropping systems. Weed Science (1997). , 45, 337-342.

[6] Harper, J. L. Populations biology of plants. Academic Pres (1977).

[7] Senseman, S. A, & Oliver, L. R. Flowering patterns, seed production, and somatic polymorphism of three weed species. Weed Science (1993). , 41, 418-425.

[8] Pitelli, R. Plantas daninhas no sistema plantio direto de culturas anuais. R. Plantio Direto (1998). , 4, 13-18.

[9] Paes JMVREZENDE AM. Manejo de plantas daninhas no sistema plantio direto na palha. Inf Agropec (2001). , 22(208), 37-42.

[10] Silva, J. B. Plantio direto: redução dos riscos ambientais com herbicidas. In: Saturnino HS and Landers JN (Eds.). O meio ambiente e o plantio direto. Brasília: APDC; (1997). , 83-88.

[11] Babujia, L. C, Hungria, M, Franchini, J. C, & Brookes, P. C. Microbial biomass and activity at various soil depths in a Brazilian oxisol after two decades of no-tillage and conventional tillage. Soil Biology and Biochemistry (2010). , 42(12), 2174-2181.

[12] Lal, R. Constraints to adopting no-till farming in developing countries. Soil & Tillage Research (2007). , 94(1), 1-3.

[13] Federação Brasileira de Plantio Direto na PalhaFEBRAPDP: Evolução da Área Culti-vada no Sistema de Plantio Direto na Palha e Brasil. Avaible from: <http://febrapdp.org.br/arquivos/EvolucaoAreaPDBr72A06.pdf (accessed 02 February (2010).

[14] Derpsch, R. Historical review of no-tillage cultivation of crops. In: Proceedings of the 1st JIRCAS Seminar on Soybean Research: No-Tillage Cultivation and Future Re-search Needs. Foz do Iguaçu, Paraná, Brazil. Tsukuba, Japan: JIRCAS (1998). and18. Avaible from: http://www.rolf-derpsch.com/notill.htm#5Working Report 13)., 1.

[15] Mascarenhas HAA; Esteves JAFWutke EB, Leão PCL. Nitrogênio residual da soja na produtividade de gramíneas e do algodão. Nucleus (2011).

[16] Liu, A. Q, Xu, Y. L, & Han, X. Z. Investigation and control of Soybean Monoculture in Heilongjiang Province. Liaoning Agric Sci (2001). , 3, 51-52.

[17] He, Z. H, Liu, Z. T, Xu, Y. L, Han, X. Z, & Xu, Y. H. Study on the reason reducing production of soybeans planted continuously and the way to get more output-yield and quality. Heilongjiang Agric Sci (2003). , 3, 1-4.

[18] Silva, A. P, Babujia, L. C, Franchini, J. C, Souza, R. A, & Hungria, M. Microbial bio-mass under various soil- and crop-management systems in short- and long-term ex-periments in Brazil. Field Crops Research (2010). , 119(1), 20-26.

[19] Restovich, S. B, Andriulo, A. E, & Portela, S. I. Introduction of cover crops in a maize-soybean rotation of the Humid Pampas: Effect on nitrogen and water dynamics. Field Crops Research (2012). , 128, 62-70.

[20] Karasawa, T, & Takebe, M. Temporal or spatial arrangements of cover crops to pro-mote arbuscular mycorrhizal colonization and P uptake of upland crops grown after nonmycorrhizal crops. Plant And Soil (2012).

[21] Syswerda, S. P, Basso, B, Hamilton, S. K, Tausig, J. B, & Robertson, G. P. Long-term nitrate loss along an agricultural intensity gradient in the Upper Midwest USA Agri-culture, Ecosystems and Environment (2012). , 149, 10-19.

[22] Djigal, D, Chabrier, C, Duyck, P. F, Achard, R, Quénéhervé, P, & Tixier, P. Cover crops alter the soil nematode food web in banana agroecosystems. Soil Biology and Biochemistry (2012). , 48, 142-150.

[23] Midegaa CAOKhana ZR, Amudaviab DM, Pittchara J, Pickettc JA. Integrated man-agement of *Striga hermonthica* and cereal stemborers in finger millet (*Eleusine coracana* (L.) Gaertn.) through intercropping with *Desmodium intortum*. International Journal of Pest Management (2010). , 56(2), 145-151.

[24] Green, J. M, Hale, T, Pagano, M. A, Andreassi, J. A, & Gutteridge, S. A. Response of 98140 corn with gat4621 and hra transgenes to glyphosate and ALS-inhibiting herbi-cides. Weed Sci (2009). , 57, 142-148.

[25] Green, J. M, & Owen, M. D. Herbicide-resistant crops: utilities and limitations for herbicide-resistant weed management. J Agric Food Chem (2011). , 59(11), 5819-5829.

[26] Norris, R. F. Ecological implications of using thresholds for weed management. In: BUHLER DD. Expanding the context of weed management. New York: Food Prod-ucts Press; (1999). , 31-58.

[27] Gunsolus, J. L, & Buhler, D. D. A risk management perspective on integrated weed management. In: D. D. Buhler (ed). Expanding the Context of Weed Management. New York: Haworth; (1999). , 167-187.

[28] Duke, S. O, & Powles, S. B. Glyphosate resistant crops and weeds: Now and in the future. AgBioForum (2009). , 12, 346-357.

[29] Duke, S. O, Baerson, S. R, & Rimando, A. M. Herbicides: glyphosate. In Encyclopedia of Agrochemicals. Plimmer JR, Gammon DW, Ragsdale NN. (Eds.). New York: Wiley; (2003). Avaible from: http://www.mrw.interscience.wiley.com/eoa/articles/agr119/frame.html.

[30] Cerdeira, A. L, Gazziero, D. L, Duke, S. O, & Matallo, M. B. Agricultural impacts of glyphosate-resistant soybean cultivation in South America. J Agric Food Chem (2011). , 59(11), 5799-5807.

[31] Neumann, G, et al. Relevance of glyphosate transfer to non-target plants via the rhizosphere. J. Plant Dis. Protect (2006). special edition) , 963-969.

[32] Fernandez, M. R, Selles, F, Gehl, D, & Depaw, R. M. Zentner RP: Crop production factors associated with *Fusarium* head blight in spring wheat in Eastern Saskatchewan. Crop Science (2005). , 45, 1908-1916.

[33] Sanogo, S, Yang, X. B, & Lundeen, P. Field response of glyphosate-tolerant soybean to herbicides and sudden death syndrome. Plant Dis. (2001). , 85, 773-779.

[34] Njiti, V. N, Myers, O, Schroeder, D, & Lightfoot, D. A. Roundup ready soybean: Glyphosate effects on *Fusarium solani* root colonization and sudden death syndrome. Agronomy Journal (2003). , 95, 1140-1145.

[35] Feng PCCBaley, GJ, Clinton WP, Bunkers GJ, Alibhai MF, Paulitz TC, Kidwell KK. Glyphosate inhibits rust diseases in glyphosate-resistant wheat and soybean. Proc. Natl. Acad. Sci. U.S.A. (2005). , 102, 17290-17295.

[36] Santos, J. B, et al. Avaliação de formulações de glyphosate sobre soja Roundup Ready. Planta Daninha (2007). , 25(1), 165-171.

[37] Santos, J. B, et al. Tolerance of Bradyrhizobium strains to glyphosate formulations. Crop Prot (2005). , 24(6), 543-547.

[38] Zablotowicz, R. M, & Reddy, K. N. Impact of glyphosate on the *Bradyrhizobium japonicum* symbiosis with glyphosate-resistant transgenic soybean: A minireview. Journal Environmental Quality (2004). , 33, 825-831.

[39] Dvoranen, E. C, et al. GR Glycine max nodulation and growth under glyphosate, fluazifop-p-butyl and fomesafen aplication. Planta Daninha (2008). , 26(3), 619-625.

[40] Zilli, J. E, et al. Efeito de glyphosate e imazaquin na comunidade bacteriana do rizoplano de soja (*Glycine max* (L.) Merrill) e em características microbiológicas do solo. R. Bras.Ci. Solo (2008). , 32(2), 633-642.

[41] Pakdaman, B. S, & Goltapeh, E. M. In vitro studies on the integrated control of Rapessed White Stem Rot disease through the application of herbicidas and Trichoderma species. Pakistan J. of Biology Science (2007). , 10(1), 7-12.

[42] Amna AliM Saleem Haider, Shabnam Javed, Ibatsam Khokhar, Irum Mukhtar and Sobia Mushatq. In vitro comparative screening of antibacterial and antigungal activities of some commom weeds extracts. Pakistan Journal of Weed Science Research (2012).

[43] Boydston, R. A, Mojtahedi, H, Crosslin, J. M, Thomas, E. E, Anderson, T, & Riga, E. Evidence for the Influence of Weeds on Corky Ringspot Persistence in Alfalfa and Scotch Spearmint Rotations. American Journal of Potato (2004). , 81, 215-225.

[44] Gustafson, T. C, Knezevic, S. Z, Hunt, T. E, & Lindquist, J. L. Early-season insect defoliation influences the critical time for weed removal in soybean. Weed Science (2006). , 54, 509-515.

[45] Collins, F. L, & Johson, S. J. Reproductive response of caged adult velvetbean caterpillar and soybean looper to the presence of weeds. Agr Ecosyst Envirom (1985). , 14, 139-149.

[46] Shelton, M. D, & Edwards, C. R. Effects of weeds on the diversity and abundance of insects in soybeans. Environ Entomol (1983). , 12, 296-298.

[47] Matioli, A. L. Ácaros predadores no controle biológico de ácaros-pragas. (2009). http://www.infobibos.com/Artigos/2009_3/acaros/index.htm.accessed 02 feb 2012).

[48] Roggia, S. Caracterização de fatores determinantes dos aumentos populacionais de ácaros tetraniquídeos em soja. Doctorate thesis. Escola Superior de Agricultura "Luiz de Queiroz", Universidade de São Paulo; (2010).

[49] Guedes JVC, Návia D, Lofego AC, Dequech STB. Ácaros associados à cultura da soja no Rio Grande do Sul, Brasil. Neotropical Entomology (2007). , 36(2), 228-293.

[50] Roggia, S. , Guedes JVC, Kuss RCR, Arnemann Já, Návia D. Spider mites associated to soybean in Rio Grande do Sul, Brazil. Pesquisa Agropecuária Brasileira (2008). , 43(3), 295-301.

[51] Meyer, S. J. Peterson RKD. Predicting movement of stalk borer (Lepidoptera: Noctuidae) larvae in corn. Crop Prot (1998). , 17, 609-612.

[52] Robbins, J. T, Snodgrass, G. L, & Harris, F. A. A review of wild hosts and their management for control of tarnished plant bug in cotton in the Southwestern U.S. Southwest. Entomol (2000). , 23, 21-25.

[53] Stadelbacher, E. A. Role of early-season wild and naturalized host plants in the buildup of the F1 generation of *Heliothis zea* and *H. virescens* in the delta of Mississippi. Environ. Entomol (1981). , 10, 766-770.

[54] Hilje, L, Costa, H. S, & Stansly, P. A. Cultural practices for managing *Bemisia tabaci* and associated viral diseases. Crop Prot (2001). , 20, 801-812.

[55] Altieri, M. A. Sustainable agricultural development in Latin America: exploring the possibilities. Agric Ecos and Environ (1992). , 39, 1-21.

[56] Suzuki, D. T. Griffiths AJF, Miller JH, Lewontin RC. Introdução à genética 4 edition. Rio de Janeiro: Guanabara Koogan; (1992).

[57] Gunsolus, J. L. Herbicide resistant weeds. Extension service. University of Minnesota (1999). p. Http://www.extension.umn.edu/documents/d/c/dc6077.htmLaccessed 09 June 2012).

[58] Betts, K. J, Ehlke, N. J, Wyse, D. L, Gronwald, J. W, & Somers, D. A. Mechanism of inheritance of diclofop resistance in italian ryegrass (Lolium multiflorum). Weed science (1992). , 40(2), 184-189.

[59] Saari, L. L, Cotterman, J. C, & Thill, D. C. Resistance to acetolactate synthase inhibiting herbicides. In: Powles SB and Holtum JAM (ed). Herbicide resistance in plants: biology and biochemistry. Boca Raton; (1994). , 83-139.

[60] Kissmann, K. G. Resistência de plantas a herbicidas. São Paulo: Basf Brasileira S.A.; (1996). p

[61] Maxwell, B. D, & Mortimer, A. M. Selection for herbicide resistance. In: Powles SB and Holtum JAM (eds). Herbicide resistance in plants: biology and biochemistry. Boca Raton; (1994). , 1-25.

[62] Heap, I. International survey of herbicide resistant weeds. http:///www.weedscience.com/(2003). accessed 24 May 2012).

[63] Mallory-smith, C. A, Thill, D. C, & Dial, M. J. Identification of sulfonylurea herbicide-resistance prickly lettuce (Lactuca serriola). Weed technology (1990). , 4(1), 163-168.

[64] Weed ScienceHerbicide-resistat weeds by year. 1998; Http://www.weedscience.com/byyear/year.htmaccessed 03 April (2012). , 17.

[65] Mortimer, A. M. Review of graminicide resistance. 1998; 32 p. Avaible from Http://ipmwww.ncsu.edu/orgs/hrac/monograph1.htmaccessed 22 May (2012).

[66] Jasieniuk, M, Brule-babel, A. L, & Morrison, I. N. The evolution and genetics of herbicides resistance in weeds. Weed science (1996). , 44(1), 176-193.

[67] Mulugeta, D, Maxwell, B. D, Fay, P. K, & Dyer, W. E. Kochia (Kochia scoparia) pollen dispersion, viability and germination. Weed science (1994). , 42(4), 548-552.

[68] Stallings, G. P, Thill, D. C, Mallory-smith, C. A, & Shafii, B. Pollen-mediated gene flow of sulfonylurea-resistant kochia (Kochia scoparia). Weed science (1995). , 43(1), 95-102.

[69] Holt, J. S, & Lebaron, H. M. Significance and distribution of herbicide resistance. Weed technology (1990). , 4(1), 141-149.

[70] Powles, S. B. Evolved glyphosate-resistant weeds around the world: lessons to be learnt. Pest Management Science (2008). , 64, 360-365.

[71] Marzocca, A, & Marsico, O. J. Del Puerto O. Manual de Malezas 3 edition. Editorial Hemisferio Sur; (1976).

[72] INTA – El cultivo de la soja en Argentina. In: Laura Giorda and Héctor Baigorri. Instituto Nacional de Tecnología Agropecuaria, Secretaría de Agricultura, Ganadería, Pesca y Alimentación(1997).

[73] Faccini, D, & Puricelli, E. Eficacia de herbicidas según la dosis y el estado de crecimiento de Malezas presentes en un suelo en barbecho. Agriscientia (2007). , 24, 29-35.

[74] Faccini, D, Tuesca, D, Puricelli, E, Nisendohn, L, Merindol, D, & Ruggeri, L. Control químico de Malezas de invierno. Revista Agromensajes (2009).

[75] Pereira, E. S, Velini, E. D, & Carvalho, L. R. Rodella RCSM. Quantitative and qualitative weed evaluation of soybean crop in no-tillage and conventional tillage systems. Planta Daninha (2000). , 18(2), 207-216.

[76] Pacheco, L P, Leandro, W M, Machado, P, Assis, O A, Cobucci, R L, Madari, T, & Petter, B E. F A. Produção de fitomassa e acúmulo e liberação de nutrientes por plantas de cobertura na safrinha. Pesq agropec bras (2011). , 46(1), 17-25.

[77] Brookes, G, Barfoot, P, & Crops, G. M. global socio-economic and environmental impacts UK :PG Economics Ltd.; (2011). , 1996-2010.

[78] Gazziero, D, Adegas, P, Voll, F S, Vargas, E, Karam, L, Matallo, D, Cerdeira, M B, Fornaroli, A L, Osipe, D A, Spengler, R, Zoia, A N, & In, L. Interferência da buva em areas cultivadas com soja [CD-ROM]; XXVII Congresso Brasileiro da Ciência das Plantas Daninhas: Ribeirao Preto, SP, Brazil; 2010. In press: Brazilian Weed Science Society: Londrina, Brazil; (2010).

[79] Oliveira Neto AMConstantin J, Oliveira Jr RS, Guerra N, Dan HA, Alonso DG, Blainski E, Santos G. Winter and summer management strategies for *Conyza bonariensis* and *Bidens pilosa* control. Planta Daninha (2010). special edition).

[80] Heatherly, L G, Reddy, K N, & Spurlock, S R. Weed management in glyphosate-resistant and non-glyphosate-resistant soybean grown continuously and in rotation. Agronomy Journal (2005). , 97, 568-577.

[81] Gibson, L, & Benson, G. O. History, and Uses of Soybean (*Glycine max*). Iowa State University, Department of Agronomy. Available from: http://www.agron.iastate.edu/courses/agron212/Readings/Soy_history.htmaccessed 06 June (2012).

[82] Talbert, R. E, & Burgos, N. R. History and Management of Herbicide-resistant Barnyardgrass (*Echinochloa Crus-galli*) in Arkansas Rice. Weed Technology (2007). , 21(2), 324-331.

[83] Bonny, S. Herbicide-tolerant Transgenic Soybean over 15 Years of Cultivation: Pesticide Use, Weed Resistance, and Some Economic Issues. The Case of the USA. Sustainability (2011). , 3(9), 1302-1322.

[84] U.D. Department of Agriculture- National Agricultural Statistics Service. USDA-NASS: Acreage 2009. Washington, DC. Available from: http://usda.mannlib.cornell.edu/usda/current/Acre/Acre-06-30-2009.pdf.

[85] Owen, M. D K. Midwest experiences with herbicide resistant crops. Proceedings of the Western Society of Weed Science (1997). , 9-10.

[86] Mortensen, D. A, Egan, J. F, Maxwell, B. D, Ryan, M. R, & Smith, R. G. Navigating a critical juncture for sustainable weed management. Bioscience (2012). , 62(1), 75-84.

[87] Usncc-ipm- U, S. National IPM Coordinating Committee. Integrated pest management: a national plan for future direction. U.S. National IPM Coordinating Committee (1988). , 12.

[88] Sanyal, D, Prasanta, C. B, Randy, L. A, & Anil, S. Revisiting the Perspective and Progress of Integrated Weed Management. Weed Science (2008). , 56(1), 161-167.

[89] Liebman, M, Mohler, C. L, & Staver, C. P. Ecological Management of Agricultural Weeds. Cambridge, UK: Cambridge University Press; (2001). p.

[90] Malone, S. Herbert JrDA, Pheasant S. Determining adoption of integrated pest management practices by grains farmers in Virginia. Journal of Extension (2004).

[91] Hammond, C. M, Luschei, E. C, Boerboom, C. M, & Nowak, P. J. Adoption of integrated pest management tactics by Wisconsin farmers. Weed Technology (2006). , 20, 756-767.

[92] Fuglie, K. O, & Kascak, C. A. Adoption and diffusion of natura lresource-conserving agricultural technology. Review of Agricultural Economics (2011). , 23, 386-403.

[93] Monjardino, M, Pannell, D. J, & Powles, S. B. Multispecies resistance and integrated management: a bioeconomic model for integrated management of rigid ryegrass (*Lolium rigidum*) and wild radish (*Raphanus raphanistrum*). Weed Science (2003). , 51(5), 798-809.

[94] Boyette, C. D, Hoagland, R. E, & Weaver, M A. Interaction of a bioherbicide and glyphosate for controlling hemp sesbania in glyphosate-resistant soybean. Weed Biology and Management (2008).

[95] Cook, J. C, Raghavan, C, Thomas, W. Z, Erin, N. R, & William, M. S. Gregory EMD. Effects of *Alternaria destruens*, Glyphosate, and Ammonium Sulfate Individually and Integrated for Control of Dodder (*Cuscuta pentagona*). Weed Technology (2009). , 23(4), 550-555.

[96] Eric, V. H, Nicholas, J, Jianhua, Z, & Donald, L. W. Integrated cultural and biological control of Canada thistle in conservation tillage Soybean. Weed Science (2001). , 49(5), 642-646.

[97] Thomas, A. G, Legere, A, Leeson, J. Y, Stevenson, F. C, Holm, F. A, & Gradin, B. Weed community response to contrasting integrated weed management systems for cool dryland annual crops. Weed Research (2011). , 51, 41-50.

[98] Oliveira, T K, & Carvalho, G. J. Moraes RNS. Plantas de cobertura e seus efeitos sobre o feijoeiro em plantio direto. Pesquisa Agropecuária Brasileira (2002). , 37(8), 1079-1087.

[99] Puente, M, Darnall, N, & Forkner, R. E. Assessing Integrated Pest Management Adoption: Measurement Problems and Policy Implications. Springer, Environmental Management, Forthcoming (2011). Available from: SSRN: http://ssrn.com/abstract=1911509

[100] Davis, V. M, Gibson, K. D, Bauman, T. T, Weller, S. C, & Johnson, W. G. Influence of weed management practices and crop rotation on glyphosate- resistant horseweed (*Conyza canadensis*) population dynamics and crop yield-years III and IV. Weed Science (2009). , 57, 417-426.

[101] Culpepper, A S, Sosnoskie, L M, Kichler, J, & Steckel, L E. Impact of Cover Crop Residue and Tillage on the Control of glyphosate-resistant Palmer Amaranth. Paper presented at the 2011 Weed Science Society of America Annual Meeting; February (2011). Portland, Oregon2011.., 7-10.

[102] Zalucki, M. P, Adamson, D, & Furlong, M. J. The future of IPM: whither or wither? Australian Journal of Entomology (2009). , 48, 85-96.

[103] Young, S. L. True Integrated Weed Management. Weed Research (2012). , 52, 107-111.

Chapter 5

Engineered Soybean Cyst Nematode Resistance

Vincent P. Klink, Prachi D. Matsye,
Katheryn S. Lawrence and Gary W. Lawrence

Additional information is available at the end of the chapter

1. Introduction

A variety of plant parasitic nematodes (PPNs), including the soybean cyst nematode (SCN), elicit the initiation, development and maintenance of a specialized nurse cell from which they derive their nutriment (Figure 1). Remarkably, during parasitism by the PPN, the nurse cell survives the apparently significant resource drain on the root cell that would be expected to detrimentally impact normal physiological processes of the cell. This outcome indicates that the nematode has developed a well tuned apparatus to ensure that the root cell does not collapse and die during parasitism. In contrast, in the soybean-SCN pathosystem, the nurse cell and sometimes the surrounding cells are the sites of the defense response to the parasite (Ross, 1958; Endo, 1965). Therefore, plants have in place a mechanism to overcome the influence of the activities of the nematode. Identifying the factor(s) is of utmost importance in developing resistance to PPNs.

1.2. History

Documented accounts reveal that soybean has been in cultivation for thousands of years (Hymowitz et al. 1970), beginning in Asia perhaps as early as 3,500 B.C. (Liu et al. 1997). While the natural range of soybean is East Asia, after thousands of years of cultivation a true understanding of its native range is complicated at best. However, the extensive range of wild soybean and obvious differences in its growth habit indicates that while environmental cues may be responsible for changes in soybean and plant growth habit in general (Garner Allard 1930; Chapin III et al. 1985; Day et al. 1999), genetic variation that exists in wild populations is of significant benefit to agriculture for production purposes and developing resistance to its many pathogens. This assessment is particularly true for soybean and its most significant pathogen, SCN, as many ecological collections have resulted in the identification of naturally occurring resistance (Ross and Brim, 1957; Ross, 1958; Epps and Hartwig, 1972; Concibido et al. 2004; Ma et al. 2006; Li et al. 2011; Matsye et al. 2012).

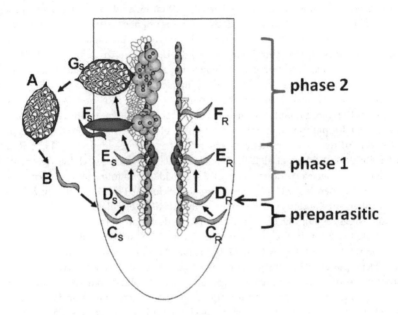

Figure 1. SCN life cycle during susceptible and resistant reactions Fig. 1A, cysts with eggs (white). Fig. 1B, second stage pre-infective juveniles (pi-J2) (gray) migrate toward the root. SUSCEPTIBLE REACTION: Fig. 1C$_s$, the infective-J2 (i-J2) nematodes (light gray) burrow in and migrate toward the stele. Fig. 1D$_s$, feeding site selection (yellow). Fig 1E$_s$, p-J2 nematodes molt into J3, then J4. During this time, the original feeding site is incorporating adjacent cells (purple) via cell wall degradation events. Meanwhile, the male discontinues feeding at the end of its J3 stage. Fig 1F$_s$, After maturation, the male and female nematodes copulate. Fig. 1G$_s$, the female at ~30 dpi. RESISTANT REACTION: Fig. 1C$_R$, Like the susceptible reaction, the infective-J2 (i-J2) nematodes (dark gray) burrow into the root and migrate toward the root stele. Fig. 1D$_R$, feeding site selection by the parasitic J2 (p-J2). Fig 1E$_R$, the syncytium begins to develop. Fig. 1F$_R$, the syncytium has collapsed resulting in nematode mortality. The right panel shows the initiation phase of infection (black arrow). Phase 1 is the development period. Phase 2 is the maintenance period (adapted from Klink et al. 2011b).

While knowledge of soybean's cultivation is long and extensive, scientific information on its dominant pathogen, the SCN, began with its description (Ichinohe, 1952). However, reports going back as early as the 1880's (Noel 1992) and late 1930s (Ichinohe, 1961) reveal knowledge of the nematode and appreciation of its pathogenic capacity. The SCN is a devastating pathogen that causes approximately 7-10% production loss, worldwide, annually and suppresses seed yield more than any other single soybean pathogen (Wrather et al. 1995, 2001a, b, 2003, 2006; Pratt and Wrather, 1998). In contrast, in some fields, as much as a 15% loss in yield has been observed with no visible signs of disease on soybean (Wang et al. 2003). Observations such as these could complicate SCN management since the disease can occur without knowledge of it being present in a particular field.

The near perfect overlap of the agricultural range of soybean with the distribution of SCN, infection creates a scenario where there is a high probability of a widespread and significant effect on yield. Conservative reports have shown that there is approximately 1.1-1.5 billion dollars in agronomic losses, annually, worldwide (Wrather et al. 2001). The overlapping distribution of SCN with that of soybean production was not always the case. Historically, SCN was not found in the U.S., or likely even North America or the New World. That situation changed when the SCN was first identified in the U.S. in North Carolina in 1954 by Winstead et al. and published a year later (Winstead et al. 1955). Unfortunately for agriculture, SCN is readily transmissible as evidenced by its identification in localities as far away as Mississippi only a few years later by 1957 (Spears, 1957). The SCN now is a registered invasive species in the U.S. Notably, in the U.S., SCN causes more agricultural loss to soybean than the rest of its pathogens combined (Wrather et al. 2001, 2006). Making the problem worse for agriculture is the genetic diversity of the SCN (Golden et al. 1970; Riggs and Schmidt 1988; 1991; Niblack et al. 2002; Bekal et al. 2008). SCN research has determined that the nematode is a species complex originally subdivided into four races (Golden et al. 1970) which later was expanded into 16 races (Riggs and Schmidt 1988, 1991) that have been reorganized, further subdivided and reclassified into distinct populations (Niblack et al. 2002; Niblack and Riggs, 2004). The term population was designated since genetically pure clones are impossible to obtain in the sexually reproducing SCN system (Niblack et al. 2002). The classification scheme of Niblack et al. (2002) is based on the varying ability of SCN populations to infect a panel of 7 soybean genotypes that can resist infection to varying levels. It is noted that some of these designated populations are "strains" that are maintained in the greenhouse and genetically purified through hundreds of generations of single cyst descent (Niblack et al. 1993). Therefore, the genetic background in these "strains" may not resemble the original field-extracted population since allelic forms of the parasitism genes would likely be lost through this purification process.

The genetic diversity found in SCN (Golden et al. 1970; Riggs and Schmidt 1988; 1991; Niblack et al. 2002; Bekal et al. 2008) likely aids in its ability to infect and reproduce on plants other than soybean. Thus, from an ecological standpoint, SCN could pose a threat to plants that grow outside of production areas. This potential problem would be exasperated if those plant species are listed as endangered or threatened species or are a significant component of the ecological community. A number of studies have shown that the SCN reproduces on at least, but certainly is not limited to, 97 legume and 63 non-legume hosts (Epps and Chambers, 1958; Riggs and Hamblen, 1962, 1966a, b) and new SCN hosts are determined on a regular basis (Creech et al. 2006). It has been many years since species range tests have been performed for SCN so it is likely that these lists of hosts are not comprehensive. This virulence capability of SCN poses a problem in terms of its management since SCN populations could be maintained by weedy plants that grow or overwinter in fallow fields or along the boundaries of acreage that is in production (Creech et al. 2006). In addition to these problems, SCN does not even have to reproduce in the plant to still cause damage to the plant. While the genetic diversity of both soybean and SCN may appear to complicate an understanding of the process of infection and the development of resistant cultivars, the natural variation in both the germplasm of soybean (Doyle et al. 1999) and SCN (Bekal et al. 2003;

2008) presents many opportunities to understand the basic machinery of infection of the SCN and the genotype-specific nuances that regulate both susceptibility and defense. These features make the soybean-SCN pathosystem an extremely valuable experimental model (Barker et al. 1993; Opperman and Bird, 1998; Niblack et al. 2006; Klink et al. 2010a).

1.3. Methods to control SCN infection

Historically, SCN has been managed through a combination of chemical control, cropping systems, biological control and the identification and use of resistant germplasm. Chemical control for pathogens using methyl bromide first occurred in France (Schneider et al. 2003; Rosskopf et al. 2005) and had subsequently been used for decades for both pre- and post plant nematode control. However, methyl bromide is a chemical that has been phased out of use in 2005 in the U.S. (Schneider et al. 2003; Rosskopf et al. 2005) because it has been classi-fied as a Class 1 (Group VI) stratospheric ozone depletor by the Environmental Protection Agency. Because of the loss of this major control agent for the SCN, even in developing countries by the year 2015 (Rosskopf et al. 2005), it was important to identify other strategies that could be included in the SCN management plan. Biocontrol measures that include bac-teria, fungi or even their proteins are feasible (Chen and Dickson, 1996; Kim and Riggs, 1991, 1995; Liu and Chen, 2001; Meyer and Huettel, 1996; Meyer and Meyer, 1996; Timper et al., 1999). Other control methods that had already been used extensively for decades include the time honored crop rotation strategy. This strategy has reduced SCN populations below damaging levels (Francl and Dropkin, 1986; Sasser and Uzzell, 1991; Koenning et al. 1993). Rotating with nonhosts over a 2-3 year period mitigated the undesirable levels of SCN in the field (Ross, 1962; Francl and Dropkin, 1986). Other cropping systems that have had success in SCN control are the use of blending, resistant cultivars and cropping sequence, among others (Niblack and Chen, 2004). While successful, a problem with cropping strategies is that the interval is not long enough to compete with the 9 year cycle that cysts can remain viable, but dormant in production fields (Inagake and Tsutsumi, 1971). With these strategies in place it is possible to develop a tightly managed regime, incorporating some or all of these technologies and principles to mitigate SCN damage. Lastly, genetic engineering has begun to take root with potential as a method to generate resistance (Steeves et al. 2006; McLean et al. 2007; Klink et al. 2009a; Matsye et al. 2012). However, for genetic engineering to be successful, it is first required that candidate genes be identified. The identification of these genes has happened through a series of RNA gene expression studies, employing soy-bean germplasm that exhibits resistance to SCN.

1.3.1. Available resistant germplasm

Once SCN was identified in the U.S. (Winstead et al. 1955), a very large need existed to deter-mine if soybean germplasm existed that could resist infection. The vast and expansive range of soybean (Morse, 1927) and visually obvious variations in growth form in its various ecological habitats provided the possibility that germplasm that was resistant to SCN existed in its wild populations. Established in 1898, the development of a substantial and publically available seed bank was initiated that is maintained by the USDA-National Plant Germplasm System (USDA-

NPGS) (Morse, 1927; Bernard et al. 1987). It now contains approximately 20,000 varieties (accessions) with each accession classified as a plant introduction (PI) through a numbering system. Many of the 7,867 PIs that were already available by 1944, just 10 years before the identification of SCN in the U.S., had been collected in trips to China, Japan, India and Korea in a small window of time between 1924 and 1932 (Bernard et al. 1987). The public availability of the germplasm allowed it to be used in a series of trials to determine if any of the available accessions was resistant to SCN. A number of accessions were determined to be resistant to SCN through two large trials that studied about 5,700 accessions (Ross and Brim, 1957; Ross, 1958; Epps and Hartwig, 1972). Research on these accessions has resulted in the identification of approximately 118 sources of resistance (Concibido et al. 2004). However, only approximately seven sources are used for cultivar development in the U.S. (Shannon et al. 2004). These accessions include the *G. max* PIs known as Peking (*G. max*$_{[Peking]}$) and *G. max*$_{[PI 88788]}$. Currently, *G. max*$_{[Peking]}$ and *G. max*$_{[PI 88788]}$ resistance germplasm is present in >97% of all commercial cultivars in the U.S. (Concibido et al. 2004). In addition to these PIs, hundreds of additional accessions that can resist SCN infection have been identified in China (Ma et al. 2006; Li et al. 2011). These banks of germplasm provide an important and substantial genetic resource for understanding the process of parasitism in soybean at the cellular level. This is important to understand because the infection of soybean involves very specific cell types that react in very specific ways to SCN parasitism.

1.4. Cytological reaction during resistance

The SCN can remain viable in the soil in eggs ensheathed within the carcass of the dead mother (cyst wall) for up to 9 years (Inagake and Tsutsumi, 1971). However, the devastating interaction of the SCN with soybean begins when it burrows into the root through the epidermal and cortical cells. This has been shown both by cytological studies and gene expression studies of time points collected before the formation of syncytia (Alkharouf et al. 2006; Klink et al. 2009b). The interaction continues through the initiation and subsequent formation a multinucleate nurse cell known as a syncytium from pericycle or neighboring cells (Ross, 1958; Endo, 1964, 1992). The formation of the syncytium is likely to be a very coordinated process, occurring through the injection and subsequent activity of nematode parasitism proteins (Atkinson and Harris, 1989; Smant et al. 1999; Lambert et al. 1999; De Boer et al. 1999, 2002; Wang et al. 1999, 2001, 2003; Gao et al. 2001, 2003; Bekal et al. 2003). These substances likely orchestrate successive waves of interference of the root cell's normal physiological processes and initiate various cell wall dissolving events (Atkinson and Harris, 1989; Smant et al. 1999; Wang et al. 1999, 2001, 2003; Lambert et al. 1999; De Boer et al. 1999, 2002; Gao et al. 2001, 2003; Bekal et al. 2003). The parasitism process merges approximately 200-250 root cells into a common cytoplasm containing as many nuclei, the definition of a syncytium (Jones and Northcote, 1972; Jones, 1981). Additional nematode activities alter the plant cell's physiology (Klink et al. 2005, 2007a; Ithal et al. 2007). The activities benefit the nematode during the sedentary period of its life cycle as they feed and mature (Edens et al. 1995; Hermsmeier et al. 1998; Mahalingam et al. 1999; Vaghchhipawala et al. 2001; Klink et al. 2005, 2007a; Alkharouf et al. 2006; Ithal et al. 2007; Matsye et al. 2011).

Cytological studies of the SCN infection process (Figure 1) have shown that the cellular response of soybean to SCN infection can be divided into an earlier phase (phase 1) and a later phase (phase 2) (Ross, 1958; Endo, 1964, 1965, 1991; Riggs et al. 1973; Kim et al. 1987; Mahalingam and Skorpska, 1996). Phase 1 and 2 span the periods including the initiation, development and maintenance of the syncytium (Figure 1). These observations are not unique to SCN since similar observations have also been made for the cyst nematode *Rotylechulus reniformis* (Robinson et al. 1997), indicating that a basic level of conservation may exist for the process of defense at the site of infection while genotype-specific gene activities also exist (Klink et al. 2011a; Matsye et al. 2011, 2012).

During phase 1, the cellular reactions leading to susceptibility or defense appear the same at the cytological level. The cellular events occurring during the earlier stages of syncytium development include hypertrophy, the dissolution of cell walls, the development of dense cytoplasm, an enlargement of nuclei and an increase in endoplasmic reticulum (ER) and ribosome content (Endo, 1964, 1965; Riggs et al. 1973; Kim et al. 1987; Kim and Riggs, 1992; Mahalingam and Skorpska, 1996). The enlargement of nuclei and increase in ER and ribosome content indicate an increase in gene expression and protein synthesis accompanies the activity of the nematode within the parasitized cells. Therefore, it is likely that the plant cell is being programmed to make specific materials to benefit the nutritional needs of the nematode. It is known that plant parasitic nematodes lack the ability to make materials such as sterols (Chitwood and Lusby, 1990). Therefore, altering the metabolism of the parasitized cell probably would involve the induction of metabolic activity that relates to these processes. Cell fate mapping experiments have demonstrated the metabolism that occurs during these stages of parasitism and some of it relates to enhanced plant sterol production (Klink et al. 2011a; Matsye et al. 2011).

After these earlier events, the cytology of susceptibility and defense become apparent and is referred to as phase 2. Phase 2 of the susceptible reaction is characterized by hypertrophy of nuclei and nucleoli. This process is accompanied by the reduction and dissolution of the vacuole. The reduction and dissolution of the vacuole suggests important events or structural features involved in membrane fusion and/or maintenance are perturbed. This topic will be described in a later section. Other cellular events that have been identified during the susceptible reaction include cell expansion as it incorporates and fuses with adjacent cells (Endo and Veech 1970; Gipson et al 1971; Jones and Northcote, 1972; Riggs et al. 1973; Jones, 1981). Additional activities include the proliferation of cytoplasmic organelles.

In contrast, the cellular aspects of the defense responses occurring during phase 2 depend on the soybean genotype being infected. Information that has been generated through a number of cytological studies have resulted in the development of a system that divides the PIs into cohorts having similar cytological reactions that is based on the cellular characteristics associated with how SCN responds during resistance (Colgrove and Niblack, 2008). Currently, the PIs have been categorized into those genotypes having the *G. max*$_{[Peking]}$ and *G. max*$_{[PI 88788]}$-types of defense responses (Colgrove and Niblack, 2008). Much more work in this area of research is required for a comprehensive understanding of the different forms of the defense response. Such knowledge would allow the commonalities of the cytological features to be correlated with the molecular events that are occurring in the parasitized cell types. By doing so, it would allow for the identification of genes that always correlate to resistance, regardless of the cytology or genotype of soybean. It would be likely that these genes are central to all forms of the defense response (Klink et al. 2011a; Matsye et al. 2011, 2012). Among the characteristics that define these cohorts, the *G. max*$_{[Peking]}$-type of defense includes the development of a necrotic layer that surrounds the head of the nematode (Kim et al. 1987; Endo, 1991). This process is followed by necrosis of the initial cell that the nematode had parasitized. In contrast, in the *G. max*$_{[PI\ 88788]}$-type of defense response, the necrotic layer that surrounds the head of the nematode is lacking and the initial cell that the nematode parasitized first experiences necrosis (Kim et al. 1987; Endo, 1991). In addition to these cytological characteristics found in the *G. max*$_{[Peking]}$ and *G. max*$_{[PI\ 88788]}$-types of defense responses are the presence or absence of cell wall appositions (CWAs). CWAs are structures defined as physical and chemical barriers that are designed to prevent cell penetration (Aist et al. 1976, Schmelzer, 2002; An et al. 2006a, b; Hardham et al. 2008). CWAs have been found and studied in other plant-organism pathosystems (Collins et al. 2003; Assaad et al. 2004; Kalde et al. 2007). However, CWAs are not a defining characteristic of all types of defense responses in soybean. CWAs have been found in the *G. max*$_{[Peking]}$-type of resistant reaction and are found in the *G. max*$_{[PI\ 437654]}$ genotype (Mahalingam and Skorpska, 1996). This makes the placement of *G. max*$_{[PI\ 437654]}$ in the *G. max*$_{[Peking]}$ cohort logical (Colgrove and Niblack et al. 2008). In contrast, CWAs are lacking in *G. max*$_{[PI\ 88788]}$. More work is required in the understanding the role(s) that CWAs play, if any, during defense of soybean to SCN. However, the significance and role of CWAs during defense were first demonstrated by Collins et al. (2003), and followed by additional studies performed by Assaad et al. (2004) and Kalde et al. (2007). In those studies, it was shown at the molecular level that CWA formation involves the vesicular transport machinery protein component known as syntaxin. This was a striking discovery since the process of vesicular transport is a conserved cellular process, meaning it has been found in other organisms. The syntaxin gene was first identified in animal systems (Inoue et al. 1992; Bennett et al. 1992) and through a number studies performed in animal and model genetic systems it was shown that syntaxin interacts with other proteins to accomplish specific cellular functions. Unfortunately, while the role of syntaxin in plant defense has been studied (Collins et al. 2003; Assaad et al. 2004; Kalde et al. 2007), the examination of other components of the vesicular transport machinery has received little attention. Until very recently (Matsye et al. 2012), no information existed on how these proteins function or interact with syntaxin during the defense of plants to pathogens. The demonstration that syntaxin plays a role in the defense of plants to pathogens, implicates that other proteins that interact directly or indirectly with syntaxin probably are also involved in the process of defense. A genetic pathway, involving *PEN1*, the β-glycosyl hydrolase *PEN2* and the ABC transporter *PEN3* transports and delivers antimicrobial compounds across the cell membrane to sites where the fungus is attempting to enter (Collins et al. 2003; Lipka et al. 2005; Stein et al. 2006). Other proteins that interact directly with syntaxin have been studied in other experimental systems and include the ATPase known as N-ethylmaleimide-sensitive factor attachment protein (NSF) (Malhotra et al. 1988), the soluble N-ethylmaleimide-sensitive factor attachment receptor protein (SNARE) complex and synaptosomal-associated protein 25 (SNAP25) (Oyler et al. 1989), the soluble N-ethylmaleimide-sensitive factor attachment protein (SNAP) (Weidman et al. 1989; Clary et al, 1990; Collins et al. 2003; Assaad et al. 2004; Kalde et al. 2007), among other proteins. Since these numerous studies have shown very specifically how the protein complex is assembled, it was then possible to determine how specific components of the CWA assembly process that are present during defense of soybean to SCN function (Matsye et al. 2012). However, even though CWAs are lacking in genotypes like G. max$_{[PI 88788]}$, it does not mean that the proteins are not involved in defense through related activities. Membrane fusion has been shown to play a role in defense through a process known as autophagy. Autophagy is a process known in plants to play crucial roles in defense (Patel and Dinesh-Kumar 2008; Hofius et al. 2009; Lenz et al. 2011; Lai et al. 2011). This knowledge has allowed for a targeted approach in understanding the protein machinery that is involved in defense (Matsye et al. 2012).

1.5. Genomics-based studies of SCN

A number of "omics" studies in the soybean-SCN pathosystem have been performed to understand both plant and nematode gene expression at the organismal level. Many of the gene expression studies that relied on the microarray technology were modeled after earlier experiments that were performed in the model plant *Arabidopsis thaliana* that were infected with the beet cyst nematode, *Heterodera schactii* (Puthoff et al. 2003). Studying SCN in *A. thaliana* is complicated by the fact that it is a nonhost to SCN infection so studies investigating the susceptible reaction elicited by SCN in *A. thaliana* cannot be done until suitable mutants or susceptible ecotypes are identified. A number of microarray studies using whole infected soybean roots as a source for the RNA samples have identified genes that are expressed during a susceptible reaction (Alkharouf et al. 2006; Ithal et al. 2007; Klink et al. 2007b).

The parasitism of soybean by SCN begins with and is sustained through the injection of materials that are synthesized in subventral and esophageal glands into the root cell. It is a costly life strategy since typically only about 10% of the nematodes ever make it to maturity in a fully susceptible genotype like G. max$_{[Williams 82/PI 518671]}$. Identifying the genes involved in parasitism would likely occur through collecting the cytoplasm of the cells composing the subventral and esophageal glands. It was hypothesized that these genes would be important for the events of parasitism and would be involved in altering the metabolic processes of the soybean to benefit the nematode. The experiments were performed by microaspirating the cytoplasm of the gland cells, constructing cDNA libraries and sequencing the genes, allowing for downstream bioinformatics analyses to help elucidate what the genes could actually be (Smant et al. 1998). The experiments were then repeated for the SCN, identifying a number of putative parasitism genes (Wang et al. 1999; Gao et al. 2001, 2003). With the development of the Affymetrix® Soybean GeneChip, it was possible to examine the expression of thousands of SCN genes simultaneously. This was made possible because 7,539 H. *glycines* probe sets representing 7,431 transcripts (genes) were printed onto the array. One analysis examined the expression of SCN genes that were expressed specifically during infection of the G. max$_{[Williams 82/PI 518671]}$ genotype that lacks a functional defense response (Ithal et al. 2007). This means that gene expression occurring during a susceptible reaction was monitored. The work examined the expression of previously identified (Wang et al. 1999; Gao et al.

2001, 2003) and analyzed (Bakhetia et al. 2007) putative parasitism genes (Ithal et al. 2007). The remaining genes that were fabricated onto the array were not a focus of the analysis. The experiments confirmed the expression pattern of dozens of putative parasitism genes (Ithal et al. 2007). A gap in the knowledge was that the experiments were not designed to determine what genes were expressed as the nematode experienced a resistant reaction in a soybean genotype that was capable of a defense reaction. This information would be important because it would provide knowledge on the metabolic pathways that may be sensitive to genetically-based control measures. That gap in knowledge was filled in experiments that performed population-specific analyses of gene expression, comparing the susceptible and resistant reactions experienced by SCN as they infected the $G.$ $max_{[Peking/PI 548402]}$ genotype that has a functional defense response to some populations of SCN (Klink et al. 2007a). Thus, from the experiments of Ithal et al. (2007) and Klink et al. (2007a), specific knowledge of gene expression occurring in genotypes both lacking and having functional resistance genes was obtained. It is noted that additional gene expression profiling experiments have also been performed (Elling et al. 2009). In earlier studies, Alkharouf et al. (2007) annotated all of the SCN genes that were available in Genbank and compared them to the genetic model free living nematode, *Caenorhabditis elegans*. The advantage of these comparisons was that the genome of *C. elegans* had been sequenced (C. elegans Sequencing Consortium, 1998), allowing for a substantial annotation process to be executed. In addition, there was a massive amount of functional data obtained through genetic and reverse genetic experiments (Fire et al. 1998; Piano et al., 2000; Kamath et al., 2003; Sonnichsen et al., 2005) that was available for essentially every gene in the genome housed in the *C. elegans* database at http://www.wormbase.org. Since the genomic sequence of *C. elegans* is known, it is possible to find highly conserved and related genes in SCN. The working hypothesis was that if the genes in *C. elegans* and SCN are nearly identical in primary sequence, it would be likely that they have similar function. If the genes have similar function, for example an essential function for survival in *C. elegans*, knocking out that gene in SCN would probably result in lethality for those nematodes if the gene could be knocked out. The annotation of the SCN genes was driven by a homology criterion whereby the SCN genes were pooled into six bins referred to as Groups 1-6 (Alkharouf et al. 2007). The six bins were based on the level of homology the sequence had to *C. elegans* genes. Group 1 had the highest level of homology and Group 6 had the lowest level. For example, Group 1 had E-values between 0 and $1E$-100; Group 2 had E-values between $1E$-100 and $1E$-80; Group 3 had E-values between $1E$-80 and $1E$-60; Group 4 had E-values between $1E$-60 and $1E$-40; Group 5 has E-values between $1E$-40 and $1E$-20 while Group 6 has E-values > $1E$-20 (Alkharouf et al. 2007). The gene annotation process resulted in taking the nearly 8,334 conserved genes between $H.$ *glycines* and *C. elegans* and identifying 1,508 that have been shown to have lethal phenotypes/phenocopies in *C. elegans* (Alkharouf et al. 2007). The research then was poised to test the function of the 1,508 genes, but it was an unmanageable number of genes. To narrow down the 1,508 genes to a manageable number for functional studies, the genes underwent further annotation procedures (Alkharouf et al. 2007). To do this annotation procedure, firstly, a pool of 150 highly conserved, Group 1, $H.$ *glycines* homologs of genes having lethal mutant phenotypes or phenocopies from the free living nematode *C. elegans* were identified from the pool of 1,508 genes that were fabricated onto the Affymetrix® microarray. Secondly, it was determined that of those 150 genes on the Affymetrix® soybean GeneChip, a subset of 131 genes could have their expression monitored during the parasitic phase of their life cycle. Thirdly, a microarray analyses identified a core set of 32 genes with induced expression occurring during the parasitic stages of infection. The identification of 32 genes that had known expression during the parasitic stages of infection provided a small, but feasible, core set of genes that could be targeted in RNAi-based, reverse genetic screens (Table 1).

1.6. Reverse genetic screens to identify essential SCN genes

Unlike *C. elegans*, SCN is not an ideal system for genetic studies because of its obligate endo-parasitic life cycle. However, from information learned in *C. elegans*, gene function can be studied by an mRNA nuclease process called RNA interference (RNAi) (Fire et al. 1998). Through this process, a specific mRNA is targeted through a ribonucleoprotein complex for degradation (Hammond et al. 2001; Caudy et al. 2003). The challenge then became demonstrating whether RNAi was functional and reliable in the SCN since the approach does not work in some organisms. However, there are two demonstrated ways that RNAi–based experiments can be performed for SCN, allowing gene function experiments to be performed through the a reverse genetic manner allowed by the RNAi technology.

The first demonstration of RNAi in SCN accomplished the experiment by taking cDNAs for the gene of interest, synthesizing double stranded RNA (dsRNA) *in vitro* and soaking the nematodes in the dsRNA cocktail (Urwin and Atkinson, 2002; Alkharouf et al. 2007). Urwin et al. (2002) examined how the SCN actin gene could be knocked down in its expression. The experiments resembled those performed in *C. elegans* whereby soaking the nematodes in dsRNA resulted in a phenocopy of the normal phenotype generated by the hypomorphic null mutant (Fire et al. 1998; Timmons et al. 2001). Similar experiments that relied on an extensive, but simple, gene annotation pipeline that identified 1,508 candidate genes (Alkharouf et al. 2007) used the cloned genes to synthesize dsRNA from *H. glycines* homologs of small ribosomal protein 3a. Experiments that soaked the SCN with dsRNA resulted in nematode mortality that was demonstrated by vital fluorescent dyes and a phenocopy where the nematodes appeared stiff (Alkharouf et al. 2007). Therefore, RNAi would work in the SCN system. The experiments were then taken a step further in experiments that used the RNAi-soaked nematodes to infect soybean plants to see if the nematodes were impaired in their ability to parasitize soybean. Modeled after the earlier experiments of Urwin et al. (2002), in experiments that used this approach for parasitism genes, it was shown that SCN infection could be suppressed (Bakhetia et al. 2007, 2008). The problem with these experiments, from a nematode biocontrol perspective, is that it would be virtually impossible to synthesize, apply and deliver enough dsRNA to nematodes that are living in the environment to obtain a positive effect even though crude dsRNA extracts can be used (Tenllado et al. 2003). Other problems would be whether the dsRNA remained residually in the soil. Therefore, a second method would be needed that could express the genes as dsRNA in soybean, allowing greater control over the delivery of the dsRNA to SCN.

ProbeSetID	Afx H.g. Accn	Best C.e. Hit	Brief ID
HgAffx.13360.1.S1_at	CB374622	pes-9	Yeast hypothetical 52.9 KD protein
HgAffx.10986.1.S1_at	CA939315	gpd-3	NULL
HgAffx.19591.1.S1_at	CB278666	cpf-1	cleavage stimulation factor like
HgAffx.16755.1.S1_at	CA939427	phb-1	NULL
HgAffx.6532.1.S1_at	CK350603	hsp-6	heat shock 70 protein
HgAffx.19651.1.S1_at	CD748666	phb-2	Prohibitin
HgAffx.17567.1.S1_at	CB825030	ben-1	tubulin
HgAffx.20551.1.S1_at	CB281634	E02H1.1	rRNA methyltransferase
HgAffx.19055.1.S1_at	CB826041	T21B10.2	enolase
HgAffx.24001.2.S1_at	CK351582	ftt-2	14-3-3 protein
HgAffx.22771.2.S1_at	BI749139	uaf-1	NULL
HgAffx.21332.1.S1_at	BI748882	daf-21	heat shock protein (HSP90)
HgAffx.10691.1.S1_at	CD748651	K04D7.1	guanine nucleotide-binding protein
HgAffx.10821.1.S1_at	CB935592	T21B10.7	t-complex protein 1
HgAffx.13633.1.S1_at	CD748017	pyp-1	inorganic pyrophosphatase
HgAffx.20969.1.S1_at	CB379877	rps-1	Ribosomal protein S3a homolog
HgAffx.24120.1.S1_at	CB935135	eft-2	Elongation factor Tu family
HgAffx.22597.1.S1_at	CB826306	kin-19	casein kinase I
HgAffx.11150.1.S1_at	CB378957	D1005.1	ATP citrate lyase
HgAffx.17961.1.S1_at	CB281421	F01G10.1	transketolase
HgAffx.18740.2.S1_at	CA940055	act-4	actin
HgAffx.14431.1.S1_at	CB935363	mdh-1	malate dehydrogenase
HgAffx.19636.2.S1_at	CA939544	rps-4	NULL
HgAffx.18811.1.S1_at	CA940369	F43G9.5	NULL
HgAffx.5490.1.S1_at	CD747934	gpi-1	glucose-6-phosphate isomerase
HgAffx.22868.1.S1_at	BG310682	cpl-1	cathepsin-like protease
HgAffx.13283.1.S1_at	CD748764	K10D6.2	NULL
HgAffx.17866.1.S1_at	CB824474	M03C11.7	NULL
HgAffx.20065.1.S1_at	AF318605	hsp-1	HSP-1 heat shock 70kd protein A
HgAffx.15252.1.S1_at	CK348813	rho-1	p21 ras-related rho (RhoA)
HgAffx.16942.1.S1_at	CK349264	ruvb-2	NULL
HgAffx.23555.2.S1_at	CD748675	Y54E10BR.6	NULL

Table 1. Annotation of the Affymetrix ® soybean GeneChip in relation to gene pathway analyses.

The second way to perform RNAi experiments for SCN control would be to express the genes in transgenic soybean roots, allowing the nematodes to feed on the genetically engineered roots. The hypothesis is that if the SCN was able to ingest the double stranded RNA manufactured in the plant cells through its stylet in high enough concentrations and if the RNAi metabolic process occurred in SCN, there was a chance that nematode development could be controlled. Prior experiments already demonstrated that the RNAi pathway functioned in SCN (Urwin et al. 2002; Alkharouf et al. 2007). The original experiments that performed host-mediated expression of SCN genes as inverted tandemly duplicated copies for RNAi control in soybean to examine SCN biology were done by Steeves et al. (2006), examining the major sperm protein. Huang et al. (2006) demonstrated the same effect for root knot nematode in the model plant *A. thaliana* so the approach would have broad applicability for PPN control. The experiments were followed by Klink et al. (2009a) that identified many genes from microarray studies that would serve as candidates for RNAi control during parasitism.

Figure 2. Soybean plants with transgenic roots. The transgenic soybean roots are expressing the enhanced green fluorescent protein (eGFP) (Haseloff et al. 1997) found in the pRAP vectors (Klink et al. 2009c; Matsye et al. 2012). Bar = 10 cm.

The problem with the transgenic approach is that soybean is a difficult to genetically engineer. However, strategies whereby composite plants (Collier et al. 2005) that are chimeras having nontransformed aerial stocks having transgenic root stocks can be readily made in soybean (Klink et al. 2008, 2009a). The simplicity of the approach is evident because the transgenic plants can be made in non-axenic conditions with the use of fluorescent reporter (Collier et al. 2005) (Figure 2). The development of vectors that work in soybean (Klink et al. 2008, 2009a; Ibrahim et al. 2011; Matsye et al. 2012) have made the experiments possible. Further improvements whereby the plant expression vectors are Gateway®-compatible (Klink et al. 2009a; Ibrahim et al. 2011; Matsye et al. 2012) allows for semi-large reverse genetic screens to be performed. In such experiments, SCN homologs of the small ribosomal protein 3a (Hg-rps-3a) and Hg-rps-4, synaptobrevin (Hg-snb-1) and a spliceosomal SR protein (Hg-spk-1) were tested for functionality in host mediated expression, RNAi-based studies (Klink et al. 2009a). After 8 days of infection, the experiments demonstrated that 81–93% fewer females developed on transgenic roots containing the genes engineered as tandem inverted repeats. Those experiments resulted in lethality for SCN feeding on plants that were expressing the genes as tandemly duplicated inverted repeats (Klink et al. 2009a). The same outcome was shown for root knot nematode in soybean using the same plant expression vector system (Ibrahim et al. 2011). These observations demonstrated that broad spectrum resistance for PPNs in soybean was probable.

1.7. Proteomic studies of SCN

The prior experiments have discussed gene expression in SCN at the RNA level. These experiments are technologically simplistic to perform, because of major advances in sequencing and detection technologies. However, in these experiments using hybridization to study gene expression, little to no information is obtained as to how much protein is actually synthesized from the RNA or modifications that are known to exist on the protein molecules. Recently, the proteome of SCN was investigated (Chen et al. 2011), resulting in a reference map of protein expression. These experiments add to the already extensive gene expression databases that are available for SCN (Ithal et al. 2007; Klink et al. 2007a, 2009a; Elling et al. 2009). The advantage of the proteomic studies is that it allows for the identification of the relative amounts of the studied proteins to be known. This is in contrast to microarray-based experiments where only different levels of expression can be inferred, but their absolute amounts are not known. Chen et al. (2011) performed experiments using LC-MS/MS on preinfective J2 SCN. The nematodes were highly pure samples since they had not yet infected soybean roots. The experiments were able to discern 803 spots on 2-D gels (Chen et al. 2011). Of those spots, 426 proteins were identified (Chen et al. 2011). Gene Ontology analyses allowed for the identification of a number of different functional groups, including secreted proteins that may act during parasitism (Chen et al. 2011). While it is likely that the protein list is not comprehensive, the work provides a solid foundation for future work to examine the proteome of SCN and compare with the gene expression studies based on the RNA.

1.8. Soybean gene expression

To understand how soybean was reacting to infection, it was going to be imperative to develop ways to monitor gene expression during infection. Unlike the model system, *A. thali-*

ana, where a number of gene expression microarrays existed (Mussig et al. 2002; Tao et al. 2003) no commercially available microarrays were in place for soybean. The fabrication of soybean microarrays from cDNAs isolated from uninfected and SCN-infected tissues resulted in the identification of genes that are expressed during parasitism by SCN (Alkharouf et al. 2006). Subsequently, after the availability of the Affymetrix soybean GeneChip, a number of gene expression studies have been performed on whole infected soybean roots (Klink et al. 2007b; Ithal et al. 2007). Some of the studies have focused in on expression occurring during the susceptible reaction (Alkharouf et al. 2006; Klink et al. 2007b; Ithal et al. 2007). These studies have resulted in the identification of genes that are highly expressed during the susceptible reaction. Alkharouf et al. (2006) performed experiments that examined the preparasitic stages of infection of the compatible reaction. The experiments identified defense-related genes such as Kunitz trypsin inhibitor (KTI), germin, peroxidase, phospholipase D, 12-oxyphytodienoate reductase (OPR), pathogenesis related-1 (PR1), phospholipase C, lipoxygenase, WRKY6 transcription factor and calmodulin. The experiments demonstrated that multiple defense pathways were induced even early (by 6 hours post infection) during the compatible reaction. This is important to note because the time point at which the sample was collected occurred before the nematode initiated the formation of the syncytium. This meant that soybean was responding in important ways to the presence of the nematode within it s root tissues. Similar lists of genes were identified by Ithal et al. 2007, demonstrating a commonality of expression even though the experiments used different soybean genotypes and populations of SCN. Unfortunately, since only the susceptible reaction was studied, it was unclear whether the expressed genes were specific to the susceptible reaction or would also be differentially expressed in roots if they were undergoing a resistant reaction. This knowledge would be important to identify actively expressed genes that relate specifically to defense.

To distinguish between expression of genes during the susceptible and resistant reactions, an experiment was performed whereby both susceptible and resistant reactions could be obtained in the same soybean genotype (*G. max*[Peking/PI 548402]) (Klink et al. 2007b). The importance in the way the experiment was designed was that it allowed gene expression that pertained specifically to the susceptible or resistant reaction to be identified. Thus, differences in plant genotype could not introduce error into the experiment. The experiments were set up whereby the *G. max*[Peking/PI 548402] genotype was infected with one of two SCN populations that would result in a susceptible or a resistant reaction. Another important feature of the experiment was that the gene expression that occurred as *G. max*[Peking/PI 548402], a genotype with functional resistance genes, failed in its effort to defend itself from SCN would be identified. The *G. max*[Peking/PI 548402] genotype was infected with *H. glycines*[NL1-RHg/HG-type 7] (originally called race 3) that obtained a resistant reaction and *H. glycines*[TN8/HG-type 1.3.6.7] (originally called race 14) that obtained a susceptible reaction (Klink et al. 2007b). The experiments revealed induced levels of some genes during different points of the susceptible reaction as compared to the resistant reaction. Some of the genes that were induced in their expression during the susceptible reaction at 12 hours post infection (hpi) were an expansin, peroxidase, plasma membrane intrinsic protein 1C (PIP1C), germin-like protein (GER) 1, beta-Ig-H3 domain-containing protein and chorismate mutase (Klink et al. 2007b). Genes induced during the susceptible reaction at 3 dpi included 4-coumarateCoA ligase family protein, expansin, LTP1, transketolase and a cytochrome P450 (Klink et al. 2007b). Related experiments showing genes that were induced specifically during a susceptible reaction at 8 dpi included 4-coumarate CoA ligase family protein, peroxidase, expansin, matrix metalloproteinase, matrixin family protein and a lipid transfer protein (LTP) (Klink et al. 2007b). All of these proteins were suppressed in their activity during the resistant reaction. However, the problem with the vast amounts of data that was being generated at the time was in obtaining a meaningful annotation that would provide an understanding of the global events occurring in the sample types.

1.9. Improvements in annotation

The described experiments resulted in the generation of a massive amount of gene expression data and gene lists for the 38,099 genes fabricated on the Affymetrix® soybean Gene-Chip. Annotated gene lists for soybean genes are very useful because no information is lost from the analysis (Table 1). However, the gene lists do not provide higher order knowledge of how the many genes are functioning during a process under study. It is possible that various metabolic pathways that pertain to a specific process could be identified if the data could be organized into a higher order structure. Since the aforementioned work was done in soybean, often considered a non-model organism, it was difficult to translate the information into gene pathway analyses applications in a manner that would reveal how the gene expression is orchestrated during the process under study. However, an investigation that had been done in *A. thaliana* infected with *Pseudomonas syringae* pv. tomato did show how useful the higher order gene expression knowledge could be in allowing for a visualization of the switch in metabolism from housekeeping to pathogen defense during infection (Scheideler et al. 2001). The development and presentation of gene pathway information, a procedure that merged the Kyoto Encyclopedia of Genes and Genomes (KEGG) framework (http://www.genome.jp/kegg/ catalog/org_list.html) (Goto et al. 1997) with the gene expression data was accomplished through the development of a computer application called Pathway Analysis and Integrated Coloring of Experiments (PAICE) (Paice_v2_90.jar) (http://sourceforge.net/projects/paice/) (Hosseini et al. unpublished; Klink et al. 2011a). This allowed for obtaining higher order cell fate mapping studies to be performed (Klink et al. 2011a; Matsye et al. 2011). Moreover, the sequencing of the soybean genome (Schmutz et al. 2010) made transcriptional mapping experiments that relate to resistance loci possible (Matsye et al. 2011).

1.10. Genomics of the syncytium

While strides were being made in obtaining a deep analysis of the physiological processes occurring in whole infected roots, the greater challenge would be to identify gene expression that occurred within the syncytium because it would require either drawing the cytoplasm out of the syncytium or a way to physically isolate the cells. The original studies that attempted to determine gene expression in nematode nurse cells was done by Bird et al. (1994) and Wilson et al. (1994). The hypothesis was that by extracting the cytoplasm of the cells that are specifically undergoing the parasitism, it would be possible to determine the gene expression that pertains specifically to parasitism. However, it is noted that gene expression in the cells surrounding the syncytium probably plays some role in the maintenance and development of the susceptible and resistant reactions. This approach to isolate the cytoplasm (Bird et al. 1994; Wilson et al. 1994) would be more challenging for syncytia because it is virtually impossible to determine what cells are infected by SCN. Therefore, instead of collecting the cytoplasm, the collection of the cells would have to occur and it would have to be done through their physical isolation.

The physical isolation of syncytia undergoing a susceptible reaction to the SCN was first described by Klink et al. (2005). The study collected syncytia by a procedure called laser microdissection (Isenberg et al. 1976; Meier Ruge et al. 1976; Emmert-Buck et al. 1996) (Figure 3). The experiments obtained RNA of suitable quality for making cDNA libraries, cloning and sequencing full length genes greater than 1,000 base pairs, making probes for *in situ* hybridization and quantitative PCR (qPCR) and immunocytochemistry which would allow for the visualization of gene expression inside of the infected cells (Klink et al. 2005). These results made it possible to study gene expression occurring within the syncytium at the genome-wide level.

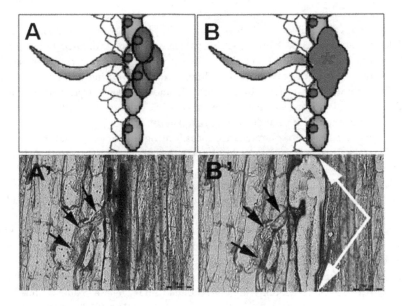

Figure 3. Laser microdissection (LM) of nematode feeding sites. A, cartoon of a nematode (gray) parasitizing a pericycle cell (yellow) that previously was uninfected (green). The parasitism process is resulting in the incorporation of neighboring cells by dissolving their cell walls, forming a syncytium. B, the syncytium (asterisk) was collected after LM. A', an actual root used for LM. The black arrows point to a nematode infecting a root cell (red outline). B', white arrows point to the feeding cell that was collected by LM.

Genomics approaches to syncytium biology resulted in a series of investigations that have focused in on gene expression that occurs during a susceptible reaction in the syncytium (Klink et al. 2005, 2007a, 2009b, 2010b, c, 2011a; Ithal et al. 2007; Matsye et al. 2011, 2012; Kandoth et al. 2011). In these studies, a number of genes were identified. However, to understand gene expression as it specifically pertains to defense, it would be required to study the cells undergoing the defense response. The main obstacle in performing studies with this goal in mind was determining whether the cells undergoing the defense response were already dead at the time of cell collection. The prediction is that cells that were dead would have already halted their physiological processes that pertained to defense and also may not provide RNA of suitable quality for microarray studies. However, it was unlikely that the cells progressing through the earlier stages of defense were dead (Figure 1) since the EM studies revealed very specific progression of cellular architecture during the defense response, suggesting that the cells had to be alive to progress through this developmental process (Endo, 1965; Kim et al. 1987; Endo, 1991). The initial collection of syncytia undergoing the developmental process that leads to their eventual collapse and death were then performed (Klink et al. 2007a). These experiments demonstrated that the cells would be a suitable source for RNA collection and genomics-based analyses. The first set of experiments to use laser microdissected cells undergoing an incompatible reaction for genomics studies determined that the expression of lipoxygenase (LOX), arabinogalactan-protein (AGP18), annexin, a thioesterase family protein heat shock protein (HSP) 70 and superoxidase dismutase (SOD) (Klink et al. 2007a). Many of the genes have very well known roles in plant defense. Subsequent experiments examined more time points occurring during the defense response, spanning phase 1 and phase 2 (Klink et al. 2009b, 2010b, c). The experiments identified a number a genes that were very highly expressed during the resistant reaction, specifically within the syncytium. In contrast, a number of genes were very highly suppressed (>1,000 fold) during the resistant reaction (Klink et al. 2009b, 2010b, c). The experiments were repeated later by Kandoth et al. (2011) in the G. $max_{[PI\ 209322]}$ genetic background that either has or lacks the $rhg1$ resistance locus. During this time, studies were also performed that examined and compared multiple forms of the resistant reaction that were found in the G. $max_{[Peking/PI\ 548402]}$ and G. $max_{[PI\ 88788]}$ genotypes (Klink et al. 2009b, 2010b, c, 2011a; Matsye et al. 2011, 2012). These studies were important because the G. $max_{[Peking/PI\ 548402]}$ and G. $max_{[PI\ 88788]}$ genotypes are well known to undergo different forms of the resistant reaction at the cellular level (Kim et al. 1987; Endo, 1991; Mahalingham and Skorupska, 1996). The G. $max_{[Peking/PI\ 548402]}$ and G. $max_{[PI\ 88788]}$ PIs are also important genotypes to obtain knowledge from because they are the source of >97% of the resistance germplasm used in commercial breeding programs (Concibido et al. 2004). In some of the earlier studies (Klink et al. 2009b), a number of genes were identified that were induced preferentially in their expression during the resistant reaction. The genes included lipoxygenase, S-adenosylmethionine synthetase, a dnaK domain-containing protein, GRF2 GENERAL REGULATORY FACTOR 2, ACT7 (actin 7), major latex protein-related protein, xyloglucan endotransglucosylase/hydrolase protein 26, cytochrome P450 monooxygenase CYP93D1, pyruvate dehydrogenase E1 beta subunit isoform 2, nitrate transporter (NTP2), endo-1,4-beta-glucancase that were all expressed preferentially between 100 to 383-fold higher in syncytia undergoing the defense response as compared to syncytia undergoing the early stages of a susceptible reaction (Klink et al. 2009b). Additional experiments aided by Illumina® deep sequencing technology which is a sequence by synthesis procedure much like quantitative PCR, but for every gene in the genome simultaneously, identified genes that were expressed only in syncytia undergoing the defense response (Matsye et al. 2011). Some of the genes were expressed at all times during the defense response. Importantly, the Illumina® deep sequencing technology revealed that some of the transcripts that are genes known to be important in defense responses represented between 1 and 17% of the sequenced tags from RNA isolated from the syncytia undergoing the defense response (Table 2) (Matsye et al. 2011). The knowledge gained from these gene expression experiments was then used to select candidate genes whose function during infection could be tested. The cross-comparison of data obtained from the Illumina® sequencing platform with the Affymetrix® microarrays determined the genes within the *rhg1* locus that were expressed specifically during defense (Matsye et al. 2011). Experimentation of these genes in functional tests determined that some of these genes play a role in defense to SCN (Matsye et al. 2012). It was shown that one gene, an α-SNAP allele isolated from the resistant *G. max*$_{[Peking/PI\ 548402]}$, provided resistance when genetically engineered into the susceptible *G. max*$_{[Williams\ 82/PI\ 518671]}$ (Matsye et al. 2012). Gene expression and functional studies will be further expanded on in a subsequent section.

category	total probe sets	percent
Affymetrix® soybean GeneChip® probe sets (PS)	38,099	
PS with matches to *Arabidopsis thaliana* accessions	23,583	62%
PS with enzyme commission (E.C.) numbers	9,717	29%
PS matching both *A. thaliana* accessions and having E.C. numbers	4,156	11%
PS with chromosomal coordinates	31,188	82%

Table 2. Group 1 SCN genes expressed during parasitism and used in RNAi studies (Klink et al. 2009)

1.10.1. Soybean resistance clusters

The major SCN resistance trait, *rhg1*, was first identified by Caldwell et al. (1960). In and around the same time, four other major loci, the recessive *rhg2*, *rhg3* (Caldwell et al. 1960) and the dominant *Rhg4* (Matson and Williams, 1965) and *Rhg5* (Rao Arelli 1994) have been identified. In all, there are approximately 61 QTLs that associate with resistance to SCN (Kim et al. 2010). Many of the details of the numerous mapping studies can be found in a review by Concibido et al. (2004). Of all of the loci that associate with resistance to SCN, the best studied is *rhg1*. It is a major resistance locus and has been fine mapped to a region defined in a span of approximately 611,794 nucleotides between the molecular markers ss107914244 and Satt038 on chromosome 18 (Concibido et al. 1994; Mudge et al, 1997; Cregan et al. 1999a; Hyten et al. 2010). It is important to note that the *rhg1* loci found in the different genotypes that exhibit resistance are not the same (Cregan et al. 1999b; Brucker et al.

2005; Matsye et al. 2012). For example, due to the variation in how soybean responds to infection by the SCN, the *rhg1* resistance allele in *G. max*$_{[PI\ 88788]}$ is designated *rhg1-b* (Kim et al. 2010). Work by Kim et al. (2010) has resulted in the fine-mapping of the *rhg1-b* locus to within a region of approximately 67 kb. This development was important for the SCN research field because the locus contains approximately 9 genes. However, work in understanding the biological nature of the genes within the locus was not the focus of the Kim et al. (2010) study because the investigation was a genetic mapping analysis. Other resistance loci that are not as well mapped, such as *Rhg4* (Matson and Williams 1965), while providing resistance, account for about a 30% of the resistance of soybean to SCN. In addition to this feature, it functions only against certain populations of SCN.

1.11. Gene expression found during defense at the *rhg1* locus

Knowing how and when genes are expressed in syncytia specifically during defense would likely provide knowledge of the genes that regulate or contribute to the process. Matsye et al. (2012) demonstrated in complimentary studies, that an amino acid transporter (AAT) (Glyma18g02580) and an α soluble NSF attachment protein (α-SNAP) (Glyma18g02590) found in the *rhg1* locus, undergo expression specifically in syncytia undergoing defense in both the *G. max*$_{[Peking/PI\ 548402]}$ and *G. max*$_{[PI\ 88788]}$ genotypes (Matsye et al. 2011). What was notable about the analysis was that AAT and α-SNAP were shown to be expressed throughout the defense response in experiments that sampled time points at 3, 6 and 9 days post infection (dpi), spanning phase 1 and 2 (Matsye et al. 2011). The AAT and α-SNAP genes did not appear to be expressed in syncytia undergoing the susceptible reaction. This difference in expression that was occurring between the resistant and susceptible reaction made it possible that the genes could be involved in the defense response. However, this would only be determined in functional studies that tested how the gene acted during infection (Matsye et al. 2012).

1.12. Genetic engineering as a solution for SCN

A number of approaches like conventional breeding programs have been shown for decades to generate resistance to SCN (Brim and Ross, 1966). The resistant cultivars have been shown to result in savings of hundreds of millions of dollars (Bradley and Duffy, 1982). One drawback of conventional breeding programs is that along with the resistance genes that are bred in, a number of genes are also introgressed that could have undesirable characteristics. This is especially a problem when is desirable traits are tightly linked to the undesirable traits. To circumvent this problem, it is possible to genetically engineer in genes of interest. A number of strategies that have been described in this chapter have shown promise in disrupting the soybean-SCN interaction. As noted earlier, RNAi of nematode parasitism genes has been shown in the *Arabidopsis thaliana-Meloidogyne* sp. system to perturb giant cell formation (Huang et al. 2006). This was also shown to work in the soybean-SCN pathosystem (Steeves et al. 2006). Later work that identified highly conserved SCN genes that were expressed during parasitism could be knocked down by RNAi and suppress infection (Klink et al. 2009a; Li et al. 2010). Due to the duplicated nature of the soybean genome (Doyle et al. 1999; Schmutz et al. 2010), RNAi studies of soybean genes may be met with complications and may require methodologies that can knock down entire gene families (Alvarez et al. 2006).

Another procedure to modulate gene expression in soybean to engineer resistance involves the engineering of soybean genes as overexpression constructs (Matsye et al. 2012). To do the studies, genes that are highly expressed during a resistant reaction, identified in accessions of little agronomic value can be expressed to high levels in a soybean genotype that is normally susceptible, but of great economic value. The hypothesis is that if the gene is important in the defense response, the overexpression of that gene in a genotype that is normally susceptible would result in suppressed nematode infection. Such a result was obtained by Matsye et al. (2012) with the overexpression of a naturally occurring truncated allele of an α-SNAP gene. When the α-SNAP gene that was identified in the G. $max_{[Peking/PI\ 548402]}$ accession was overexpressed in the normally susceptible G. $max_{[Williams\ 82/PI\ 518671]}$ genotype, nematode infection was suppressed (Figure 4). The experiments demonstrated the efficacy of the approach, opening up the possibility for large scale reverse genetic screens since the plasmid vectors used to engineer the genes into soybean through the hairy root procedure (Tepfer et al. 1984) was designed with an enhanced green fluorescent reporter (eGFP) (Collins et al. 2005; Klink et al. 2008) was designed using the Gateway® technology for both RNAi and overexpression studies (Klink et al. 2009a; Matsye et al. 2012).

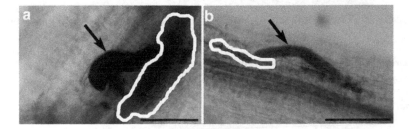

Figure 4. An overexpressed gene affects nematode development. A, a nematode, stained with acid fuchsin for visualization, developing in an experimental control plant. The boundary of the nematode feeding site is encircled in blue. B, a nematode failing to develop in a plant overexpressing a gene identified in the gene expression studies of the syncytium. The boundary of the nematode feeding site is encircled in blue.

2. Conclusion

The soybean-SCN pathosystem has been under study for over 60 years. Through a massive amount of basic studies involving agricultural production practices, genomics and genetic engineering, solutions to the chronic and global SCN problem are emerging. The difficulty of studying the system has been met with many improvements in technology that are allowing for basic features of the pathosystem to be exploited so that agricultural practices and economic returns are improved. The basic knowledge gained in this system can now be applied as a model for understanding other recalcitrant pathogens affecting soybean, to obtain a comprehensive understanding of infection and defense.

Acknowledgements

VPK is thankful for start-up support provided by Mississippi State University and the Department of Biological Sciences; funds in the forms of a competitive Research Improvement Grant; support from the Mississippi Soybean Promotion Board. GWL is thankful to the department of Department of Biochemistry, Molecular Biology, Entomology and Plant Pathology, Mississippi State University, KSL is thankful to the Department of Entomology and Plant Pathology, Auburn University.

Author details

Vincent P. Klink[1]*, Prachi D. Matsye[1], Katheryn S. Lawrence[2] and Gary W. Lawrence[3]

*Address all correspondence to: vklink@biology.msstate.edu

1 Department of Biological Sciences, Mississippi State University, Mississippi State, MS, U.S.A.

2 Department of Biochemistry, Molecular Biology, Entomology and Plant Pathology, Mississippi State University, Mississippi State, MS,, U.S.A.

3 Department of Entomology and Plant Pathology, Auburn University, Auburn, AL, U.S.A.

References

[1] Aist JR. 1976. Papillae and related wound plugs of plant cells. Annu Rev Phytopathol 14: 145–163

[2] Alkharouf NW, Klink VP, Chouikha IB, Beard HS, MacDonald MH, Meyer S, Knap HT, Khan R, Matthews BF. 2006. Timecourse microarray analyses reveals global changes in gene expression of susceptible *Glycine max* (soybean) roots during infection by *Heterodera glycines* (soybean cyst nematode). Planta 224: 838-852

[3] Alkharouf N, Klink VP, Matthews BF. 2007. Identification of *Heterodera glycines* (soybean cyst nematode [SCN]) DNA sequences with high similarity to those of *Caenorhabditis elegans* having lethal mutant or RNAi phenotypes. *Exp Parasitol* 115: 247-258

[4] Alvarez JP, Pekker I, Goldshmidt A, Blum E, Amsellem Z, Eshed Y. 2006. Endogenous and synthetic microRNAs stimulate simultaneous, efficient, and localized regulation of multiple targets in diverse species. Plant Cell 18: 1134-1151

[5] An Q, Ehlers K, Kogel KH, van Bel AJ, Hückelhoven R. 2006a. Multivesicular compartments proliferate in susceptible and resistant MLA12-barley leaves in response to infection by the biotrophic powdery mildew fungus. New Phytol 172: 563-57

[6] An Q, Hückelhoven R, Kogel KH, van Bel AJ. 2006b. Multivesicular bodies participate in a cell wall-associated defence response in barley leaves attacked by the pathogenic powdery mildew fungus. Cell Microbiol 8: 1009-1019

[7] Assaad FF, Qiu JL, Youngs H, Ehrhardt D, Zimmerli L, Kalde M, Wanner G, Peck SC, Edwards H, Ramonell K, Somerville CR, Thordal-Christensen H. 2004. The PEN1 syntaxin defines a novel cellular compartment upon fungal attack and is required for the timely assembly of papillae. Mol Biol Cell 15: 5118-5129

[8] Atkinson HJ, Harris PD. 1989. Changes in nematode antigens recognized by monoclonal antibodies during early infections of soya bean with cyst nematode Heterodera glycines. Parasitology 98: 479-487

[9] Bakhetia M, Urwin PE, Atkinson HJ. 2007. QPCR analysis and RNAi define pharyngeal gland cell-expressed genes of Heterodera glycines required for initial interactions with the host. *Mol Plant Microbe Interact* 20: 306-312

[10] Bakhetia M, Urwin PE, Atkinson HJ. 2008. Characterisation by RNAi of pioneer genes expressed in the dorsal pharyngeal gland cell of Heterodera glycines and the effects of combinatorial RNAi. *Int J Parasitol* 38: 1589-1597

[11] Barker KR, Koenning SR, Huber SC, Huang JS. 1993. Physiological and structural responses of plants to nematode parasitism with Glycine max-Heterodera glycines as a model system. Pp. 761-771 in DR Buxon R Shibles RA Forsberg BL Blad KH Asay GM Paulsen and RF Wilson, Eds. International Crop Science I: Madison, WI: Crop Science Society of America

[12] Bekal S, Niblack TL, Lambert KN. 2003. A chorismate mutase from the soybean cyst nematode Heterodera glycines shows polymorphisms that correlate with virulence. Molecular Plant-Microbe Interactions 16: 439-446

[13] Bekal S, Craig JP, Hudson ME, Niblack TL, Domier LL, Lambert KN. 2008. Genomic DNA sequence comparison between two inbred soybean cyst nematode biotypes facilitated by massively parallel 454 micro-bead sequencing. Mol Genet Genomics 279:535-543

[14] Bennett MK, Calakos N, Scheller RH. 1992. Syntaxin: a synaptic protein implicated in docking of synaptic vesicles at presynaptic active zones. Science 257: 255-259

[15] Bernard RL, Juvik GA, Nelson RL. 1987. USDA Soybean Germplasm Collection Inventory, Vol. 2. INTSOY Series Number 31. IL: International Agriculture Publications, University of Illinois

[16] Bird D McK, Wilson MA. 1994. DNA sequence and expression analysis of root-knot nematode-elicited giant cell transcripts. MPMI 7:419-424

[17] Bradley EB, Duffy M. 1982. The value of plant resistance to soybean cyst nematode: a case study of Forrest soybeans. National Resource Economics Staff Report No. AGES820929, USDA. Washington DC: U.S. Government Printing Office

[18] Brucker E, Carlson S, Wright E, Niblack T, Diers B. 2005. Rhg1 alleles from soybean PI 437654 and PI 88788 respond differently to isolates of Heterodera glycines in the greenhouse. Theor Appl Genet 111: 44-49

[19] C. elegans Sequencing Consortium. 1998. Genome sequence of the nematode C. elegans: a platform for investigating biology. Science 282: 2012-2018

[20] Caldwell BE, Brim CA, Ross JP. 1960. Inheritance of resistance of soybeans to the soybean cyst nematode, *Heterodera glycines*. Agron J 52: 635-636

[21] Caudy AA, Ketting RF, *Hammond* SM, Denli AM, Bathoorn AM, Tops BB, Silva JM, Myers MM, Hannon GJ, Plasterk RH. 2003. A micrococcal nuclease homologue in *RNAi* effector complexes. Nature 425: 411-414

[22] Chapin III FS, Shaver GR. 1985. Individualistic growth response of tundra plant species to environmental manipulations in the field. Ecology 66: 564–576

[23] Chen W, Chao G, Singh KB. 1996. The promoter of a H2O2-inducible, Arabidopsis glutathione S-transferase gene contains closely linked OBF- and OBP1-binding sites. Plant J 10: 955-966

[24] Chen SY, Dickson DW. 1996. Pathogenicity of fungi to eggs of *Heterodera glycines*. Journal of Nematology 28: 148-158

[25] Chen X, MacDonald MH. Khan F, Garrett WM, Matthews BF, Natarajan SS. 2011. Two-dimentional proteome reference maps for the soybean cyst nematode Heterodera glycines. Proteomics 11: 4742-4746

[26] Chitwood DG, Lusby WR: Metabolism of plant sterols by nematodes. *Lipids* 1990, 26: 619-627

[27] Clary DO, Griff IC, Rothman JE. 1990. SNAPs, a family of NSF attachment proteins involved in intracellular membrane fusion in animals and yeast. Cell 61: 709–721

[28] Colgrove AL, Niblack TL. 2008. Correlation of female indices from virulence assays on inbred lines and field populations of Heterodera glycines. J Nematol 40: 39–45

[29] Collier R, Fuchs B, Walter N, Kevin Lutke W, Taylor CG. 2005. Ex vitro composite plants: an inexpensive, rapid method for root biology. Plant J 43: 449-457

[30] Collins NC, Thordal-Christensen H, Lipka V, Bau S, Kombrink E, Qiu JL, Hückelhoven R, Stein M, Freialdenhoven A, Somerville SC, Schulze-Lefert P. 2003. SNARE-protein mediated disease resistance at the plant cell wall. Nature 425: 973–977

[31] Concibido VC, Denny RL, Boutin SR, Hautea R, Orf JH, Young ND. 1994. DNA Marker analysis of loci underlying resistance to soybean cyst nematode (*Heterodera glycines* Ichinohe). Crop Sci 34: 240–246

[32] Concibido VC, Diers BW, Arelli PR. 2004. A decade of QTL mapping for cyst nematode resistance in soybean. Crop Sci. 44: 1121-1131

[33] Creech JE, Johnson WG. 2006. Survey of broadleaf winter weeds in Indiana production fields infested with soybean cyst nematode (*Heterodera glycines*). Weed Technol. 20: 1066-1075

[34] Cregan PB, Mudge J, Fickus EW, Danesh D, Denny R, Young ND. 1999a. Two simple sequence repeat markers to select for soybean cyst nematode resistance conditioned by the rhg1 locus. Theor Appl Genet 99: 811–818

[35] Cregan PB, Mudge J, Fickus EW, Marek LF, Danesh D, Denny R, Shoemaker RC, Matthews BF, Jarvik T, Young ND. 1999b. Targeted isolation of simple sequence repeat markers through the use of bacterial artificial chromosomes. Theor Appl Genet 98:919–928

[36] Day TA, Ruhland CT, Grobe CW, Xiong F. 1999. Growth and reproduction of Antarctic vascular plants in response to warming and UV radiation reductions in the field. Oecologia 119: 24–35

[37] De Boer JM, Yan Y, Wang X, Smant G, ussey RS, Davis EL. 1999. Developmentla expression of secretory beta 1, 4-endonucleases in the subventral esophageal glands of Heterodera glycines. Molecular Plant-Microbe Interactions 12: 663-669

[38] De Boer JM, Mc Dermott JP, Davis EL; Husses RS, Popeijus H, Smant G, Baum TJ. 2002. Cloning of a putative pectate lyase gene expressed in the subventral esophageal glands of Heterodera glycines. J. Nematology 34: 9-11

[39] Doyle JJ, Doyle JL, Brown AH. 1999. Origins, colonization, and lineage recombination in a widespread perennial soybean polyploid complex. Proc Natl Acad Sci 96: 10741-10745

[40] Edens RM, Anand SC, Bolla RI. 1995. Enzymes of the phenylpropanoid pathway in soybean infected with Meloidogyne incognita or Heterodera glycines. J Nematology. 27: 292-303

[41] Elling AA, Mitreva M, Gai X, Martin J, Recknor J, Davis EL, Hussey RS, Nettleton D,McCarter JP, Baum TJ. 2009. Sequence mining and transcript profiling to explore cyst nematode parasitism. BMC Genomics 10: 58

[42] Emmert-Buck MR, Bonner RF, Smith PD, Chuaqui RF, Zhuang Z, Goldstein SR, Weiss RA, Liotta LA. 1996. Laser capture microdissection. Science 274: 998–1001

[43] Endo BY. 1964. Penetration and development of *Heterodera glycines* in soybean roots and related and related anatomical changes. Phytopathology 54: 79–88

[44] Endo BY. 1965. Histological responses of resistant and susceptible soybean varieties, and backcross progeny to entry development of *Heterodera glycines*. Phytopathology 55: 375–381

[45] Endo BY. 1991. Ultrastructure of initial responses of susceptible and resistant soybean roots to infection by *Heterodera glycines*. Revue Nématol 14: 73-84

[46] Endo BY, Veech JA. 1970. Morphology and histochemistry of soybean roots infected with *Heterodera glycines*. Phytopathology 60: 1493–1498

[47] Epps JM, Chambers AY. 1958. New host records for Heterodera glycines including one in the Labiate. Plant Disease Reporter 42: 194

[48] Epps JM, Hartwig EE. 1972. Reaction of soybean varieties and strains to soybean cyst nematode. J Nematol 4: 222

[49] Fire A, Xu S, Montgomery MK, Kostas SA, Driver SE, Mello CC. 1998. Potent and specific genetic interference by doublestranded RNA in Caenohrabditis elegans. Nature 391:806–811

[50] Francl LJ, Dropkin VH. 1986. Heterodera glycines population dynamics and relation of initial population density tp soybean yield. Plant Disease: 70: 791-795

[51] Gao B, Allen R, Maier T, Davis EL, Baum TJ, Hussey RS. 2001. Identification of putative parasitism genes expressed in the esophageal gland cells of the soybean cyst nematode Heterodera glycines. *Mol Plant Microbe Interact* 2001, 14:1247-1254.

[52] Gao B, Allen R, Maier T, Davis EL, Baum TJ, Hussey RS. 2003. The parasitome of the phytonematode Heterodera glycines. Mol Plant Microbe Interact 16: 720-726

[53] Garner WW, Allard HA. 1930. Photoperiodic response of soybeans in relation to temperature and other environmental factors. J. Agric. Res. 41:719-735

[54] Gipson I, Kim KS, Riggs RD. 1971. An ultrastructural study of syncytium development in soybean roots infected with *Heterodera glycines*. Phytopathology 61: 347-353

[55] Golden AM, Epps JM, Riggs RD, Duclos LA, Fox JA, Bernard RL. 1970. Terminology and identity of infraspecific forms of the soybean cyst nematode (Heterodera glycines). Plant Dis Rep 54: 544–546

[56] Goto S, Bono H, Ogata H, Fujibuchi W, Nishioka T, Sato K, Kanehisa M. 1997. Organizing and computing metabolic pathway data in terms of binary relations. Pac Symp Biocomput. 1997: 175-186

[57] Hammond SM, Boettcher S, Caudy AA, Kobayashi R, Hannon GJ. 2001. Argonaute2, a link between genetic and biochemical analyses of RNAi. Science. 293: 1146-1150

[58] Hardham AR, Takemoto D, White RG. 2008. Rapid and dynamic subcellular reorganization following mechanical stimulation of Arabidopsis epidermal cells mimics responses to fungal and oomycete attack. BMC Plant Biol 8: 63

[59] Haseloff J, Siemering KR, Prasher DC, Hodge S. 1997. Removal of a cryptic intron and subcellular localization of green fluorescent protein are required to mark transgenic Arabidopsis plants brightly. Proc Natl Acad Sci U S A 94: 2122-2127

[60] Hermsmeier D, Mazarei M, Baum TJ. 1998. Differential display analysis of the early compatible interaction between soybean and the soybean cyst nematode. Molecular Plant-Microbe Interactions 11: 1258-1263

[61] Hofius D, Schultz-Larsen T, Joensen J, Tsitsigiannis DI, Petersen NH, Mattsson O, Jørgensen LB, Jones JD, Mundy J, Petersen M. 2009. Autophagic components contribute to hypersensitive cell death in Arabidopsis. Cell 137: 773-783

[62] Huang G, Allen R, Davis EL, Baum TJ, Hussey RS. 2006. Engineering broad root-knot resistance in transgenic plants by RNAi silencing of a conserved and essential root-knot nematode parasitism gene. Proc Natl Acad Sci USA 103: 14302-14306

[63] Hymowitz t. 1970. On the domestication of soybean. Economic Botany 24: 408-421

[64] Hyten DL, Choi IY, Song Q, Shoemaker RC, Nelson RI, Costa JM, Specht JE, Cregan PB. 2010. Highly variable patterns of linkage disequilibrium in multiple soybean populations. Genetics 175: 1937-1944

[65] Ichinohe M. 1952. On the soybean nematode, Heterodera glycines n. sp., from Japan. Magazine of Applied Zoology 17: 1-4

[66] Ichinohe M. 1961. Studies on the soybean cyst nematode, *Heterodera glycines* Hakkaido National Experiment Station Report no. 56

[67] Inagaki H, Tsutsumi M. 1971. Survival of the soybean cyst nematode, Heterodera glycines Ichinohe (Tylenchida: Heteroderidae) under certain storage conditions. Appl Entomol Zool (Jpn) 8: 53–63

[68] Inoue A, Obata K, Akagawa K. 1992. Cloning and sequence analysis of cDNA for a neuronal cell membrane antigen, HPC-1. J Biol Chem 267: 10613-10619

[69] Isenberg G, Bielser W, Meier-Ruge W, Remy E. 1976. Cell surgery by laser micro-dissection: a preparative method. J. Microsc 107: 19–24

[70] Ithal N, Recknor J, Nettleston D, Hearne L, Maier T, Baum TJ, Mitchum MG. 2007. Developmental transcript profiling of cyst nematode feeding cells in soybean roots. Molecular Plant-Microbe Interactions 20: 293-305

[71] Jones MGK. 1981. The development and function of plant cells modified by endoparasitic nematodes. Pages 255-279 in: Plant Parasitic Nematodes, Vol. III. B. M. Zuckerman and R. A. Rohde, eds. Academic Press, New York, U.S.A.

[72] Jones MGK, Northcote DH. 1972. Nematode-induced syncytium-a multinucleate transfer cell. J Cell Sci 10: 789–809

[73] Kandoth PK, Ithal N, Recknor J, Maier T, Nettleton D, Baum TJ, Mitchum MG. 2011. The Soybean Rhg1 locus for resistance to the soybean cyst nematode Heterodera gly-

cines regulates the expression of a large number of stress- and defense-related genes in degenerating feeding cells. Plant Physiol 155: 1960-1975

[74] Kalde M, Nühse TS, Findlay K, Peck SC. 2007. The syntaxin SYP132 contributes to plant resistance against bacteria and secretion of pathogenesis-related protein 1. Proc Natl Acad Sci U S A 104: 11850-11855

[75] Kamath RS, Fraser AG, Dong Y, Poulin G, Durbin R, Gotta M, Kanapin A, Le Bot N, Moreno S, Sohrmann M, Welchman DP, Zipperlen P, Ahringer J. 2003. Systematic functional analysis of the *Caenorhabditis elegans* genome using RNAi. Nature 421: 231–237

[76] Kim DG, Riggs RD. 1991. Characteristics and efficacy of a sterile hyphomycete (ARF18), a new biocontrol agent for *Heterodera glycines* and other nematodes. Journal of Nematology 23: 275–282

[77] Kim DG, Riggs RD. 1995. Efficacy of the nematophagous fungus ARF18 in alginate-clay pellet formulation against *Heterodera glycines*. Journal of Nematology 23:275-282

[78] Kim KS, Riggs RD. 1992. Cytopathological reactions of resistant soybean plants to nematode invasion. Pp. 157–168 *in* J. A. Wrather and R. D. Riggs, eds. Biology and Management of the Soybean Cyst Nematode. St. Paul: APS Press

[79] Kim M, Hyten DL, Bent AF, Diers BW. 2010. Fine mapping of the SCN resistance locus *rhg1-b* from PI 88788. The Plant Genome 3: 81-89

[80] Kim YH, Riggs RD, Kim KS. 1987. Structural changes associated with resistance of soybean to *Heterodera glycines*. J Nematol 19: 177–187

[81] Klink VP, MacDonald M, Alkharouf N, Matthews BF. 2005. Laser capture microdissection (LCM) and expression analyses of *Glycine max* (soybean) syncytium containing root regions formed by the plant pathogen *Heterodera glycines* (soybean cyst nematode). Plant Mol Bio 59: 969-983

[82] Klink VP, Overall CC, Alkharouf N, MacDonald MH, Matthews BF. 2007a. Laser capture microdissection (LCM) and comparative microarray expression analysis of syncytial cells isolated from incompatible and compatible soybean roots infected by soybean cyst nematode (*Heterodera glycines*). Planta 226: 1389-1409

[83] Klink VP, Overall CC, Alkharouf N, MacDonald MH, Matthews BF. 2007b. A comparative microarray analysis of an incompatible and compatible disease response by soybean (Glycine max) to soybean cyst nematode (Heterodera glycines) infection. Planta 226: 1423-1447

[84] Klink VP, MacDonald MH, Martins VE, Park S-C, Kim K-H, Baek S-H, Matthews BF. 2008. MiniMax, a new diminutive Glycine max variety, with a rapid life cycle, embryogenic potential and transformation capabilities. Plant Cell, Tissue and Organ Culture 92: 183-195

[85] Klink VP, Kim K-H, Martins VE, MacDonald MH, Beard HS, Alkharouf NW, Lee S-K, Park S-C, Matthews BF. 2009a. A correlation between host-mediated expression of

parasite genes as tandem inverted repeats and abrogation of the formation of female Heterodera glycines cysts during infection of Glycine max. Planta 230: 53-71

[86] Klink VP, Hosseini P, Matsye P, Alkharouf N, Matthews BF. 2009b. A gene expression analysis of syncytia laser microdissected from the roots of the *Glycine max* (soybean) genotype PI 548402 (Peking) undergoing a resistant reaction after infection by *Heterodera glycines* (soybean cyst nematode) Plant Mol Bio: 71: 525-567

[87] Klink VP, Hosseini P, MacDonald MH, Alkharouf N, Matthews BF. 2009c. Population-specific gene expression in the plant pathogenic nematode Heterodera glycines exists prior to infection and during the onset of a resistant or susceptible reaction in the roots of the Glycine max genotype Peking. BMC-Genomics 10: 111

[88] Klink VP, Matsye PD, Lawrence GW. 2010a. Developmental Genomics of the Resistant Reaction of Soybean to the Soybean Cyst nematode, Pp. 249-270, In Plant Tissue Culture and Applied Biotechnology. Eds. Kumar A., Roy S. Aavishkar Publishers, Distributors, India.

[89] Klink VP, Hosseini P, Matsye P, Alkharouf N, Matthews BF. 2010b. Syncytium gene expression in *Glycine max*[PI 88788] roots undergoing a resistant reaction to the parasitic nematode *Heterodera glycines* Plant Physiology and Biochemistry 48: 176-193

[90] Klink VP, Overall CC, Alkharouf N, MacDonald MH, Matthews BF. 2010c. Microarray detection calls as a means to compare transcripts expressed within syncytial cells isolated from incompatible and compatible soybean (Glycine max) roots infected by the soybean cyst nematode (Heterodera glycines). Journal of Biomedicine and Biotechnology 1-30

[91] Klink VP, Hosseini P, Matsye PD, Alkharouf N, Matthews BF. 2011a. Differences in gene expression amplitude overlie a conserved transcriptomic program occurring between the rapid and potent localized resistant reaction at the syncytium of the *Glycine max* genotype Peking (PI 548402) as compared to the prolonged and potent resistant reaction of PI 88788. Plant Mol Bio 75: 141-165

[92] Klink VP, Matsye PD, Lawrence GW. 2011b. Cell-specific studies of soybean resistance to its major pathogen, the soybean cyst nematode as revealed by laser capture microdissection, gene pathway analyses and functional studies. in *Soybean - Molecular Aspects of Breeding* pp. 397-428. Ed. Aleksandra Sudaric. Intech Publishers

[93] Koenning SR, Schmitt DP, Barker KR. 1993. Effects of cropping systems on population density of Heterodera glycines and soybean yield. Plant Disease 77: 780-786

[94] Lai Z, Wang F, Zheng Z, Fan B, Chen Z. 2011. A critical role of autophagy in plant resistance to necrotrophic fungal pathogens. Plant J 66: 953-968

[95] Lambert KN, Allen KD, Sussex IM. 1999. Cloning and characterization of an esophageal-gland specific chorismate mutase from the phytopathogenic nematode Meloidogyne javanica. Molecular Plant-Microbe Interactions 12: 328-336

[96] Lenz HD, Haller E, Melzer E, Kober K, Wurster K, Stahl M, Bassham DC, Vierstra RD, Parker JE, Bautor J, Molina A, Escudero V, Shindo T, van der Hoorn RA, Gust AA, Nürnberger T. 2011. Autophagy differentially controls plant basal immunity to biotrophic and necrotrophic pathogens. Plant J 66: 818-830

[97] Li J, Todd TC, Oakley TR, Lee J and Trick HN. 2010. Host derived suppression of nematode reproductive and fitness genes decreases fecundity of *Heterodera glycines*. Planta 232: 775-785

[98] Li Y-H, Qi X-T, Chang R and Qiu L-J. 2011. Evaluation and Utilization of Soybean Germplasm for Resistance to Cyst Nematode in China. in *Soybean - Molecular Aspects of Breeding* pp. 373-396. Ed. Aleksandra Sudaric. Intech Publishers

[99] Lipka V, Dittgen J, Bednarek P, Bhat R, Wiermer M, Stein M, Landtag J, Brandt W, Rosahl S, Scheel D, Llorente F, Molina A, Parker J, Somerville S, Schulze-Lefert P. 2005. Pre- and postinvasion defenses both contribute to nonhost resistance in Arabidopsis. Science 310: 1180-1183

[100] Liu X, Z, Chen SY. 2001. Screening isolates of *Hirsutella* species for biocontrol of *Heterodera glycines*. Biocontrol Science and Technology 11:151-160

[101] Liu XZ, Li JQ, Zhang DS. 1997. History and status of soybean cyst nematode in China. International Journal of Nematology 7: 18-25

[102] Ma Y, Wang W, Liu X, Ma F, Wang P, Chang R, Qiu L. 2006. Characteristics of soybean genetic diversity and establishment of applied core collection for Chinese soybean cyst nematode resistance. Journal of Intergrative Biology 48: 722-731

[103] Mahalingham R, Skorupska HT. 1996. Cytological expression of early response to infection by Heterodera glycines Ichinohe in resistant PI 437654 soybean. Genome 39: 986–998

[104] Mahalingam R, Wang G, Knap HT. 1999. Polygalacturonidase and polygalacturonidase inhibitor protein: gene isolation and transcription in Glycine max-Heterodera glycines interactions. Molecular Plant-Microbe Interactions 12: 490-498

[105] Malhotra V, Orci L, Glick BS, Block MR, Rothman JE. 1988. Role of an N-ethylmaleimide-sensitive transport component in promoting fusion of transport vesicles with cisternae of the Golgi stack. Cell 54: 221–227

[106] Matson AL, Williams LF. 1965. Evidence of a fourth gene for resistance to the soybean cyst nematode. Crop Sci. 5: 477

[107] Matsye PD, Kumar R, Hosseini P, Jones CM, Alkharouf N, Matthews BF, Klink VP. 2011. Mapping cell fate decisions that occur during soybean defense responses. Plant Mol Bio 77: 513-528

[108] Matsye PD, Lawrence GW, Youssef RM, Kim K-H, Lawrence KS, Matthews BF, Klink VP. 2012. The expression of a naturally occurring mutant of an alpha soluble NSF at-

tachment protein gene in *Glycine max* (soybean) partially suppresses infection by the plant parasitic nematode *Heterodera glycines*. Plant Molecular Biology (in press)

[109] McLean MD, Hoover GJ, Bancroft B, Makhmoudova A, Clark SM, Welacky T, Simmonds DH, Shelp BJ. 2007. Identification of the full-length $Hs1^{pro-1}$ coding sequence and preliminary evaluation of soybean cyst nematode resistance in soybean transformed with $Hs1^{pro-1}$ cDNA. Canadian Journal of Botany 85: 437-441

[110] Meier-Ruge W, Bielser W, Remy E, Hillenkamp F, Nitsche R, Unsold R. 1976. The laser in the Lowry technique for microdissection of freeze-dried tissue slices. Histochem J 8: 387-401

[111] Meyer SLF, Huettel RN. 1996. Application of a sex pheromone, pheromone analogs, and *Verticillum lecanii* for management of Heterodera glycines. J. Nematology 28: 36-42

[112] Meyer SLF, Meyer RJ. 1996. Greenhouse studies comparing strains of the fungus *Verticillium lecanii* for activity against the nematode *Heterodera glycines*. Fundamentals of Applied Nematology 19: 305-308

[113] Morse WJ. 1927. Soybeans: culture and varieties. Farmer's bulletin NO. 1520. Washington, D.C.: U.S. Dept. of Agriculture. 38 pp

[114] Mudge J, Cregan PB, Kenworthy JP, Kenworthy WJ, Orf JH, Young ND. 1997. Two microsatellite markers that flank the major soybean cyst nematode resistance locus. Crop Sci 37: 1611-1615

[115] Müssig C, Fischer S, Altmann T. 2002. Brassinosteroid-regulated gene expression. Plant Physiol 129: 1241-1251

[116] Niblack TL, Chen SY. 2004. Cropping systems and crop management practices. Breeding for resistance and tolerance. Pp. 181-206. *in* D. P. Schmitt, J. A. Wrather, and R. D. Riggs, eds. Biology and management of soybean cyst nematode, 2nd ed. Marceline, MO: Schmitt & Associates of Marceline

[117] Niblack TL, Heinz RD, Smith GS, Donald PA (1993) Distribution, density, and diversity of Heterodera glycines in Missouri. J Nematol 25:880–886

[118] Niblack TL, Arelli PR, Noel GR, Opperman CH, Orf JH, Schmitt DP, Shannon JG, Tylka GL. 2002. A revised classification scheme for genetically diverse populations of *Heterodera glycines*. J Nematol 34: 279-288

[119] Niblack TL, Lambert KN, Tylka GL. 2006. A model plant pathogen from the kingdom animalia: Heterodera glycines, the Soybean Cyst Nematode. Annu Rev Phytopathol 44: 283-303

[120] Niblack TL, Riggs RD. 2004. Variation in virulence phenotypes. Breeding for resistance and tolerance. Pp. 57-71. *in* D. P. Schmitt, J. A. Wrather, and R. D. Riggs, eds. Biology and management of soybean cyst nematode, 2nd ed. Marceline, MO: Schmitt & Associates of Marceline

[121] Noel GR. 1992. History, distribution and economics. Pp 1-13 in RD Riggs and JA Wrather, editors. Biology and Management of the soybean cyst nematode. St. Paul, MN: APS Press

[122] Opperman CH, Bird D McK. 1998. The soybean cyst nematode, *Heterodera glycines:* a genetic model system for the study of plant-parasitic nematodes. Current Opinion in Plant Biology 1: 1342-1346

[123] Oyler GA, Higgins GA, Hart RA, Battenberg E, Billingsley M, Bloom FE, Wilson MC. 1989. The identification of a novel synaptosomal-associated protein, SNAP-25, differentially expressed by neuronal subpopulations. J Cell Biol 109: 3039-3052

[124] Patel S, Dinesh-Kumar SP. 2008. Arabidopsis ATG6 is required to limit the pathogen-associated cell death response. Autophagy 4: 20-27

[125] Piano F, Schetter AJ, Mangone M, Stein L, Kemphues KJ. 2000. RNAi analysis of genes expressed in the ovary of *Caenorhabditis elegans.* Current Biology 10: 1619–1622

[126] Pratt PW, Wrather JA. 1998. Soybean disease loss estimates for the southern United States, 1994-1996. Plant Disease 82: 114-116

[127] Puthoff DP, Nettleton D, Rodermel SR, Baum TJ. 2003. *Arabidopsis* gene expression changes during cyst nematode parasitism revealed by statistical analyses of microarray expression profiles. Plant J 33: 911–921

[128] Rao-Arelli AP. 1994. Inheritance of resistance to *Heterodera glycines* race 3 in soybean accessions. Plant Dis. 78: 898-900.

[129] Riggs RD, Hamblen ML. 1962. Soybean-cyst nematode host studies in the Leguminosae. Ark Agric Exp Stn Rep Series 110 Fayetteville AR 17p

[130] Riggs RD, Hamblen ML. 1966. Additional weed hosts of *Heterodera glycines.* Plant Dis Rep 50: 15-16

[131] Riggs RD, Hamblen ML. 1966. Further studies on the host range of the soybean-cyst nematode. Ark Agric Exp Stn Bulletin 718 Fayetteville AR 19p

[132] Riggs RD, Schmitt DP. 1988. Complete characterization of the race scheme for Heterodera glycines. J Nematol 20: 392-395

[133] Riggs RD, Schmitt DP. 1991. Optimization of the *Heterodera glycines* race test procedure. J Nematol 23: 149-154

[134] Riggs RD, Kim KS, Gipson I. 1973. Ultrastructural changes in Peking soybeans infected with *Heterodera glycines.* Phytopathology 63: 76–84

[135] Robinson AF, Inserra RN, Caswell-Chen EP, Vovlas N, Troccoli A. 1997. *Rotylenchulus* species: Identification, distribution, host ranges, and crop plant resistance. Nematropica 27: 127-180

[136] Ross JP, Brim CA. 1957. Resistance of soybeans to the soybean cyst nematode as determined by a double-row method. Plant Dis Rep 41: 923–924

[137] Ross JP. 1958. Host-Parasite relationship of the soybean cyst nematode in resistant soybean roots. Phytopathology 48: 578-579

[138] Ross JP. 1962. Crop rotation effects on the soybean cyst nematode population and soybean yields. Phytopathology 52: 815-818

[139] Brim CA, Ross JP. 1966. Registration of Pickett soybeans. Crop Science 6: 305

[140] Rosskopf EN, Chellemi DO, Kokalis-Burelle N, Church GT. 2005. Alternatives to Methyl Bromide: A Florida Perspective. American Phytopathological Society. APSnet feature, http://www.apsnet.org/publications/apsnetfeatures/Documents/2005/MethylBromideAlternatives.pdf

[141] Sasser JN, Uzzell G, Jr. 1991. Control of the soybean cyst nematode by crop rotation in combination with nematicide . J. Nematology 23: 344-347

[142] Scheideler M, Schlaich NL, Fellenberg K, Beissbarth T, Hauser NC, Vingron M, Slusarenko AJ, Hoheisel JD. 2001. Monitoring the switch from housekeeping to pathogen defense metabolism in *Arabidopsis thaliana* using cDNA arrays. J Biol Chem 277: 10555–10561

[143] Schmutz J, Cannon SB, Schlueter J, Ma J, Mitros T, Nelson W, Hyten DL, Song Q, Thelen JJ, Cheng J, Xu D, Hellsten U, May GD, Yu Y, Sakurai T, Umezawa T, Bhattacharyya MK, Sandhu D, Valliyodan B, Lindquist E, Peto M, Grant D, Shu S, Goodstein D, Barry K, Futrell-Griggs M, Abernathy B, Du J, Tian Z, Zhu L, Gill N, Joshi T, Libault M, Sethuraman A, Zhang XC, Shinozaki K, Nguyen HT, Wing RA, Cregan P, Specht J, Grimwood J, Rokhsar D, Stacey G, Shoemaker RC, Jackson SA. 2010. Genome sequence of the palaeopolyploid soybean. Nature 463: 178-183

[144] Schmelzer E. 2002. Cell polarization, a crucial process in fungal defence. Trends Plant Sci 7: 411-415

[145] Schneider SM, Rosskopf EN, Leesch JG, Chellemi DO, Bull CT, Mazzola M. 2003. United States Department of Agriculture-Agricultural Research Service research on alternatives to methyl bromide: pre-plant and post-harvest. Pest Manag Sci. 59: 814-826

[146] Shannon JG, Arelli PR, Young LD. 2004. Breeding for resistance and tolerance. Pp. 155-180. *in* D. P. Schmitt, J. A. Wrather, and R. D. Riggs, eds. Biology and management of soybean cyst nematode, 2nd ed. Marceline, MO: Schmitt & Associates of Marceline

[147] Smant GA, Stokkermans JPWG, Yan Y, De Boer JM, Baum TJ, Wang X, Hussey RS, Gommers FJ, Henrissat B, Davis EL, Helder J, Schots A, Bakker J. 1998. Endogenous cellulases in animals: isolation of 1,4-endoglucanase genes from two species of plant-parasitic nematodes. PNAS USA 95: 4906-4911

[148] Sönnichsen B, Koski LB, Walsh A, Marschall P, Neumann B, Brehm M, Alleaume AM, Artelt J, Bettencourt P, Cassin E, Hewitson M, Holz C, Khan M, Lazik S, Martin

C, Nitzsche B, Ruer M, Stamford J, Winzi M, Heinkel R, Röder M, Finell J, Häntsch H, Jones SJ, Jones M, Piano F, Gunsalus KC, Oegema K, Gönczy P, Coulson A, Hyman AA, Echeverri CJ. 2005. Full-genome RNAi profiling of early embryogenesis in Caenorhabditis elegans. Nature 434: 462-429

[149] Spears JF. 1957. Review of soybean cyst nematode situation for presentation at public hearing on the need for Federal Domestic Plant Quarantine, July 24, 1957

[150] Steeves RM, Todd TC, Essig JS, Trick HN. 2006. Transgenic soybeans expressing siRNAs specific to a major sperm protein gene suppress Heterodera glycines reproduction. Funct Plant Biol 33: 991–999

[151] Stein M, Dittgen J, Sánchez-Rodríguez C, Hou BH, Molina A, Schulze-Lefert P, Lipka V, Somerville S. 2006. Arabidopsis PEN3/PDR8, an ATP binding cassette transporter, contributes to nonhost resistance to inappropriate pathogens that enter by direct penetration. Plant Cell 18: 731-746

[152] Tao Y, Xie Z, Chen W, Glazebrook J, Chang HS, Han B, Zhu T, Zou G, Katagiri F. 2003. Quantitative nature of Arabidopsis responses during compatible and incompatible interactions with the bacterial pathogen Pseudomonas syringae. Plant Cell 15: 317-330

[153] Tenllado F, Martı´nez-Garcı´a B, Vargas M, Dı´az-Ruı´z JR. 2003. Crude extracts of bacterially expressed dsRNA can be used to protect plants against virus infections. BMC Biotechnol 3:3

[154] Tepfer D. 1984. Transformation of several species of higher plants by Agrobacterium rhizogenes: sexual transmission of the transformed genotype and phenotype. Cell 37: 959-967

[155] Timmons L, Donald LC, Andrew F (2001) Ingestion of bacterially expressed dsRNAs can produce specific and potent genetic interference in Caenorhabditis elegans. Gene 263:103–112

[156] Timper P, Riggs RD, Crippen DL. 1999. Parasitism of sedentary stages of Heterodera glycines by isolates of a sterile nematophagous fungus. Phytopathology 89: 1193-1199

[157] Urwin PE, Lilley CJ, Atkinson HJ. 2002. Ingestion of double-stranded RNA by preparasitic juvenile cyst nematodes leads to RNA interference. Mol Plant Microbe Interact 15: 747-752

[158] Vaghchhipawala Z, Bassuner R, Clayton K, Lewers K, Shoemaker R,m Mackenzie S. 2001. Modulations in gene expression and mapping of genes associated with cyst nematode infection of soybean. Molecular Plant-Microbe Interactions 14: 42-54

[159] Wang D, Stravopodis D, Teglund S, Kitazawa J, Ihle JN. 1996. Naturally occurring dominant negative variants of Stat5. Mol Cell Biol 16: 6141-6148

[160] Wang X, Allen R, Ding X, Goellner M, Maier T, DeBoer JM, Baum TJ, Hussey RS, Davis EL. 2001. Signal peptide-selection of cDNA cloned directly from the esophageal

gland cells of the soybean cyst nematode Heterodera glycines. Molecular Plant-Microbe Interactions 14: 536-544

[161] Wang J, Meyers D, Yan Y, Baum T, Smant G, Hussey R. 2003. The soybean cyst nematode reduces soybean yield without causing obvious symptoms. Plant Disease 87: 623-628

[162] Wang X, Myers D, Yan Y, Baum T, Smant G, Hussey R, Davis E. 1999. In planta localization of a 1, 4-endoglucanase secreted by Heterodera glycines. Molecular Plant-Microbe Interactions 12: 64-67

[163] Weidman PJ, Melançon P, Block MR, Rothman JE. 1989. Binding of an N-ethylmaleimide-sensitive fusion protein to Golgi membranes requires both a soluble protein(s) and an integral membrane receptor. J Cell Biol 108: 1589-1596

[164] Wilson MA, Bird D McK van der Knaap E. 1994. A comprehensive subtractive cDNA cloning approach to identify nematode-induced transcripts in tomato. Phytopathology 84: 299-303

[165] Winstead NN, Skotland CB Sasser JN. 1955. Soybean cyst nematodes in North Carolina. Plant Disease Reporter 39: 9-11.

[166] Wrather JA, Chambers AY, Fox JA, Moore WF, Sciumbato GL. 1995. Soybean disease loss estimates for the southern United States, 1974-1994. Plant Disease 79: 1076-1079

[167] Wrather JA, Anderson TR, Arsyad DM, Tan Y, Ploper LD, Porta-Puglia A, Ram HH, Yorinori JT. 2001a. Soybean disease loss estimates for the top ten soybean producing countries in 1998. Canadian Journal of Plant Pathology 23: 115-121

[168] Wrather JA, Steinstra WC, Koenning SR. 2001b. Soybean disease loss estimates for the United States from 1996-1998. Canadian Journal of Plant Pathology 23: 122-131

[169] Wrather JA, Koenning SR, Anderson TR. 2003. Effect of diseases on soybean yields in the United States and Ontario (1999-2002). Online. Plant Health Progress doi: 10.1094/PHP-2003-0325-01-RV, http://www.plantmanagement network.org/sub/php/review/2003/soybean/>(19 November 2003)

[170] Wrather JA, Koenning SR. 2006. Estimates of disease effects on soybean yields in the United States 2003-2005. J Nematology 38: 173-180

Chapter 6

Arthropod Fauna Associated to Soybean in Croatia

Renata Bažok, Maja Čačija, Ana Gajger and
Tomislav Kos

Additional information is available at the end of the chapter

1. Introduction

The importance of soybean (*Glycine max* (L.) Merr.), as today's world leading oil and protein crop, is increasing in Croatia. As a plant species, soybean was registered for the first time in Croatia in 1876. Soybean is relatively new field crop for Croatia. It was grown for the first time in 1910 but, starting with 1970s it became important field crop [1]. In 1981, soybean was cultivated on an area of 3.381 ha. Since that time the area cultivated by soybean has increased considerably, and productivity has also risen steadily. Figure 1 presents the trends in soybean production in Croatia in the period 1993-2010 [2].

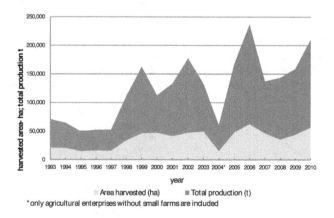

Figure 1. Harvested area and total production of soybean in Croatia, 1993-2010

Today's the area on which soybean is cultivated in Croatia varies, depending on the year, from 30.000 to 50.000 ha. Croatian government statistics [2] show gains in average yielding ability, from 2.160 to 3.000 kg/ha, between 1993 and 2010. Comparing to data from USA [3] on the average yield between 2.197 and 2.896 kg/ha, soybean yields obtained in "regular" years in Croatia are satisfactory. The exceptions in registered yield quantity were observed in extremely dry and warm years 2000, 2003 and 2007 in which yield was between 1.380 and 1.900 kg/ha. Therefore, the main problem of soybean yielding stability is related to vulnerability of soybean production in extreme climatic conditions in which pest outbreaks influence yields negatively. Global climate changes are often discussed by numerous scientists. Besides the increase of global mean temperature [4], the incidence of the years in which extreme conditions are present vs. "regular" years is increasing. This is proved by the fact that in the period from 2000 to 2009, three years with extremely dry and warm conditions were observed. Consequently, to mitigate the negative consequences of pest outbreaks and improve profits soybean growers, in these extreme years, attempt to control the pests which can reduce crop productivity.

Comparing to weeds and diseases, in "regular years" pests are of somewhat less importance for soybean production in Croatia. In different agro-ecosystems, the arthropod fauna of soybean contains a great number of damaging species [5-8]. Soybean pests have not been investigated completely in Croatia. It was reported [8] that in the region where Croatia belongs, soybean crops are attacked by over 180 pests (150 insects and 30 species from other animal classes) among which approximately 25 pest species are the most important.

Some investigations or observations on arthropod fauna of soybean were conducted in the past on the territory of Croatia [9 - 13], and in neighboring countries [14 - 24]. Additionally, some of the species were registered recently as the pests which could cause significant yield damage on soybean [25 - 29].

The most comprehensive overview of the potential arthropod pests' fauna of soybean in Croatia is given by Maceljski [9]. This overview is a result of the literature review and author's long time work experience in entomology. On the other hand, investigations carried out by other scientists in Croatia [10 - 13] and neighboring countries [6-8, 14-24] reported on the presence or harmfulness of some additional species. In the Table 1 arthropod species that are reported as soybean pests both, in Croatia and in neighboring countries are listed.

Besides arthropod species, nematodes are established as potential pests on soybeans in Croatia [11, 12] and in neighboring countries [19]. Jelić [12] established 43 species of phytoparasitic nematodes on 18 localities distributed in east Croatia (region of Slavonia). Identified species belonged to the genera *Ditylenchus* Filipjev, *Meloidogyne* Goeldi, *Paratylenchus* Micoletzky, *Pratylenchus* Filipjev, *Rotylenchus* Filipjev, and *Tylenchorhynchus* Cobb. However significant damages caused by nematodes haven't been recorded jet. Besides mentioned pests, some authors [8, 30] reported that significant damage on soybean crops in Serbia could be caused by other animal species as are *Cricetus cricetus* L., *Microtus arvalis* Pallas and *Lepus europaeus* Pallas as well.

In regular farming practice in Croatia soybean seed is not treated with insecticides. Among the arthorpod pests, mites (*Tetranychus urticae* Koch and *Tetranychus atlanticus* = *T. turkestani*

Ugarov & Nikolskii) could be controlled by the use of acaricides if their populations reach economic threshold (usually in warm and dry years). Other pest species are controlled only occasionally if pest outbreaks occur.

Order	Suborder	Family	Species	Literature source	Croatia	Neighboring countries
Collembola		Smynthuridae	*Sminthurus* sp. Latreille 1802	21		+
Thysanoptera		Thripidae	*Frankliniella intonsa* (Trybom 1985)	21		+
Hemiptera	Heteroptera	Miridae	*Lygus* sp. Hahn 1833	9	+	
			Lygus gemellatus (Heerrich-Schaeffer 1835)	21		+
			Lygus pratensis (Linneaus 1758)	24		+
			Lygus rugulipennis Poppius 1911	8		+
			Halticus apterus (Linnaeus 1758)	9	+	
			Apolygus lucorum (Meyer-Dur 1843)	24		+
		Pentatomidae	*Dolycoris bacarrum* (Linnaeus 1758)	21, 24		+
			Eurydema oleracea (Linnaeus 1758)	24		+
			Nezara viridula (Linnaeus 1758)	9, 26	+	
			Piezodorus sp. Fieber 1861	9	+	
		Anthocoridae	*Anthocoris* sp. Fallen 1814	9	+	
			Orius niger Woolf 1811	6		+
		Nabidae	*Nabis (Nabis) ferus* Linnaeus 1758	6, 9	+	+
			Nabis feroides Wagner 1967	6		+
			Nabis pseudoferus Remane 1949	6		+
	Homoptera	Membracidae	*Sctictocephala bisonia* Koop & Yonke	9, 22, 24	+	+
			Cicadella viridis (Linnaeus 1758)	9	+	
		Aphididae	*Aphis craccivora* Koch 1854	21		+

Order	Suborder	Family	Species	Literature source	Croatia	Neighboring countries
		Diaspididae	*Lepidosaphes* sp. Shimer 1898	21		+
Coleoptera		Elateridae	*Agriotes ustulatus* Schaller 1793	15		
			Agriotes sp. Eschscholtz 1829	15		+
		Scarabaeidae	*Anomala* sp. Schoenherr 1817	21		+
		Anobiidae	*Stegobium paniceum* (Linnaeus 1758)	21		+
		Cocinelidae	*Subcocinella vigintiquatuorpunctata* (Linnaeus 1758)	9, 21	+	+
		Chrysomelidae	*Longitarsus* sp. Berthold 1827	9	+	
			Phylotreta undulata Kutschera 1860	9	+	
			Haltica oleracea Linnaeus 1758	9	+	
		Lathiridae	*Corticaria* sp. Marsham 1802	9	+	
		Curculionidae	*Phyllobius* sp. Germar 1824	9	+	
			Sitophillus sp. Schnherr, 1838	9	+	
Lepidoptera		Gracilariidae	*Phylonorycter insignitella* Zeller 1846	24		+
		Pyralidae	*Etiella zinckenella* (Treitschke 1832)	8, 9, 14, 21	+	+
		Crambidae	*Udea ferrugalis* Hubner 1796	21		+
		Tortricidae	*Olethreutes lacunana* Freeman 1941	21		+
			Grapholita compositella Fabricius 1775	24		+
		Lymanthridae	*Orgya gonostigma* L.	21		+
		Geometridae	*Ascotis selenaria* Dennis & Schiffermuller 1775	21		+
		Nymphalidae	*Vanessa cardui* Linnaeus 1758	9, 20, 23, 24, 25, 27, 28, 29	+	+
		Noctuidae	*Acronicta (Viminia) rumicis (Linnaeus 1758)*	21		+

Order	Suborder	Family	Species	Literature source	Croatia	Neighboring countries
			Chloridea dipsacea L. (*Heliothis viriplaca* (Hufnagel, 1766))	21		+
			Phragmatobia fuliginosa (Linnaeus 1758)	21		+
			Autographa gamma (Linnaeus 1758)	7, 8		+
			Helicoverpa armigera (Hubner 1808)	7, 8		+
			Mamestra sp. Ochensheimer 1816	7, 8		+
Diptera		Cecidomyidae	*Clinodiplosis trotteri = Anabremia trotteri* (Kieffer 1909)	21		+
			Acarolestes tetranychorum (Kiefer 1909)	21		+
		Anthomyiidae	*Delia platura* (Meigen 1826)	8		+
		Agromyzidae	*Lyriomyza congesta* (Becker 1903)	21		+
Prostigmata		Tetranychidae	*Tetranychus urticae* Koch 1836	7, 8, 9, 10, 12, 13, 16, 17, 19	+	+
			Tetranicuhus atlanticus = T. turkestani Ugarov & Nikolskii 1937	7, 8, 9, 10, 16, 17, 19, 24	+	+
			Tetranychus tumidus Banks 1900	24		+

Table 1. Arthropod species established to damage soybean in Croatia and neighbouring countries

In only one investigation which was carried out in Serbia [21] beneficial fauna on soybean was recorded. Only three predatory species were established, *Coccinella septempunctata* L. (Coleoptera: Coccinelidae), *Chrysopa carnea* Stephens (Neuroptera: Chrysopidae) and *Acarolestes tetranychorum* Kief. (Diptera: Cecidomyidae). There are no similar investigations conducted in Croatia but, out of all species listed [9] as potential members of entomofauna of soybean, two species (*Anthocoris sp.* Fallen and *Nabis (Nabis) ferus* L.) are listed as potential beneficial insects.

The subject of pest control is rarely discussed without the reference to the concept of integrated pest management (IPM). IPM is essentially a holistic approach to pest control that seeks to optimize the use of a combination of methods to manage whole spectrum of pests within particular cropping system. IPM relies heavily on biological controls with a perspective chemical input only as a last resort. For effective control, there needs to be an understanding of a pest's interaction with its environment. This is so called concept of "life system" which was initially conceived by Clark et al. [31] to reinforce the idea that population cannot be considered apart from the ecosystem with which it interacts. The life system consists of the pest population plus its "effective environment". Most ecological pest management concentrates on the agro-ecosystem, defined as "effective environment" at the crop level [32]. Monitoring in insect pest management can be used to determine the geographical distribution of pests, to assess the effectiveness of control measures, but in its widest sense monitoring is the process of measuring the variables required for the development and use of forecast to predict pest outbreaks [33]. Such forecasts are an important component of pest management strategies because a warning of the timing and extent of pest attack can improve the efficiency of control measures. For successful pest control according to the principles of IPM it is of great importance to have deep knowledge in harmful and beneficial arthoropods in particular agro-ecological conditions.

The study was conducted to determine the harmful and beneficial arthropod fauna during the soybean growing season, and based on their dynamic of occurrence and abundance to identify the harmful and beneficial species of greater importance for soybean production in Croatia.

2. Materials and methods

Research was conducted on experimental field located in Zagreb. The soybean variety Zlata (BC Institute Zagreb, Croatia) was planted on April 27th 2010 on an experimental area of 162 m². The average plant density was 630.000 plants/ha. Soybean variety Zlata belongs to the maturity group "0" and according to the information given by producers [34] it has a "good" tolerance to pests and diseases. In order to control weeds gyphosate (pre-sowing), metribuzin, metholachlor and clomazone (in the phase of the first trifoliate - V1, according to [35]) and bentazon (in the phase of the third trifoliate - V3) were applied.

Sweep net sampling consisted of making a set of 50 sweeps across three rows of soybeans while walking down the row [36]. A 30 cm diameter sweep net was used. Sampling began when soybeans were in the beginning of flowering (R1) on June 24th 2010 and continued through September 9th 2010 when plants reached physiological maturity (R7). Weekly sampling was done on the same day each week in late morning. It was performed for 12 weeks. At each sampling date four samples were collected.

Whole plant counts were conducted on 10 plants per each of four replicates. As it was proposed by Kogan and Pitre [36] randomly selected plants were initially scanned for large, often fast moving species. After the initial scan, both sides of each leaf on the plant were searched, as were petioles, axils and stems. Additionally, one leaf per plant was collected at each whole plant count date to establish mite population by leaf inspection. Therefore, four samples each containing 10 leaves were transported to laboratory to be examined under the stereomicroscope and all life stages of mites were counted [37]. Whole plant counts and leaf collection began one week later than sweep net sampling i.e. on July 1st 2010 and continued through September 9th 2010. It was performed for 11 weeks.

All collected insects were identified to the family or genus and species (if possible). For identifying insects identification keys were used [38-42].

Based on the number of all individuals, cenological characteristics (dominance and frequency) of the insect orders and families (where appropriate) were determined [43].

The dominance was calculated by Balogh formula:

$$D1 = \frac{a_1}{\sum a_i} \times 100$$

Where: a_1 = number of identified specimens of one species;

Σ_{a1} = total number of all collected specimens.

The frequency was calculated by Balogh formula:

$$C_{a1} = \frac{U_{a1}}{\sum U_i} \times 100$$

Where: U_{a1} = number of samples with identified species;

$\sum U_i$ = total number of samples.

3. Results and discussion

The total catch was 1357 specimens which belong to six orders: Thysanoptera, Hemiptera, Coleoptera, Lepidoptera, Diptera and Prostigmata (Table 2).

Out of 1357 specimens, only 73 individuals (5.37%) belong to beneficial fauna (mostly predators), while all other collected specimens are herbivorous and therefore potential pests on soybean. All found beneficials belonged to predators and majority of them (70 individuals) belong to Hemiptera what confirms the statement of Ketzschmar [44] that predaceous Hemiptera are usually more abundant in soybean fields than all other insect predators combined. In earlier investigations [21] conducted in Serbia no predaceous Hemiptera have been found while more recent investigations in Serbia [6] and in Croatia [9] stated that they are present in soybean crops. All predaceous Hemiptera feed on a wide range of hosts and may extend this polyphagy to plant feeding to some extent [45]. Such plant feeding causes no damage to row crops but almost certainly has survival value for the predators by maintaining populations where prey are scarce or absent. Some of the species which belong to family Pentatomidae are also recognized as predators [45]. Since some of the individuals collected in our investigation were classified as family Pentatomidae but, were not identified to the species, it is possible that some of them are predaceous as well.

ORDER	FAMILY	GENUS	SPECIES	TOTAL NUMBER OF INDIVIDUALS CAPTURED BY		
				SWEEP NET	WHOLE PLANT COUNTS	LEAF INSPECTION
Thysanoptera		*		52	12	
Hemiptera	Miridae	*Halticus*	apterus	44	3	
		Lygus Hahn 1833	sp.	4		
	Nabidae	*Nabis*	ferus Linnaeus 1758	55		
	Anthocoridae	*		3	8	
	Pentatomidae	*		12		
		Nezara	viridula Linnaeus 1758	472	181	
		Piezodorus	sp. Fieber 1861	28		
	Membracidae	*Stictocephala*	bisonia Kopp & Yonke 1977	2		
	Cicadellidae	*Cicadella*	viridis Linnaeus 1758	3		
Coleoptera	Coccinelidae	*		3		
	Chrysomelidae	*Phyllotreta*	undulata Kutschera 1860	4		
		Haltica	oleracea Linnaeus 1758	7		
		Longitarsus Berthold 1827	sp.	2	1	
	Latridiidae	*Corticaria* Marsham 1802	sp.	21		

ORDER	FAMILY	GENUS	SPECIES	TOTAL NUMBER OF INDIVIDUALS CAPTURED BY		
				SWEEP NET	WHOLE PLANT COUNTS	LEAF INSPECTION
	Curculionidae	*			1	
		Phyllobius Germar 1824	sp.	5		
		Sitophillus Schnherr, 1838	sp.	1		
Lepidoptera		*			2	
	Nymphalidae	*Vanessa*	*cardui* Linnaeus 1758	7		
	Noctuidae			15		
Diptera	Nematocera	*			1	
		*		16		
Prostigmata	Tetranychidae	*Tetranychus*	*urticae* Koch 1836			387
TOTAL				759	211	387

* Identification was not possible; Beneficial species are marked in grey;

Table 2. Arthropod species established during the soybean vegetation in 2010 by three different methods

Using the entomological net, 759 individuals were collected, whereas 211 individuals were gathered by whole plant counts and 387 individuals by leaf inspection.

3.1. Sweep net sampling

Number of arthropod individuals collected by sweep net sampling was the highest among the three methods applied. Using these methods, species belonging to 20 different systematic categories were collected. The collected individuals belonged to five insect orders, Thysanoptera, Hemiptera, Coleoptera, Lepidoptera and Diptera. The abundance of insect orders established by sweep net sampling is shown in Figure 2.

Order Hemiptera was present in the sweep net sampling in the highest abundance (82.48%). The same order was the most frequent. It was present in 87.5% of all samples obtained by sweep net sampling. Order Coleoptera was present in 57.5% of all samples and was designated as constant. Other orders (Thysanoptera, Lepidoptera and Diptera) were less frequent; they were present in 30-37.5% of all samples. Investigations conducted in different agro-eco-systems showed that the sweep net sampling is the most effective method to collect different

■ Thysanoptera □ Hemiptera Coleoptera ▩ Lepidoptera ■ Diptera

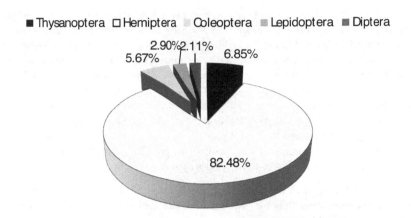

Figure 2. Abundance of insect orders collected by sweep net sampling of soybean crop, Zagreb, 2010

leaf feeding pests as are leafhoppers [46], lepidopterous larvae [46, 47], leaf feeding Coleoptera [48] and phytophagous Pentatomidae [49] as well as predaceous Hemiptera [45].

3.2. Whole plant count

The lowest number of individuals was established by whole plant count method. The majority of established individuals belonged to order Hemiptera. Only few Thysanoptera and Lepidoptera were established by this method. Some authors stated that this method is suitable for larvae of Lepidoptera [46, 47] and for phytophagous thryps [50]. There are no data on any damage caused by any phytophagous thrips in Croatia while *Vanessa cardui* L. was recorded in previous investigations as important pest.

3.3. Leaf sampling and inspection

The only species established by leaf inspection was *T. urticae*. It was proved that this method is good for establishing population of phytophagous thrips [48], whiteflies [49] and mites [35]. It is obviously that out of these three groups, only phytophagus mites were present in experimental field.

3.4. Collected species: abundance and importance

Sampling arthropod populations is a cornerstone of basic research on agricultural ecosystems and the principal tool for building and implementing pest management programs. The purpose of sampling is dual, it is a research method for defining the nature and dynamics of communities in agricultural ecosystems and it is also a mean for providing pest management decision. The purpose of sampling in our investigation was to get deep knowledge on pest and beneficial species present in soybean crop. Conducted investigation encompassed three most common sampling methods for investigations of soybean arthropod fauna. The need to encompass all three methods is confirmed later by the fact that species identified by each particular method differ. By employed methods we were able to get all relevant data on above ground arthropod fauna that could be found on soybean canopy. We did not aim to collect information on underground soybean arthropods and ground predators in soybean fields. To collect this information we should use common methods for sampling soil arthropods, soil samples and extraction [51] or pitfall trapping [48]. Some of earlier researches on soybean arthropod fauna in the region collected information on underground soybean arthropods but, no research did pay attention on ground predators in soybean field. No researches among all conducted [6-29] did pay attention to abundance and frequency of particular orders, genus or species as well, so it is not possible to compare if there are some discrepancies with the results of previous researches.

Individuals that belonged to the order Thysanoptera have been found by sweep net sampling in highest abundance than by whole plan counts, and they haven't been found by leaf inspection at all. In Serbia one phytophagous thrips species (*Frankliniella intonsa* Trybom) on soybean has been identified [21]. Since in our investigation thrips were not established by leaf sampling and inspection, it could be concluded that they did not feed and develop on soybean. It might be that those species are predaceous because it is reported [52] that thrips are natural enemies of different pests. Important predaceous genera of thrips are *Aelothrips* Haliday, *Franklinothrips* Back, *Scolothrips* Hinds, *Leptothrips* Hood, *Karnyothrips* Watson and *Podothrips* Hood. Within the genus *Aeolothrips*, the species *Aeolothrips intermedius* Bagnall is distributed throughout western and eastern Europe [53], the middle East and India but now it can be considered cosmopolitan [54]. Comparative tests by many authors [53, 55, 56] using different types of prey (including various species of Thysanoptera), suggested that both the larvae and the adult females are generic predators, even though they present marked dietary preferences. In Italy [57] *A. intermedius* was detected in association with various different phytophagous Thysanoptera, which included *T. tabaci* but also frequently *F. occidentalis* on many different plant species including legume species *Medicago sativa* L. Predaceous thyrips belonging to the genus *Aelothips* are reported as important predators of *T. urticae* in soyben crops in north-eastern Italy [58]. *Franklinothrips* sp. adults and larvae are generalist predators and attack a wide variety of arthropod pests including two spotted spider mite (*T. urticae*) [59]. Genus *Scolothrips* is counting six species in Europe [60], and one of them, *Scolothrips longicornis* Priesner is a predator of *T. urticae* [61] and *T. turkestani* [62]. Both pest species are registered as soybean pests in Croatia. Genus *Leptothrips* is not present in Europe [60]. Genus *Karnyothrips* is counting three species in Europe [60]. Some species are reported as predators of scale insects [63]. Genus *Podothrips* is known as grass-living genus. It counts only two species in Europe present only in Italy and Cyprus [60]. Since identification to the species was not possible, we cannot state which species of Thysanoptera were present.

Individuals belonging to six families of the **order Hemiptera** were identified in our research. Four families belong to the **suborder Heteroptera** (so called typical bugs) and two families belong to the **suborder Homoptera.**

Family Miridae was presented by genus *Lygus* sp. Four individuals were captured. Identification to species was not possible. Species belonged to the genus *Lygus* were reported by different authors [6, 8, 21, 24] to feed on soybean crops in Bosnia and Herzegovina and Serbia as well as in Croatia [9] without causing serious damages. More numerous were individuals of *Halticus apterus*. This species is distributed through Mediterranean region [64]. It was reported to feed on soybean only in Croatia [9]. Other research showed that it feeds on some legume plants such as *Medicago sativa* L., *Lotus corniculatus* L. and *Trifolium repens* L. in Italy, and also to be able to cause damages on onions and Gallium [64]. Since it was not reported as serious pest of soybean anywhere, it should be monitored in the future but the probability for this species to become important pest of soybean is low.

Family Nabidae was represented with one species, *N. ferus*. The same species was previously reported in Croatia [9] and in Serbia [6]. Additionally two other species of this genus, *N. feroides* and *N. pseudoferus*, were reported in Serbia [6]. The density of *N. ferus* was moderate, total of 55 individuals were captured. This species was reported as common predator species in Ukraine [65]. Aphids are the principal prey insects for this species, but numerous other families are acceptable, including other bugs [65]. Because of its possible importance in soybean agro-ecosystems, the dynamic of the appearance of this species will be further analyzed.

Eleven individuals belonging to **family Anthocoridae** were captured in our investigation. Family Anthocoridae was previously reported in soybean in Croatia [9] and in Serbia [6]. This family is mentioned as one of the most important predaceous family of Heteroptera in soybean crops [45]. Within the family Anthocoridae, members of the genus *Orius* occur as predators in soybean fields all around the world [45]. The species *Orius niger* Woolf has been found in soybean fields in Serbia [6]. In some areas, species of the genus *Anthocoris* Fallen are probably also important predators in soybean [45]. Captured individuals were not identified to the species so it is not possible to discuss which genus was the most abundant in our investigations.

Among the established Hemiptera, **family Pentatomidae** was the most abundant. Altogether 512 individuals were found in sweep-net samples and 181 individuals by whole plant counts. The most abundant species was the southern green stink bug, *Nezara viridula* L. This species was reported as present in Croatia [9]. Recently the serious damages caused by this species were reported in Croatia [26]. It is not mentioned as serious pest in neighboring countries, while it was mentioned as serious threat to soybeans in Italy [58, 66]. It was reported [49] as one of the most abundant phytophagous stink bugs on soybean worldwide among of almost 40 species of stink bugs that have been found on soybean. Due to high number of captured individuals and registered damages caused by this species, it might be identified as one of the potential pests on soybean in Croatia. Therefore the dynamic of the appearance will be further analyzed. Species belonging to phytophagous genus *Piezodorus* were captured in lower number. Genus *Piezodorus* was reported as possible pest genus in Croatia [9] and in Italy [58]. The importance of the species *P. guildinii* Westwood is increasing in USA as well as in Brazil. This species was observed for the first time in southern Louisiana in 2000 and since 2002, it has been a significant pest of soybean [67]. At present, *Eustichus heros* (F.) and *P. guildinii* are more widespread and occur in greater numbers than *N. viridula,* and *P. guildinii* is principally responsible for the green bean syndrome observed in Brazilian soybean [68]. Genus *Piezodorus* is counting three species in Europe [60]. Twelve individuals, members of family Pentatomidae remained unidentified. It is possible that some of them are phytophagous. Also it is possible that some of them are predators because species which belong to family Pentatomidae are also recognized as predators [45].

Two families each represented with one species from the **suborder Homoptera** have been collected in low numbers. *Stictocephala bisonia* was reported to feed on soybean in Croatia [9], Serbia [21] and Bosnia and Herzegovina [24]. The second identified species was *Cicadella viridis*. This species was registered to feed on soybean in Croatia [9] but without significant damage. Within the USA, potato leafhopper (*Empoasca fabae* Harris) is the most important leafhopper species [69]. Even though aphids are recognized as a regular part of entomofauna of soybean, we did not record them. Several species of aphids are known to attack soybean crops. The most important species in North America is *Aphis gyicines* Matsumura [70, 71]. This species is not registered in Europe [60]. Some other species of aphids that are members of the fauna of Europe [60] and Croatia [72, 73] successfully colonize and reproduce parthenogenetically on soybean [71]: *A. craccivora* Koch, *Aulacorthum solani* (Kaltenbach) and *Aphys gossypii* Glover.

Out of four families of the order Coleoptera that were identified, one represents mainly predaceous species (family Coccinelidae). Some species of the family Coccinelidae are reported as the members of arthropod fauna on soybean in Croatia and Serbia [9, 21]. Species *Epilachna varivestris* Mulsant is known as a soybean pest in USA [48]. Three species of the phytophagous genus *Epilachna* are present in Europe, including Croatia [60] but only in Dalmatia where soybean cultivation is not common. Individuals from the genus *Corticaria*, family Latridiidae were the most numerous. Adults and larvae of this family feed on the conidia of fungi and *Myxomycetes* [74]. All found species from the order Coleoptera were previously listed as potential members of soybean fauna in Croatia [9] but, due to the low populations, their potential to be significant pests or predators is not very high. We did not employ any method for sampling soil dwelling insects or underground fauna. Therefore we did not collect the species which belong to families Elateridae and Scarabaeidae that are known as polyphagous soil pests that could cause damage on soybean crops [15, 21].

Only 22 individuals from **order Lepidoptera** were collected. Painted Lady (*Vanessa cardui*) was the only identified species. Other specimens were classified into the family Noctuidae. *V. cardui* was previously recorded to significantly damage soybean in Croatia [9, 25, 27, 29] and in neighboring countries [20, 23]. The outbreak of this pest is periodical. Higher population could be expected in weedy soybean fields because females are attracted by pollen sources and heavy plant density [25]. There are many of species from the family of Noctuidae and from the other families, members of the order Lepidoptera which could cause the damage but, until now, serious damage in Croatia was reported only by *V. cardui*. In USA, the most important lepidopterous species are *Anticarsia gemmatalis* (Hubner), *Pseudoplusia includens* (Walker), *Trichoplusia ni* (Hubner), *Platypena scabra* F., *Heliothis zea* Boddie, *Heliothis virescens* (Fabricius) and *Heliothis* (=*Helicoverpa*) *armigera* (Hubner) [46, 47, 75]. Except *T. ni* and *H. armigera*, other species are not distributed in Europe [60]. *H. armigera* is often mentioned as one of the potentially very dangerous species. Because of its invasive nature this pest is currently placed on Annex I A II of Council Directive 2000/29/EC, indicating that it is considered to be relevant for the entire EU and that phytosanitary measures are required when it is found on any plants or plant products. Some countries made pest risk analyses [76]. Damages caused by this species were reported on soybeans in Vojvodina Province of Serbia and in Montenegro in the very warm summer of 2003 [77] when 85.3% of the soybean pods were injured in August. H. *armigera* is a serious pest on outdoor crops in Portugal and Spain, predominantly on tomato crops as well as on cottom and maize. It developed resistance to many groups of insecticides [78]. We did not find caterpillars of limabean pod borer (*Etiella zinckenella* Treitschke) even though this species was reported as soybean pest in Croatia [9] and in neighboring countries [6, 14, 21]. In Southern Europe and in Central and South America E. *zinckenella* is only damaging pod borer species in soybean.

Order Diptera was represented by 17 individuals that were not indentified to the species. The pest species from the order Diptera reported in the literature are *Delia platura* Meigen [6, 58], *Liriomyza congesta* Becker and *Clinodiplosis trotteri* Kief. [21]. Larvae of D. *platura* could cause damage on soybean seed during the emergence. Larvae of L. *congesta* are damaging leaves [21] and larvae of C. *trotteri* are damaging plant stem [21]. Some of Diptera species in soybean could be natural enemies, for example predaceous species *Acarolestes tetranychorum* feed on mites [21].

We established one species from **order Prostigmata** (infraclass Acari). This was the species T. *urticae* which was established only by leaf inspection. This species is the most important pest of soybean in the whole region [6, 8-10, 13, 17, 19, 29]. The pest outbreaks are occurring in dry and warm years in which farmers must apply control measures. Besides T. *urticae*, soybean in Croatia [9] and neighboring countries [16, 17, 19] is often attacked by T. *atlanticus* (= *turkestani*). Both species have similar thermal requirements but, T. *atlanticus* prefers extremely dry conditions [9]. Some differences were established in their response to host plant nutrient status [79]. The development of T. *urticae* is positively influenced by potassium content in the plant host, while T. *atlanticus* is positively influenced by content of phosphorus. According to the data obtained from Meteorological and Hydrological Service of Republic of Croatia, somewhat lower temperatures and higher amount of rainfalls in July and August in 2010, comparing to the average were recorded. That could cause the absence of T. *atlanticus* in experimental field and relatively low population of T. *urticae*.

The dominance indices of the insect orders established in total capture are shown in Figure 3.

In total catch the eudominant orders were Hemiptera (60.46%) and Acarina (28.6%), while subdominant were orders Thysanoptera (4.73%), Coleoptera (3.33%), Lepidoptera (1.63%) and Diptera (1.26%).

3.5. Most important phytophagous species

The significant feeding on soybean was established by two species, N. *viridula* and T. *urticae*. Therefore we will further analyze their appearance with the respect of their life cycle and possible damages that they can cause.

The dynamic of the appearance of N. *viridula* is shown in Figure 4.

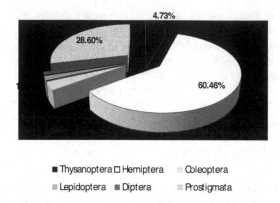

Figure 3. The dominance indices of arthropod orders established in the total capture of arthropod species on soybean in Croatia in 2010

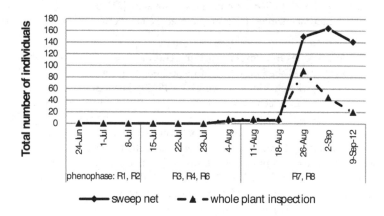

Figure 4. The dynamic of the appearance of *Nezara viridula* (L.) on soybeans in Croatia in 2010 established by two sampling methods

The southern green stink bug, *N. viridula*, is one of the most economically important soybean pests [80]. It has a worldwide distribution, occurring throughout the tropical and subtropical regions of Europa, Asia, Africa and America [49]. This pest is in constant expansion as a consequence of the increased acreage for soybean production, particularly in South America [80]. This pentatomid is highly polyphagous, attacking more than 145 species of plants (including cultivated and uncultivated species) within 32 families [49]. Life cycle of the southern green stink bug has been studied by number of authors in different parts of the world [81-85]. The biology of this pest has not been studied in Croatia jet but, some data were presented by different authors [26, 29]. From literature it is known [49] that southern green stink bug, like most pentatomids, overwinters in the adult stage under different objects that offer protection (litter, bark etc.). In the northern hemisphere [49] overwintering adults emerge in March and first generation develops on clover. The total developmental time from eggs to adults lasts between 23 days [82] and 49 days [81]. In USA, it develops 3-5 generations per year, depending on the climate. The 3rd, 4th and 5th generations attack soybean. We found it on soybean when soybeans began to mature, in August and onward what corresponds with the data presented by Todd and Herzog [49]. Probably the first two generations developed on some other plants. Stink bugs feed primarily on the seeds of soybean. Feeding results with puncture marks on the seed coat, deformation of the seed coats and reduced seed weight and size. Adults live longer, approximately 30 to 50 days [81, 82] and they cause more damage than nymphs [86]. Croatian authors [26] proposed economic threshold of 1 bug/30 m of soybean row or 8-10 bugs/10 sweep nets what seems to be too low. It is important to point out that the threshold depends on the period when insects occur. Early infestation causes more damage than late infestation [86]. Late in the season high infestation level of 2 bugs/m^2 will not result with the damage [86]. In our investigation we established infestation of 2 bugs/plant by whole plant count method and 4-5 bugs/10 sweep nets without seeing any damage on the yield. The appearance of N. *viridula* was in literature [26] connected with higher temperatures and drought, what was not the case in our investigation. Generally, in other countries the southern green stink bug is controlled with non-selective insecticides, which belong to carbamates, the organophosphate group, or the cyclodiene group, such as endosulfan or to pyrethroids [80]. Some of the mentioned insecticides are banned in Croatia and others are not allowed for that purpose. In the case of pest outbreak farmers do not have any available option to control this pest.

The second species which was recorded in high population density was T. *urticae*. The infestation with T. *urticae* started somewhat earlier that the attack of N. *viridula*. The dynamic of the appearance of T. *urticae* is shown in Figure 5.

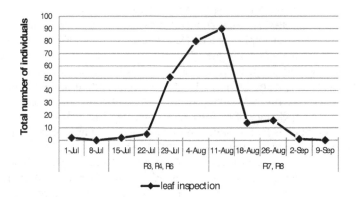

Figure 5. The dynamic of the appearance of *Tetranychus urticae* on soybeans in Croatia in 2010

The maximal infestation of *T. urticae* was recorded on August 11th and it was 2.25 mites/leaf. This infestation is considered as very weak to weak infestation [87]. After that date, the number of mites significantly decreased without the use of acaricides. The reason of the decrease of the population is the period in August in which strong rain occurred. Strong rain probably caused washing up the spider mites from the leaves as it was mentioned by some authors [29, 87]. *T. urticae* injure soybean by feeding on the green foliage and pods. By their needle-like chelicerate mouthparts that are used to puncture individual plant tissue cells and consume the entire cytoplasmic contents, they are leaving and empty irreversible damaged cell. The presence of numerous empty cells results in the yellow or brown spots on mite-injured leaves. Extensive feeding by large numbers of mites causes the leaves to appear yellow or brown [37]. Complete defoliation due to mite feeding can reduce pod set and seed yield. Under the favorable conditions mites have very short developmental time, between 8-20 days [9]. That enables them to develop several generations in a very short time and to increase population up to the economic threshold very fast. Therefore permanent monitoring of the pest population is needed. No acaricides are allowed for the control of *T. urticae* in soybean crops at the moment in Croatia. Even though there are some mite resistant cultivars in USA [88], they are not registered in Croatia.

3.6. Most important zoophagus species

Total of 73 predaceous species are collected in the investigation. Family Nabidae was represented by one specius, *N. ferus*. Members of the family Nabidae are confirmed predators of different kind of insects [89]. Most types of insect prey of nabids are plant-feeding species, but nabids sometimes attack predaceous insects, including members of their own species. The polyphagous feeding habits of the nabids make them less effective than species-specific predators against specific prey species [89]. Altogether 55 individuals of *N. ferus* were counted. The dynamic of the appearance of *N. ferus* is shown in Figure 6.

Nabis ferus is a common, widespread species in the Palearctic region [89]. It was reported as predatory species on *Trialeurodes vaporariorum* Westwood [90], *Oulema melanopus* L. [91], *Sitobion avenae* F. [92] and other aphid species [65], *N. viridula* [93] and leafhoppers in all stages [89]. Species *N. ferus* overwinters in the adult stage [65]. Adults emerge from the soil and migrate to field of various crops in April and May according to the weather. They mate, lay eggs and the nymphs appear between late May and June and are present until July [65]. There is a second generation with nymphs in July-August and adults in August-October. The dynamic of the appearance of *N. ferus* in experimental field corresponds with the data on life cycle of this species [65]. In mid-late July we collected adults of the first generation and nymphs were collected in August. Kereši [94] stated that zoophagous *Nabis* species develop one generation per year in soybeans. She mentioned adult appearance at the end of July and maximal larval appearance at the end of August. It remains unknown in which crop species is developing the first generation. Due to its preference to prey aphids which are abundant in wheat fields, it could be that the first generation is developing in wheat fields. The experimental field in our investigation was surrounded by wheat fields what could influence high population of *N. ferus*.

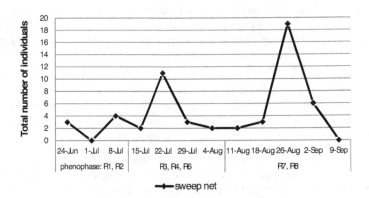

Figure 6. The dynamic of the appearance of *Nabis ferus* on soybeans in Croatia in 2010

4. Conclusions

Literature reports that soybean crops in the region where Croatia belongs (Croatia, Hungary, Serbia, Romania, Bulgaria and Bosnia and Herzegovina) are attacked by over 180 pests (150 insects and 30 species from other animal classes). However, by literature review from Croatia, Serbia and Bosnia and Herzegovina we established that 52 species (or genus) of arthropods are reported to be associated with soybean crops. Out of these 52 species, seven species are zoophagous, 44 species are phytophagous and one species is myceliophagous. Additionally, we have found data on 43 species of phytoparasitic nematodes that can be find in soybean fields but without causing significant damages and literature also reports on three species of rodents that could cause significant damage on soybean fields.

In our investigations the number of established species was lower than the number obtained by literature review. Total of 1357 individuals were collected and classified into the five orders from the class of Insects and one order from the class of Arachnida (infraclass Acari). 1232 individuals were classified in 15 species or genus, 58 individuals were classified into the six families while 67 individuals were classified into the orders. Phytophagous arthropods were more abundant than zoophagous. The ratio between phytophagous and zoophagous specimens was 94.63% : 5.37%.

Based on the results of the literature review and of the research conducted, it could be concluded that significance of the arthropod pest fauna connected with soybean has changed over the time. Nowadays, soybean production in Croatia could be endangered by four phytophagous arthropod species: *N. viridula* L., *V. cardui* L., *T. urticae* and *T. turkestani*.

N. viridula is attacking soybean pods and seed causing the loss in yield quantity and quality. The early infestation is very dangerous. The population of this pest has been increasing in the past few years. This could be connected with the increase of soybean cultivation area. Obtained results indicate that the increase in pest population has occurred recently and that one existing species became significant pest of soybean. The life cycle of this pest as well as other issues related to sampling procedure, economic threshold and control possibilities have not been studied jet in Croatian agro-ecological conditions. In the future, investigations should be carried out with the aim to collect more data on this pest and to be able to implement IPM principles in its control.

V. cardui is causing defoliation of the soybean plants. As periodical pest it appears from time to time in certain areas causing significant damage. Sampling procedure for this pest is established but, research should be conducted in order to determine economic threshold and the efficacy of environmentally acceptable insecticides (*B.t.k.*, spinosad, neem, IGRs, avermectins).

Phytophagous mites, *T. urticae* and *T. turkestani* as soybean pests are well known to Croatian farmers. Due to their feeding on soybean leaves they are causing defoliation. Their infestation is related to the climatic condition. In warm and dry years these pests cause more severe damage than in "normal" years. Sampling procedure for these pests is established but, due to the lack of registered acaricides, their control is not possible. The future research should be focused on finding appropriate ecologically acceptable acaricide for the control of this pest.

The critical period for the infestation by all four species is from flowering through maturity in which period all four pests should be monitored and sampled on a regularly basis in order to ensure the proper information about the need for control measure.

Some other pests that were found in our investigation are capable of becoming key pests if environmental conditions and population of their natural control agents are disrupted by unnecessary application of insecticides. One of these species belongs to *Piezodorus* sp. which is world widely recognized as very important soybean pest. Thus the future systematic and intensive study of the arthropod fauna associated with soybean in Croatia has to be continued. It will allow us to monitor the changes in the pest population and to prepare strategies for the control of the new pests that could arise over the time.

The main zoophagous species found on soybeans was *Nabis ferus*. The role of this species in soybean ecosystems, including its varying feeding strategies, needs much additional attention.

Acknowledgements

We thank Prof. Hrvoje Šarčević for providing us adequate experimental field conditions.

Author details

Renata Bažok, Maja Čačija, Ana Gajger and Tomislav Kos

*Address all correspondence to: rbazok@agr.hr

University of Zagreb, Faculty of Agriculture, Zagreb, Croatia

References

[1] Vratarić M, Sudarić A. Tehnologija proizvodnje soje. Osijek: Poljoprivredni institut Osijek; 2007.

[2] Statistical Yearbook of the Republic of Croatia 2011. http://www.dzs.hr/Hrv_Eng/ ljetopis/2011/SLJH2011.pdf (accessed 15 June 2012)

[3] Wilson R. Soybean: Market driven research needs. In: G. Stacey (ed.) Genetics and Genomics of Soybean. New York: Springer Science+Business Media, LLC 2008. 3-15.

[4] Solomon S, Qin D, Manning M, Chen Z, Marquis M, Averyt KB, Tignor M, Miller HL (eds) (2007) Climate change 2007: the physical science basis. In: Contribution to working group I to the fourth assessment report of the intergovernmental panel on climate change. Cambridge University Press, chap 3 and 11

[5] Higley LG, Boethel DJ. Handbook of Soybean Insect Pests. Lanham, MD: Entomological Society of America; 1994.

[6] Kereši T, Sekulić R, Čamprag D. Important insect pests in soyabean fields. Biljni Lekar (Plant Doctor) 2008;36(3-4) 259-272.

[7] Kereši T. Bug fauna (Heteroptera) on winter wheat and soybean dependent on cropping system. PhD thesis. University of Novi Sad; 1999.

[8] Čamprag D. Harmful fauna in soybean fields and integral pest management. Biljni Lekar (Plant Doctor) 2008;36(3-4) 240-246.

[9] Maceljski M. Poljoprivredna entomologija. Čakovec: Zrinski; 2002.

[10] Dimitrijević M, Valenčić Lj, Jurković D, Ivezić M, Kondić Đ. Bolesti i štetnici važnijih ratarskih kultura na području stočne Slavonije i Baranje. Poljoprivredne aktualnosti 1985; 22(1-2) 203-223.

[11] Jelić A. Proučavanje fitoparazitskih nematoda biljaka za proizvodnju ulja na području Slavonije i Baranje sa posebnim osvrtom na soju. PhD thesis, University of Josip Juraj Strossmayer Osijek; 1989.

[12] Nikolić M. Proizvodnja i zaštita soje na društvenim površinama SOUR-a PIK „Vinkovci" u 1983. Godini. Glasnik zaštite bilja 1984; 2 50-52.

[13] Pagliarini N. Rezultati ispitivanja efikasnosti nekih akaricida protiv T. urticae Koch. na soji u vegetaciji, In: Zbornik radova: Prvo jugoslovensko savetovanje o primeni pesticida u zaštiti bilja, 19-23 November 1979, Kupari, 1: 241-250.

[14] Atanasov P. Sojin moljac (Etiella zinckenella Tr.). Biljna zaštita 1964; 3 50-51.

[15] Čamprag D, Đurkić J, Sekulić R, Kereši T, Almaši R, Thalji R. Brojnost larvi Elateridae (Coleoptera) na raznim poljoprivrednim kulturama na području Vojvodine. Zaštita bilja (Plant Protection) 1985; 36(4) 399-404.

[16] Đurkić J. Pojava Tetranychus atlanticus Mc. Gregor štetočine poljoprivrednih kultura u Vojvodini u 1956. godini, Zaštita bilja, (1956): 7(1): 67-70.

[17] Đurkić J, Srečković R, Sabadin T. Zapažanja o pojavi grinje na soji u 1976. godini. Suvremena poljoprivreda 1977; 5-6 45-57.

[18] Klindić O. Proučavanje nematoda prouzrokovača pjegavosti korjena – Pratylenchus spp., Zaštita bilja 1967; 18(1) 133-142.

[19] Ratajac R, Rajković D. Praćenje dinamike populacije grinja na soji u nekim lokalitetima SAP Vojvodine. Glasnik zaštite bilja 1976; 6 191-197.

[20] Sekulić R, Thalji R, Kereši, T. Prilog proučavanju ishrane gusenica i suzbijanja stričkovog šarenjaka (Pyrameis cardui L.) na soji i boraniji. Agronomski Glasnik 1983; 1 57-63.

[21] Simova-Tošić D, Vuković M, Plazinić V, Mihajlović Lj. Pojava i identifikacija najznačajnijih štetnih insekata soje u SR Srbiji. Zaštita bilja 1988; 39(1) 17-24.

[22] Tešić T. Rogati cvrčak (C. bubalus) u Srbiji. Zaštita bilja 1964; 15(6) 593-665.

[23] Vaclav V, Batinica J. Stričkov šarenjak kao štetočina soje. Poljoprivredni pregled 1962; 11(12): 408-410.

[24] Vaclav V, Radman Lj, Batinica J, Ristanović M, Dimić N, Numić R, Beš A. Prilog poznavanju bolesti i štetočinja soje na proizvodnim površinama Bosne. Zaštita bilja 1970; 21(109) 229-236.

[25] Bažok R, Ljikar K. Stričkov šarenjak - malo poznati štetnik soje. Glasilo biljne zaštite 2007; 8(1) 44-46.

[26] Majić I, Ivezić M, Raspudić E, Vratarić M, Sudarić A, Sarajlić A, Matoša M. Pojava stjenica na soji u Osijeku. Glasilo biljne zaštite 2010; 11(1-2/dodatak) 51-51.

[27] Raspudić E; Ivezić M, Ladocki Z, Pančić S, Brmež M. Stričkov šarenjak – opasnost za soju. Glasilo biljne zaštite 2007; 8(1-dodatak) 12-12

[28] Vratarić M, Sudarić A. Soja Glycine max (L.) Merr. Osijek: Poljoprivredni institut Osijek 2008.

[29] Vratarić M, Sudarić A. Važnije bolesti i štetnici na soji u Republici Hrvatskoj. Glasnik zaštite bilja 2009;36(6) 6-23.

[30] Sekulić R, Kereši T. Major pests of soybean- mites and rodents. Biljni Lekar (Plant Doctor) 2008; 36(3-4) 247-258.

[31] Clark LR, Geier PW, Hughes RD, Morris RF. The Ecology of Insect Populations in Theory and Practice. New York: Chapman & Hall. 1967.

[32] Altieri MA.. Biodiversity and Pest Management in Agroecosystems. Binghamton, NY: Food Products Press; 1994.

[33] Conway GR, Strategic models. In: Conway GR. (ed) Pest and Pathogen Control: Strategic, Tactical and Policy Models. New York: John Willey & Sons; 1984. p15-28.

[34] Bc Institut d.d. Zagreb http://www.bc-institut.hr/s_zlata.htm (accessed 20 June 2012)

[35] Iowa State University- Soybean Extension and Research Program http://extension.agron.iastate.edu/soybean/production_growthstages.html (accesed 20 June 2012)

[36] Kogan, M., Pitre H.N. Jr. General Sampling Methods for Above-Ground Populations of Soybean Arthropods. In: Kogan, M., Herzog, D.C.(eds) Sampling Methods in Soybean Entomology. New York: Springer; 1980. p30-60.

[37] Poe, S.L. Sampling Mites on Soybean. In: Kogan, M., Herzog, D.C.(eds) Sampling Methods in Soybean Entomology. New York: Springer; 1980. p312-323.

[38] Schmidt, L: Tablice za determinaciju insekata. Zagreb: Sveučilišna naklada Liber; 1972.

[39] Villiers, A. Atlas des Hémiptères de France, I Hétéroptères Gynocérates: 108. Paris: Editions N. Boubee; 1951.

[40] Auber, L. Atlas des Coléoptères de France I, Belgique, Suisse. Paris: Editions N. Boubée; 1965.

[41] Bechyně, J. Welcher Käfer ist das? Kosmos-Naturfuhrer, Stuttgart: Balogh Scientific Books; 1974.

[42] Harde, K.W., Severa, F. Der Kosmos Käferführer . Kosmos-Naturfuhrer, Stuttgart: Balogh Scientific Books; 1984.

[43] Balarin I. Fauna Heteroptera na krmnim leguminozama i prirodnim livadama u SR Hrvatskoj. Doktorska disertacija, Zagreb: Agronomski fakultet. 1974.

[44] Kretzschmar GP. Soybean insects in Minnesota with special reference to sampling techniques. Journal of Economic Entomology 1948; 41 586-591.

[45] Irwin ME, Shepard M. Sampling Predaceous Hemiptera on Soybean. In: Kogan, M., Herzog, D.C.(eds) Sampling Methods in Soybean Entomology. New York: Springer; 1980. p505-531.

[46] Pedigo LP. Sampling Green Cloverworm on Soybean. In: Kogan, M., Herzog, D.C. (eds) Sampling Methods in Soybean Entomology. New York: Springer; 1980. p169-186.

[47] Herzog DC. Sampling Soybean Looper on Soybean. In: Kogan, M., Herzog, D.C.(eds) Sampling Methods in Soybean Entomology. New York: Springer; 1980. p141-168.

[48] Shelton MD, Edwards CR. Effects of Weeds on the Diversity and Abundance of Insects in Soybeans. Environmental Entomology 1983; 12(2) 266-298.

[49] Todd JW, Herzog DC. Sampling Phytophagous Pentatomidae on Soybean. In: Kogan, M., Herzog, D.C.(eds) Sampling Methods in Soybean Entomology. New York: Springer; 1980. p438-478.

[50] Irwin ME, Yaergan KV. Sampling Phytophagous Thrips on Soybean. In: Kogan, M., Herzog, D.C.(eds) Sampling Methods in Soybean Entomology. New York: Springer; 1980. p283-304.

[51] Eastman CE. Sampling Phytophagous Underground Soybean Arthropods. In: Kogan, M., Herzog, D.C.(eds) Sampling Methods in Soybean Entomology. New York: Springer; 1980. p327-354.

[52] Lewis T. Thrips, their biology, ecology and economic importance. London: Academic Press; 1973.

[53] Bournier A, Lacasa A, Pivot Y, Biologie d'un Thrips prédateur Aeolothrips intermedius (Thys.: Aeolothripidae). Entomophaga 1978; 23 (4) 403-410.

[54] Riudavates J. Predators of *Frankliniella occidentalis* (Perg.) and *Thrips tabaci* Lind.: a review. Wageningen Agricultural University Papers 1995;95(1) 43-87.

[55] Bournier, A., Lacasa, A., Pivot, Y. Régime alimentaire d un Thrips prédateur *Aeolothrips intermedius* (Thys.: Aeolothripidae). Entomophaga 1979;24(4) 353-361.

[56] Lacasa, A., Bournier, A., Pivot, Y. Influencia de la temperatura sobre la biologia de un trips depredador *Aeolothrips intermedius* Barnall (Thys: Aeolothripidae).- Anales del Instituto Nacional de Investigaciones Agrarias, Agricola. 1982;20 87-98.

[57] Conti B. Notes on the presence of *Aeolothrips intermedius* in northwestern Tuscany and on its development under laboratory conditions. Bulletin of Insectology 2009;62(1) 107-112.

[58] Zandigiacomo P. Pest found in soybean crops in north-eastern Italy. Informatore Agrario 1992;48(7) 57-59.

[59] Hoddle MS, Robinson L, Drescher K, Jones J. Developmental and Reproductive Biology of a Predatory *Franklinothrips* sp. (Thysanoptera: Aeolothripidae). Biological Control 2000;18 27-38.

[60] Fauna Europea. http://www.faunaeur.org/ (accessed 24 June 2012)

[61] Pakyari H, Fathipour Y, Rezapanah M, Kamali K. Temperature-dependent functional response of *Scolothrips longicornis* (Thysanoptera: Thripidae) preying on *Tetranychus urticae*. Journal of Asia-Pacific Entomology 2009; 12(1) 23-26.

[62] Gheibi M, Hesami S. Life Table and Reproductive Table Parameters of *Scolothrips longicornis* (Thysanoptera: Thripidae) as a Predator of Two-Spotted Spider Mite, *Tetranychus Turkestani* (Acari: Tetranychidae). World Academy of Science, Engineering and Technology 2011;60 262-264.

[63] Pitkin BR. A revision of the Indian species of *Haplothrips* and related genera (Thysanoptera: Phlaeothripidae). Bulletin of the British Museum of National History, Entomology 1976;34 223-280.

[64] Limonta L, Dioli P, Bonomelli N. Heteroptera on flowering spontaneous herbs in differently managed orchards. Bollettino di Zoologia agraria e di Bachicoltura 2004;Ser. II;36 (3) 355-366.

[65] Puchkov AV. Particulars of the biology of predacious Nabis spp. Zashchita Rastenii 1980; 8 1-44.

[66] Giorgi R. The defence of soybean: control of the principal diseases and insects. Terra e Sole 1992;47(596) 231-234.

[67] Baur ME, Sosa-Gomez DR, Ottea j, Leonard BJ, Corso IC, Da Silva JJ, Temple J, Boethel DJ. Susceptibility to Insecticides Used for Control of *Piezodorus guildinii* (Heteroptera: Pentatomidae) in the United States and Brazil. Journal of Economic Entomology 2010;103(3) 869-876.

[68] Sosa-Gomez, DR, Moscardi F. Different foliar retention in soyabean caused by stink bugs (Heteroptera: Pentatomidae). Anais da Sociedade Entomologica do Brasil 1995; 24(2) 401-404.

[69] Helm CG, Kogan M, Hill BG. Sampling Lefhoppers on Soybean. In: Kogan, M., Herzog, D.C.(eds) Sampling Methods in Soybean Entomology. New York: Springer; 1980. p260-282.

[70] Ragsdale DW, Landis DA, Brodeur J, Heimpel GE, Desneux N. Ecology and Management of the Soybean Aphid in North America. Annual Review of Entomology 2011;56 375–399.

[71] Irwin ME. Sampling Aphids in Soybean Fields. In: Kogan, M., Herzog, D.C.(eds) Sampling Methods in Soybean Entomology. New York: Springer; 1980. p239-259.

[72] Igrc Barčić J. Lisne uši. In: Maceljski M. (ed) Poljoprivredna entomologija. Čakovec: Zrinski; 2002. p74-126.

[73] Gotlin Čuljak T, Igrc Barčić J, Bažok R, Grubišić D. Aphid fauna (Hemiptera: Aphidoidea) in Croatia. Entomologia Croatica 2006; 9(1-2) 57-69.

[74] Andrews FG. Latridiidae Erichson 1842. In: Arnett RH Jr, Thomas MC. (eds) American Beetles Plyphaga: Scarabaeoidea through Curculionoidea. Boca Ratton:CRC; 2002. p.395-398

[75] Hillhouse TL, Pitre HN. Comparison of Sampling Techniques to Obtain Measurements of Insect Populations on Soybeans. Journal of Economic Entomology 1974;67(3) 411-414.

[76] Lammers JW, MacLeod A. Report of a Pest Risk Analysis *Helicoverpa armigera* (Hübner, 1808). Plant Protection Service (NL) and Central Science Laboratory (UK) 2007.

http://www.fera.defra.gov.uk/plants/plantHealth/pestsDiseases/documents/helico-verpa.pdf (accessed 24 June 2012)

[77] Sekulic R, Kereši T, Masirevic S, Vajgand D, Radojcic S. Incidence and damage of cotton bollworm (*Helicoverpa armigera* Hbn.) in Vojvodina Province in 2003. Biljni Lekar (Plant Doctor) 2004;32(2) 113-124.

[78] Torres Vila L M, Rodriguez Molina MC, Lacasa Plasencia A, Bielza Lino P, Rodriguez del Rincon A. Pyrethroid resistance of *Helicoverpa armigera* in Spain: current status and agroecological perspective. Agriculture Ecosystems and Environment 2002;93 55-66.

[79] Cannon WN, Connel WA. Populations of *Tetranychus atlanticus* McG. (Acarina: Tetranychidae) on soybean supplied with various levels of nithrogen, phosphorus, and potassium. Entomologia Experimentalis et Applicata 1965; 8 153-161.

[80] Werdin Gonzalez JO, Gutierrez MM, Murray AP, Ferrero AA. Composition and biological activity of essential oils from Labiatae against *Nezara viridula* (Hemiptera: Pentatomidae) soybean pest. Pest Management Science 2011; 67 948–955

[81] Werdin Gonzalez JO, Ferrero AA. Table of life and fecundity by *Nezara viridula* var. *smaragdula* (Hemiptera: Pentatomidae) feed on *Phaseolus vulgaris* L. (Fabaceae) fruits. IDESIA 2008; 26(1) 9-13.

[82] Bharathimeena T, Sudharma K. Biological studies on the southern green stink bug, *Nezara viridula* (L.) and the smaller stink bug *Piezedorus rubrofasciatus* (F.) (Pentatomidae: Hemiptera) infesting vegetable cowpea. Pest Management in Horticultural Ecosystems 2008; 14(1) 30-36.

[83] DerChien C, ChingChung C. Life history of *Nezara viridula* Linnaeus and its population fluctuations on different crops. Bulletin of Taichung District Agricultural Improvement Station 1997; 55 51-59.

[84] Mukopadhyay B, Roychoudhury N. Biology of green stink bug *Nezara viridula*. Environment and Ecology 1987; 5(2) 325-327.

[85] Cividanes FJ, Parra JRP. Biology of soybean pests with different temperatures and thermal requirements. I. *Nezara viridula* (L.) (Hemiptera: Pentatomidae). Anais da Sociedade Entomologica de Brasil 1994; 23(2) 243-250.

[86] Miller LA, Rose HA, McDonald FJD. The effect of damage by the green vegetable bug *Nezara viridula* (L.) on yield quality of soybeans. Journal of Australian Entomological Society 1977; 16(4) 421-246.

[87] Dobrinčić R. Neki štetnici soje. Glasnik zaštite bilja 1999; 4 184-187.

[88] Hildebrand D F, Rodriguez JG, Brown GC, Luu KT, Volden CS. Peroxidative Responses of Leaves in Two Soybean Genotypes Injured by Twospotted Spider Mites (Acari: Tetranychidae). Journal of Economic Entomology 1996; 79(6)1459-1465.

[89] Lattin JD. Bionomics of the Nabidae. Annual Review of Entomology 1989; 34 383-400.

[90] Perić P, Marčić D, Stamenkvić S. Natural enemies of whitefly (*Trialeurodes vaporario-rum* Westwood) in Serbia. Acta Horticulturae 2009; 830 539-544.

[91] Popov C, Malschi D, Vilau F, Stoica V. Insect pest management of *Lema melanopa* in Romania. Romanian Agricultural Research 2005; 22 47-51.

[92] Piotrowska E. Introductory studies on some food requirements of *Nabis ferus* L. and *Nabis pseudoferus* Rem. (Heteroptera: Nabidae) on cereals. Roczniki Nauk Rolniczych (Ochrona Rastlin) 1982; 12(1-2)47-56.

[93] Hokyo N, Kiritani K. Two species of egg parasites as contemporaneous mortality factors int he egg population of the southern green stink bug, *Nezara viridula*. Japanese Journal of Applied Entomology and Zoology 1963; 7(3)214-227.

[94] Kereši T. The Heteroptera fauna on soybeans in Bačka. Zaštita Bilja 1993; 44(3)189-195.

Chapter 7

Screening of Soybean (*Glycine Max* (L.) Merrill) Genotypes for Resistance to Rust, Yellow Mosaic and Pod Shattering

M. H. Khan, S. D. Tyagi and Z. A. Dar

Additional information is available at the end of the chapter

1. Introduction

Soybean (Glycine max (L.) Merrill) is known as 'Golden bean' and miracle crop of 20th century. Soybean is a native of North China, Asia belongs to family fabaceae. It is a versatile and fascinating crop with innumerable possibilities of not only improving agriculture but also supporting industries. Soybean besides having high yielding potential (40-45 q/ha) also provides cholesterol free oil (20%) and high quality protein (40%). It is a rich source of lysine (6.4%) in addition to other essential amino acids, vitamins and minerals. Its oil is also used as a raw material in manufacturing antibiotics, paints, varnishes, adhesives and lubricants etc.

Like other economically important crops soybean is also suffering from many diseases viz, rust (*Phakopsora pachyrhizi* Syd.) and yellow mosaic (Mungbean Yellow Mosaic Virus) are the major disease under Indian conditions, which causes considerable reduction in yield up to 80 per cent under severe conditions [3]. Further, another major problem in soybean is pods shattering which also reduces yield and in some varieties 100 per cent yield losses have been observed. The extent of yield loss due to pod shattering may range from negligible to significance levels depending upon the time of harvesting, environmental condition and genetic endowment of the variety [11]. Hence screening for soybean genotypes for identifying resistance to above major problems with high yielding potential will help to increase the production to a greater extent.

2. Materials and methods

The material consisted of 84 genotypes of soybean originated from different places of India and abroad. The experiment was laid in augmented design at the Research Farm of Kisan (PG) College, Simbhaoli, Ghaziabad, during *kharif*, season of 2008. In each replication the genotypes were grown in 2 m long rows with spacing of 40cm × 10cm for row to row and plant to plant, respectively. Within a row, seeds were hand dibbled 10 cm apart. Standard package of practices was followed to raise the crop. Ten competitive plants were randomly selected from each treatment in each replication and data were recorded on 3 qualitative characters namely, pod shattering resistance, rust resistance and yellow mosaic disease resistance.

2.1. Screening for pod shattering resistance

The pod shattering resistance was recorded at physiological maturity of the pod. The screening was done under laboratory condition by following the methodology adopted by IITA [4]. The results were recorded as percentage of pod shattering. IITA method of calculating pod shattering under lab conditions:

A sample of 25 pods were collected and kept in oven at 40°C for 7 days.

On the 7th day the number of shattered pods were counted and expressed in percentage as below,

Number of pods shattered

Pod shattering percentage (%) = x 100

Total number of pods

The genotypes were classified into different categories based on their reaction to pod shattering. The scoring rate was followed according to method adopted by IITA.

Sl.No	Category Resistant reaction
1.	No pod shattering Shattering resistant
2.	<25% pod shattering Shattering tolerant
3.	25-50% pod shattering Moderately shattering
4.	51-75% pod shattering Highly shattering
5.	>75% pod shattering very highly shattering

2.2. Screening for rust resistance

The scoring for rust was done just after initiation of flowering and before pod formation. The observations were taken on lower, middle and upper leaves for density of pustule and sporulating intensity. Based on the symptoms, pustule density and sporulation intensity grades were given. The genotypes were later grouped into different categories from immune to highly susceptible. The scale (0-9) used was as follows:

Sl. No.	Scale	Category
1.	0	Immune
2.	1	Resistant
3.	3	Moderately resistant
4.	5	Moderately susceptible
5.	7	Susceptible
6.	9	Highly susceptible

2.3. Screening for yellow mosaic disease resistance

84 soybean genotypes grown in natural (field) conditions at Research Farm of Kisan (PG) College, Simbhaoli, Ghaziabad during *kharif*, 2008 were screened. Number of plants showing distinct symptoms in each line was counted 60 days after sowing and per cent disease incidence was calculated by using the following formula:

Number of plants infected in a row

Per cent Disease Incidence (PDI) = x 100

Total number of plants in a row

The genotypes were later grouped into different categories from immune to highly susceptible [7]. The scale used was as follows (0-9):

Scale	Description Category
0	No symptoms of plants Immune
1	1% or less plants exhibiting symptoms resistant
3	1 to 10% plants exhibiting symptoms moderately resistant
5	11 to 20% plants exhibiting symptoms moderately susceptible
7	21 to 50% plants exhibiting symptoms susceptible
9	51% or more plants exhibiting symptoms highly susceptible

3. Experimental results

3.1. Screening for pod shattering

84 genotypes of soybean were screened for pod shattering resistance in order to identify resistant cultivars during *kharif*, 2008. The screening was done according to method adopted by IITA, Nigeria. The data presented in Table 1 revealed that pod shattering percentage ranged from 8.7 (Himsoy-1560) to 93.3 per cent (Punjab- 1). The result indicated that there is no variety, which is resistant to pod shattering. However, some of the varieties *viz.*, Bragg, CGP-76, EC-322536, EC-34092, JS 93-05, Lee, MAUS-2, NRC-7, EC-34101, EC-34092, JS 71-05, EC-34101, EC-392536, G-26, Himsoy-1560, Himsoy-1514, Pusa-16, Pusa-22, VLS-1, VLS-2, VLS-47 and the check JS-335 were found to be tolerant. Later these genotypes were grouped into different categories based on IITA, Nigeria scale and the data is presented in Table 2. The results revealed that none of the genotypes were immune or resistant to pod shattering.

Sl. No	Genotypes	Shattering %	Grade	Sl No	Genotype	Shattering %	Grade
1.	Alankar	58.7	HS	43.	EC-392536	16.0	TO
2.	Ankur	47.0	MS	44.	EC-394839	45.3	MS
3.	AGS-34	59.7	HS	45.	G-48	15.0	TO
4.	AGS-50	52.7	HS	46.	G-479	35.0	MS
5.	Bragg	15.3	TO	47.	G-482	51.7	HS
6.	Local black soybean	78.3	VHS	48.	G-7340	83.7	VHS
7.	CO-1	80.3	VHS	49.	G-26	35.7	MS
8.	CO-2	57.0	HS	50.	G-5-1	61.7	HS
9.	CGP-76	15.3	TO	51.	Hardee	46.0	MS
10.	CGP-248	62.0	HS	52.	Hara soya	17.3	MS
11.	CGP-2037	46.0	MS	53.	Himsoya-1560	8.7	TO
12.	DSb-1	46.3	MS	54.	Himsoya-1514	11.8	TO
13.	DSb-2	41.7	MS	55.	Improved pelican	83.3	VHS
14.	DSb-3-4	44.3	MS	56.	Indira Soya 9	32.0	MS
15.	DSb-5	48.7	MS	57.	C-39506	51.7	HS
16.	DSb-6-1	37.7	MS	58.	IC-49859	56.7	HS
17.	DSb-7	56.0	HS	59.	IC-104877	46.3	MS
18.	DSb-8	45.0	MS	60.	JS-2	83.0	VHS
19.	DS-17-5	35.3	MS	61.	JS-71-05	18.0	TO
20.	EC-103369	58.3	MS	62.	JS-72-280	36.7	MS

Sl. No	Genotypes	Shattering %	Grade	Sl No	Genotype	Shattering %	Grade
21.	EC-109923	75.0	VHS	63.	JS-72-44	44.3	MS
22.	EC-322536	20.7	TO	64.	JS-75-46	55.0	HS
23.	EC-241778	36.7	MS	65.	JS-76-205	31.9	MS
24.	EC-241780	63.3	HS	66.	JS-80-21	67.0	HS
25.	EC-34092	16.0	TO	67.	JS-90-41	55.7	HS
26.	EC-118420	61.3	HS	68.	JS-93-105	19.0	TO
27.	EC-34101	22.0	TO	69.	JS-87-25	31.7	MS
28.	EC-251449	72.7	HS	70.	KB-79	30.0	MS
29.	Lee	15.0	TO	71.	Pusa-20	32.7	MS
30.	MACS-13	43.0	MS	72.	Pusa-22	16.3	TO
31.	MACS-330	72.7	VHS	73.	Pusa-24	33.0	HS
32.	MACS-450	57.3	MS	74.	Pusa-37	62.3	HS
33.	MACS-57	46.0	MS	75.	Pusa-40	74.7	HS
34.	MAUS-47	85.0	VHS	76.	Samrat	31.3	MS
35.	MAUS-68	73.0	HS	77.	T-49	71.7	HS
36.	MAUS-2	19.0	TO	78.	VLS-1	12.1	TO
37.	NRC-7	19.7	TO	79.	VLS-2	25.3	TO
38.	NRC-12	26.3	MS	80.	VLS-21	31.7	MS
39.	PK-1024	80.0	VHS	81.	VLS-47	17.0	TO
40.	PK-1029	32.0	MS	82.	JS-335 (C)	10.3	TO
41.	Punjab-1	93.3	VHS	83.	KHSb-2(C)	43.5	MS
42.	Pusa-16	23.7	TO	84.	Monetta (C)	90.7	VHS

Table 1. Screening of soybean genotypes for pod shattering resistance

3.2. Screening for rust resistance

Growing resistant varieties is the most economical and safe method of controlling the rust of soybean, which is a devastating disease resulting in heavy yield loss. In order to identify the resistant cultivars 84 genotypes of soybean were screened for rust resistance during *kharif* 2008 under natural epiphytotic conditions at Dharwad. The rust incidence was recorded at physiological maturity of the genotypes and the results are presented in Table 3. Reactions of 84 genotypes to rust revealed that, none of the genotypes showed immune reaction to rust. Two genotypes *viz.*, EC 241778 and EC 241780 showed resistant reaction (1 grade), which were considered as resistant and the remaining 82 genotypes as highly susceptible (9 grade).

Sl. No.	Category	Resistant reaction	Number of genotypes	Genotypes
1	No pod shattering	Shattering resistant	00	-
2	< 25% pod shattering	Shattering tolerant	20	Bragg, CGP-76, EC-322536, EC-34092, JS 71-05, JS-93-05, Lee, MAUS-2, WEC-7, EC-34101, EC-392536, G-26, Himsoya-1560, Himsoya-1514, Pusa-16, Pusa-22, VLS-1, VLS-2, VLS-47, JS-335(C)
3	25-50% pod shattering	Moderately shattering	32	Ankur, CGP-2037, DSb-1, DSb-2, DSb-3-4, DSb-5, DSb-6-1, DSb-8, PS-17-5, EC-103369, EC-241778, IC-104877, JS 72-280, JS 72-44, JS 76-205, JS 87-25, KB-79, MACS-13, MACS-450, MACS-57, NRC-12, PK-1029, EC-394839, G-48, G-7340, Hardee, Harasoya, Indira soya, Pusa-20, Samrat, VLS-21, KHSb-2 (C).
4	51-75% pod shattering	Highly shattering	21	Alankar, AGS-34, AGS-50, CO-2, CGP-248, DSb-7, EC-241780, EC-118420, IC-39506, IC-49859, JS 75-46, JS 80-21, JS 90-41, MAUS-68, EC-251449, G-479, G 5-1, Pusa-24, Pusa-37, Pusa-40, T-49
5	>75% pod shattering	Very highly shattering	11	Local black soybean, CO-1, EC-109923, JS-2, MACS-330, MAUS-47, PK-1024, G-482, Improved pelican, Punjab-1, Monetta (C)

Table 2. Grouping of Soybean genotypes for pod shattering resistance

Sl. No.	Reaction	Grade (0-9)	Number of genotypes	Genotypes responded
1	Immune	0	00	-
2	Resistant	1	02	EC- 241778, EC- 241780
3	Moderately resistant	3	00	-
4	Susceptible	5	00	-
5	Moderately Susceptible	7	00	-
6	Highly Susceptible	9	82	Alankar, Ankur, AGS-34, AGS-50, Bragg, Local black soybean, CO-1, CO-2, CGP-76, CGP-248, CGP-2037, DSb-1, DSb-2, DSb 3-4, DSb-5, DSb 6-1, DSB-74, DSb-8, DS 17-5, EC-103369, EC-109923, EC-322536, EC-34092, EC-118420, EC-34101, EC-251449, EC-392536, EC-394839, G-48, G-479, G-482, G-7340, G-26, G-5-1, Hardee, Harasoya, Himsoya-1560, Improved pelican, Indirasoya, IC-39506,

Sl. No.	Reaction	Grade (0-9)	Number of genotypes	Genotypes responded
				IC-49859, IC-104877, JS-2, JS 71-05, JS 72-280, JS 72-44, JS 75-46, JS 76-205, JS 80-21, JS 90-41, JS 93-105, JS 87-25, KB-79, Le, MACS-13, MACS-330, MACS-450, MACS-57, MAUS-47, MAUS-68, MAUS-2, NRC-7, NRC-12, PK-1024, PK-1029, Punjab-1, Pusa-16, Pusa-20, Pusa-22, Pusa-24, Pusa-37, Pusa-40, Samrat, T-49, VLS-1, VLS-2, VLS-21, VLS-47, JS-335 (C), KHSb-2 (C), Monetta (C).

Table 3. Grouping of soybean genotypes for soybean rust resistance

3.3. Screening for yellow mosaic disease (YMD)

84 genotypes of soybean were screened for yellow mosaic disease under natural conditions at Research Farm of Kisan (PG) College, Simbhaoli, Ghaziabad during *kharif*, 2008. The data presented in Table 4 revealed that, YMD incidence ranged from 0.95 to 90.12 per cent. Among the 84 genotypes screened lowest incidence was recorded with genotype MACS 57 (0.48%), followed by EC 241778 (0.49%). Genotypes JS 90-41 (90.12) recorded highest incidence followed by JS 76-205 (89.15%) and T 49 (86.21%). All the genotypes and their percent disease incidence are tabulated in Table 5, which categorizes these genotypes based on 0-9 scale into different reaction types. It is evident from the table that none of the genotypes tested were immune or resistant.

Sl. No	Genotypes	PDI*	Reaction	Sl No	Genotype	PDI*	Reaction
1.	Alankar	9.25	MR	43.	EC-392536	27.00	S
2.	Ankur	0.75	R	44.	EC-394839	42.12	S
3.	AGS-34	8.21	MR	45.	G-48	15.25	MS
4.	AGS-50	7.68	MR	46.	G-479	17.25	MS
5.	Bragg	8.58	MR	47.	G-482	25.65	S
6.	Local black soybean	61.25	HS	48.	G-7340	32.15	S
7.	CO-1	3.58	MR	49.	G-26	9.12	MR
8.	CO-2	7.54	MR	50.	G-5-1	42.12	S
9.	CGP-76	15.61	MS	51.	Hardee	39.15	S
10.	CGP-248	19.25	MS	52.	Hara soya	33.89	S
11.	CGP-2037	9.21	MR	53.	Himsoya-1560	19.14	MS
12.	DSb-1	25.23	S	54.	Himsoya-1514	40.01	S
13.	DSb-2	32.15	S	55.	Improved pelican	42.15	S

Sl. No	Genotypes	PDI*	Reaction	Sl No	Genotype	PDI*	Reaction
14.	DSb-3-4	15.25	MR	56.	Indira Soya 9	0.45	R
15.	DSb-5	75.25	HS	57.	C-39506	7.12	MR
16.	DSb-6-1	13.25	MS	58.	IC-49859	62.15	HS
17.	DSb-7	27.85	S	59.	IC-104877	75.12	HS
18.	DSb-8	26.32	S	60.	JS-2	46.12	S
19.	DS-17-5	16.25	MR	61.	JS-71-05	78.98	HS
20.	EC-103369	0.75	R	62.	JS-72-280	46.25	S
21.	EC-109923	39.25	HS	63.	JS-72-44	29.12	S
22.	EC-322536	0.52	R	64.	JS-75-46	89.12	HS
23.	EC-241778	0.49	R	65.	JS-76-205	8.81	MR
24.	EC-241780	9.85	MR	66.	JS-80-21	90.12	HS
25.	EC-34092	45.25	S	67.	JS-90-41	11.85	MS
26.	EC-118420	37.12	S	68.	JS-93-105	8.82	MR
27.	EC-34101	13.25	MS	69.	JS-87-25	7.81	MR
28.	EC-251449	8.25	MR	70.	KB-79	6.23	MR
29.	Lee	46.3	S	71.	Pusa-20	7.15	MR
30.	MACS-13	5.12	MR	72.	Pusa-22	0.92	R
31.	MACS-330	81.21	HS	73.	Pusa-24	8.25	MR
32.	MACS-450	0.89	R	74.	Pusa-37	7.10	S
33.	MACS-57	0.48	R	75.	Pusa-40	29.12	S
34.	MAUS-47	0.56	R	76.	Samrat	36.57	S
35.	MAUS-68	85.12	HS	77.	T-49	86.21	HS
36.	MAUS-2	31.25	S	78.	VLS-1	40.25	S
37.	NRC-7	0.78	R	79.	VLS-2	79.85	HS
38.	NRC-12	16.25	MS	80.	VLS-21	33.25	S
39.	PK-1024	0.85	R	81.	VLS-47	30.5	S
40.	PK-1029	0.52	R	82.	JS-335 (C)	30.96	S
41.	Punjab-1	5.23	MR	83.	KHSb-2(C)	28.25	S
42.	Pusa-16	6.12	MR	84.	Monetta (C)	7.6	MR

* Percent disease incidence

Table 4. Screening of soybean genotypes for yellow mosaic disease resistance

Scale	Description	Category	Number of genotypes	Genotypes
0	No symptoms on plants	Immune	00	-
1	1% or less plants exhibiting symptoms	Resistant	12	Ankur, EC-103369, EC-322536, EC-241778, Indirasoya 9, MACS-450, MACS-57, MAUS-47, NRC-7, PK-1024, PK-1025, Pusa-22
3	1-10% plants exhibiting symptoms	Moderately resistant	22	Alankar, AGS-34, Bragg, AGS-50, CO-1, CO-2, CGP-2037, DSb 3-4, PS-17-5, EC-241780, EC-251449, G-26, IC-39506, JS 80-21, JS 87-25, KB-79, MACS-13, Punjab-1, Pusa-16, Pusa-20, Pusa-24, Monetta (Check).
5	11-20% plants exhibiting symptoms	Moderately susceptible	10	CGP-76, CGP-248, DSb-6-1, EC-34101, G-48, G-479, Himory-1560, IC- 49859, JS-93-105, NRC-12
7	21-50% plants exhibiting symptoms	Susceptible	28	DSb-1, DSb-2, DSb-7, DSb-8, EC-34092, EC-118420, EC-392536, EC-394839, G-7340, G-482, G-5-1, Hardee, Harasoya, Improved pelican, JS 71-05, JS 72-44, JS 75-46, Lee, MAUS-2, Pusa-37, Pusa-40, Samrat, VLS-1, Himsoya-1514, VLS-21, VLS-47, JS-335(C), KHSb-2 (Check)
9	51% or more plants exhibiting symptoms	Highly susceptible	12	Local black soybean, DSb-5, EC-109923, IC-104877, JS-2, JS 72-280, JS 76-205, JS 90-41, MACS-330, MAUS-68, T-49, VLS-2

Table 5. Grouping of genotypes into different categories for soybean yellow mosaic virus resistance

4. Discussion

4.1. Screening for pod shattering resistance

Pod shattering is one of the major constraints in soybean, which reduces the yield potential considerably. So management of pod shattering is of great importance for achieving higher productivity. Hence, the identification of resistant sources for pod shattering is one of the most important aspect in the management of pod shattering. In the present study 84 genotypes of soybean were screened for pod shattering resistance under lab condition. The pod shattering values ranged from 8.7 to 93.3 per cent. JS-335 one of the most popular variety recorded as tolerant with mean pod shattering value of 10.3 per cent. It is evident from the table that, none of the genotypes were better than the JS-335 except Himsoy-1560, which recorded 8.7 per cent mean pod shattering value. Among 84 genotypes, 20 genotypes fall under tolerant category and 32 under moderately shattering. Fifteen Indian soybean varieties were screened for pod shattering resistance and out of these three varieties viz., JS-1515, JS-1608 and JS-1625 were found resistant against pod shattering [16]. Similarly, while screening for pod shattering resistance, Bragg and JS-71-05 recorded the lowest pod shattering and Punjab-1 with highest pod shattering value [12]. Similar results were also reported [1, 13].

4.2. Screening for rust resistance

Among many of the diseases in soybean, rust is the major fungal disease which may reduce the yield drastically. So identification of resistant sources and involving them in resistant breeding forms one of the criteria in resistant breeding programme. In the present study 84 genotypes of soybean were screened for rust resistance under natural epiphytotic condition. None of the genotypes showed immune reaction. However, genotypes EC-241778 and EC-241780 showed resistance reaction. Remaining all genotypes exhibited highly susceptible reaction. In general, over all disease incidence was very high. Similar results are reported in [9], who evaluated several soybean genotypes and varieties under natural epiphytotic condition and reported EC-392530, EC-392538, EC-392539, EC-392541, SL-423, RSC-1, RSC-2, JS-80-21 and PK-1029 as moderately resistant. Hundekar (1999) also evaluated S-22, WC-12 and 92-10 as rust resistant germplasm. Among varieties PK-1162, PK-1029, JS-80-21 and PK-1024 showed moderately resistant reaction with better yield. Basavaraja (2002) identified three useful mutants which are moderately resistant to rust among 270 induced mutant families studied in M_3 generation. Similar results were also reported by various researchers [8, 10, 14, 15]

4.3. Screening for yellow mosaic disease

Yellow mosaic is one of the major viral diseases in India and it is causing major problem during *rabi/summer* in Utter Pradesh in recent years. The yield loss due to disease may range from minor to complete loss depending upon severity. So identification of resistant sources will help in optimum management and thus help in future breeding programmes. In the present study, 84 genotypes of soybean were taken for screening against yellow mosaic disease under natural conditions. None of the genotypes tested were immure to the disease. Over the entire disease incidence was high which was evident from the results as most of the genotypes fall under the category moderately susceptible to susceptible. Similar results were also reported [6, 17]. They screened 88 indigenous and exotic soybean genotypes in the field and found EC-107014, EC-107003 and EC-100777 resistant.

Author details

M. H. Khan[1], S. D. Tyagi[2] and Z. A. Dar[3]

1 Central Institute of Temperate Horticulture, Indian Council of Agriculture Research, Srinagar, India

2 Department of Plant Breeding and Genetics, Kisan (P.G) College, Simbhaoli, Ghaziabad, India

3 S.K. University of Agricultural Sciences and Technology of Kashmir, Shalimar, Srinagar, India

References

[1] Agrawal AP, Salimath PM, Patil SA. Soybean Pod Growth Analysis and its Relationship with Pod Shattering. Karnataka Journal of Agricultural Sciences 2004; 17(1) 41-45.

[2] Basavaraja GT. Studies on induced mutagenesis in soybean. PhD thesis. University of Agricultural Sciences, Dharwad; 2002.

[3] Bromfield KR, Yang CY. Soybean rust: Summary of available knowledge In: Expanding the use of Soybean. (Editors: Roberts,M.Goodman). INTSOY Series No.10, Illionis; 1976. pp.161-163.

[4] Dashell KE, Bello L. Screening for resistant to pod shattering. IITA Grain Legume Improvement Programme. Annual report for 1986.Ibadan, Nigeria; 1988. p.120.

[5] Hundekar AR. Studies on some aspects of soybean rust caused by Phakopsora pachyrizi Syd. PhD thesis, University of Agricultural Sciences, Dharwad; 1999.

[6] Koranne KD, Tyagi PC. Screening of soybean germplasm against yellow mosaic diseases. Indian Journal of Genetics and Plant Breeding 1985; 45(1) 30-33.

[7] Mayee CR, Datar VV. Phytopathametry, Maharashtra Agricultural University, Parbhani, Technical Bulletin No.1, 1986; pp: 145-146.

[8] Miles MR, Bonde MR, Nester SE, Berner DK, Frederick RD, Hartman GL. Characterizing Resistance to Phakopsora pachyrhizi in Soybean. Plant Disease 2011; 95(5) 577-581.

[9] Patil PV, Basavaraja GT. A prospective source of resistance to soybean rust. Karnataka Journal of Agricultural Sciences 1997; 10 1241-1243.

[10] Pham TA, Miles MR, Frederick RD, Hill CB, Hartman GL. Differential Responses of Resistant Soybean Entries to Isolates of Phakopsora pachyrhizi. Plant Disease 2009; 93(3) 224-228.

[11] Tiwari SP, Bhatnagar PS. Pod shattering of soybean in India. Journal of Oilseed Research 1988; 5 : 92-93.

[12] Tiwari SP, Bhatnagar PS. Consistent resistance for pod shattering in soybean (Glycine max (L.) Merrill) varieties. Indian Journal of Agricultural Sciences 1992; 63(3) 173-174.

[13] Tukamuhabwa P, Rubaihayo P, Dashiell KE. Genetic components of pod shattering in soybean. Euphytica 2002; 125(1) 29-34.

[14] Twizeyimana M, Ojiambo PS, Hartman GL, Bandyopadhyay R. Dynamics of Soybean Rust Epidemics in Sequential Plantings of Soybean Cultivars in Nigeria. Plant Disease 2011; 95(1): 43-50.

[15] Twizeyimana M, Ojiambo PS, Ikotun T, Ladipo JL, Hartman GL, Bandyopadhyay R. Evaluation of Soybean Germplasm for Resistance to Soybean Rust (Phakopsora pachyrhizi) in Nigeria. Plant Disease 2008; 92(6) 947-952.

[16] Upadhaya and Paradkar. Pod shattering in soybean (Glycine max (L.) Merrill). Journal of Oilseed Research 1991; 8 121-122.

[17] Yadav RK, Shukla RK, Chattopadhyay D. Soybean cultivar resistant to Mungbean Yellow Mosaic India Virus infection induces viral RNA degradation earlier than the susceptible cultivar. Virus Research 2009;144(1-2) 89-95.

Chapter 8

Integrated Management of
Helicoverpa armigera in Soybean Cropping Systems

Yaghoub Fathipour and Amin Sedaratian

Additional information is available at the end of the chapter

1. Introduction

In most developing countries, agriculture is the driving force for broad-based economic growth and low agricultural productivity is a major cause of poverty, food insecurity, and malnutrition. However, food production per unit of land is limited by many factors, including fertilizer, water, genetic potential of the crop and the organisms that feed on or compete with food plants. Despite the plant-protection measures adopted to protect the principal crops, 42.1% of attainable production is lost as result of attack by pests [1]. Therefore, accelerated public investments are needed to facilitate agricultural growth through high-yielding varieties with adequate resistance to biotic and abiotic stresses, environment-friendly production technologies, availability of reasonably priced inputs in time, dissemination of information, improved infrastructure and markets, and education in basic health care.

Soybean (*Glycine max* (L.) Merrill) is one of the most important and widely grown oil seed crops in the world. Successful production in soybean cropping systems is hampered due to the incidence of several insect pests such as *Etiella zinkienella* Treitschke, *Tetranychus urticae* Koch, *Thrips tabaci* Lindeman, *Spodoptera exigua* (Hübner) and *Helicoverpa armigera* (Hübner) [2-9]. Among these pests, *H. armigera* represents a significant challenge to soybean production in different soybean-growing areas around the world. *Helicoverpa armigera* is an important pest of many crops in many parts of the world and is reported to attack more than 60 plant species belonging to more than 47 families (such as soybean, cotton, sorghum, maize, sunflower, groundnuts, cowpea, tomato and green pepper) [10-12]. This noctuid pest is distributed eastwards from southern Europe and Africa through the Indian subcontinent to Southeast Asia, and thence to China, Japan, Australia and the Pacific Islands [13]. The pest status of this species can be derived from its four life history characteristics (polyphagy, high mobility, high fecundity and a facultative diapause) that enable it to survive in unstable habitats and adapt to seasonal changes. Direct damage of the larvae of this noctuid pest to flowering and fruiting structures together with extensive insecticide spraying resulted in low crop yield and high costs of production [14].

Different methods have been applied to control *H. armigera* in order to improve the quality and quantity of soybean production in cropping systems of this oil seed crop. However, synthetic insecticides including organophosphates, synthetic pyrethroids and biorational compounds are the main method for *H. armigera* control in different parts of the world. This wide use of pesticides is of environmental concern and has repeatedly led to the development of pesticide resistance in this pest. Furthermore, the deleterious effects of insecticides on nontarget organisms including natural enemies are among the major causes of pest outbreaks. It is therefore necessary to develop a novel strategy to manage population of *H. armigera* and reduce the hazardous of synthetic chemicals.

The common trend towards reducing reliance on synthetic insecticides for control of insect pests in agriculture, forestry, and human health has renewed worldwide interest in integrated pest management (IPM) programmes. IPM is the component of sustainable agriculture with the most robust ecological foundation [15]. IPM not only contributes to the sustainability of agriculture, it also serves as a model for the practical application of ecological theory and provides a paradigm for the development of other agricultural system components. The concept of IPM is becoming a practicable and acceptable approach among the entomologists in recent past all over the world and focuses on the history, concepts, and the integration of available control methods into integrated programmes. However, this approach advocates an integration of all possible or at least some of the known natural means of control with or without insecticides so that the best pest management in terms of economics and maintenance of pest population below economic injury level (EIL) is achieved.

Fundamental of effective IPM programmes is the development of appropriate pest management strategies and tactics that best interface with cropping system-pest situations. Depending on the type of pest, however, some of the primary management strategies could be selected. In the case of *H. armigera*, several management tactics should be considered to implement a comprehensive integrated management. Potential of some of the control tactics to reduce population density of *H. armigera* in different cropping systems were evaluated by several researchers and attempts have been made to develop integrated management approach for *H. armigera* using host plant resistance [2, 4, 6, 11] including transgenic Bt crops [16], biological control (predators and parasitoids) [17], interference methods including sex pheromones [18], biopesticides (especially commercial formulations of *Bacillus thuringiensis*) [19], cultural practices (including appropriate crop rotations, trap crops, planting date and habitat complexity) [20] and selective insecticides [21]. Likewise there remains a need for ongoing research to develop a suite of control tactics and integrate them into IPM systems for sustainable management of *H. armigera* in cropping systems. Keeping this in view, integration of these methods based on the ecological data especially thermal requirements of this pest and its crucial role in forecasting programme of *H. armigera* could lead a successful integrated management for this pest in soybean cropping systems.

As discussed above, integrated management is typically problematic in cropping systems, especially in the case of *H. armigera* on soybean. However, our intent in this section is not to develop an exhaustive review of all resources that may possibly contribute to more effective pest management for the future, but to select several topic areas that will make essential contributions to sustainable soybean cropping systems. Although the past research focused on developing various pest management tactics that would be packaged into an integrated pest management strategy, we have selected several types of resources for our discussions, realizing that there are other resources can be used in developing IPM programmes. Furthermore, to generate a comprehensive management programme, we present our perspectives on future research needs and directions for sustainable management of this pest in soybean cropping systems such as tri-trophic interactions [22], importance of modeling of insect population [23], crucial role of forecasting and monitoring programmes in IPM [24], interactions among different management tactics in IPM [25] and significance of biotechnology and genetically modified plants in IPM. Therefore, considering the importance of *H. armigera* in successful production of soybean, this review intends to provide an appropriate document to the scientific community for sustainable management of *H. armigera* in soybean cropping systems.

2. History of terminology and definition of IPM

Although many IPM programmes were initiated in the late 1960s and early 1970s in several parts of the world, it was only in the late 1970s that IPM gained momentum [26]. Throughout the late 19th and early 20th centuries, in the absence of powerful pesticides, crop protection specialists relied on knowledge of pest biology and cultural practices to produce multi tactical control strategies that, in some instances, were precursors of modern IPM systems [27]. That stance changed in the early 1940s with the advent of organosynthetic insecticides when protection specialists began to focus on testing chemicals, to the detriment of studying pest biology and non-insecticidal methods of control [15]. The period from the late 1940s through the mid-1960s has been called the dark ages of pest control. By the late 1950s, however, warnings about the risks of the preponderance of insecticides in pest control began to be heard. The publication of the book "Silent Spring" by Rachael Carson in 1962 ignited widespread debate on the real and potential hazards of pesticides. This still ongoing dialogue includes scientists in many disciplines, environmentalists, and policy makers. However, "Silent Spring" contributed much to the development of alternatives to pesticides for pest management purposes, augmented global interests in developing cropping systems that limit crop pests, and added much to the environmental movement [26]. In fact, widespread concerns about the detrimental impact of pesticides on the environment and related health issues were responsible in large part for the development of the concept of IPM.

The seed of the idea of integrated control appears in a paper by Hoskins *et al.* [28]. Conceivably, "integrated control" was uttered by entomologists long before formally appearing in a publication. However, it was the series of papers starting with Smith and Allen [29] that established integrated control as a new trend in economic entomology. Towards the end of the 1960s, integrated control was well entrenched both in the scientific literature and in the practice of pest control, although by then "pest management" as a sibling concept was gaining popularity [30]. However, in subsequent publications, integrated control was more narrowly defined as "applied pest control which combines and integrates biological and chemical control", a definition that stood through much of the late 1950s and the early 1960s but began to change again in the early 1960s as the concept of pest management gained acceptance among crop protection specialists [15, 26].

The concept of "protective population management", later shortened to "pest management", gained considerable exposure at the twelfth International Congress of Entomology, London [31]. The Australian ecologists who coined the expression contended that "control", as in pest control, subsumes the effect of elements that act independently of human interference. Populations are naturally controlled by biotic and abiotic factors, even if at levels intolerable to humans. Management, on the other hand, implies human interference. Although the concept of pest management rapidly captured the attention of the scientific community, in 1966 Geier seemed to minimize the semantic argument that favored "pest management" by stating that the term had no other value than that of a convenient label coined to convey the idea of intelligent manipulation of nature for humans' lasting benefit, as in "wildlife management" [32].

Not until 1972, however, were "integrated pest management" and its acronym IPM incorporated into the English literature and accepted by the scientific community. In creating the synthesis between "integrated control" and "pest management", no obvious attempt was made to advance a new paradigm. Much of the debate had been exhausted during the 1960s and by then there was substantial agreement that: (a) "integration" meant the harmonious use of multiple methods to control single pests as well as the impacts of multiple pests; (b) "pests" were any organism detrimental to humans, including invertebrate and vertebrate animals, pathogens, and weeds; (c) "management" referred to a set of decision rules based on ecological principles and economic/social considerations and (d) "IPM" was a multidisciplinary endeavor.

The search for a perfect definition of IPM has endured since integrated control was first defined. A survey recorded 65 definitions of integrated control, pest management, or integrated pest management [26]. Unfortunately, most of them perpetuate the perception of an entomological bias in IPM because of the emphasis on pest populations and economic injury levels, of which the former is not always applicable to plant pathogens, and the latter is usually attached to the notion of an action threshold often incompatible with pathogen epidemiology or many weed management systems [33]. Furthermore, most definitions stress the use of combination of multiple control methods, ignoring informed inaction that in some cases can be a better IPM option for arthropod pest management [15]. It was, however, in 1972 that the term 'integrated pest management' was accepted by the scientific community, after the publication of a report under the above title by the Council on Environmental Quality [34]. Much of the debate had already taken place during the 1960s and by then there was substantial agreement on the following issues [15]: (a) the appropriate selection of pest control methods, used singly or in combination; (b) the economic benefits to growers and to society; (c) the benefits to the environment; (d) the decision rules that guide the selection of the control action and (e) the need to consider impacts of multiple pests.

Several authors have come close to meeting the criteria for a good definition, but a consensus is yet to be reached. Accordingly, some of the IPM definitions were listed in Table 1. A broader definition was adopted by the FAO Panel of Experts [35]: "Integrated Pest Control is a pest management system that, in the context of the associated environment and the population dynamics of the pest species, utilizes all suitable techniques and methods in as compatible a manner as possible and maintains the pest population at levels below those causing economic injury." This definition has been cited frequently and has served as a template for others. However, based on an analysis of definitions spanning the past 35 years, the following is offered in an attempt to synthesize what seems to be the current thought: "IPM is a decision support system for the selection and use of pest control tactics, singly or harmoniously coordinated into a management strategy, based on cost/benefit analyses that take into account the interests of and impacts on producers, society, and the environment" [15].

Definition	Reference
IPM refers to an ecological approach in pest management in which all available necessary techniques are consolidated in a unified programme, so that pest populations can be managed in such a manner that economic damage is avoided and adverse side effects are minimized.	[36]
IPM is a multidisciplinary ecological approach to the management of pest populations, which utilizes a variety of control tactics compatibly in a single coordinated pest-management system. In its operation, integrated pest control is a multi-tactical approach that encourages the fullest use of natural mortality factors, complemented, when necessary, by artificial means of pest management.	[37]
IPM is a pest population management system utilizes all suitable techniques in a compatible manner to reduce pest populations and maintain them at levels below those causing economic injury.	[38]
IPM is a systematic approach to crop protection that uses increased information and improved decision-making paradigms to reduce purchased inputs and improve economic, social and environment conditions on the farm and in society.	[39]
IPM is a comprehensive approach to pest control that uses combined means to reduce the status of pests to tolerable levels while maintaining a quality environment.	[40]
IPM is an intelligent selection and use of pest-control tactics that will ensure favourable economic, ecological and sociological consequences.	[41]
IPM is a sustainable approach that combines the use of prevention, avoidance, monitoring and suppression strategies in a way that minimizes economic, health and environmental risks.	[42]
IPM is a decision support system for the selection and use of pest control tactics, singly or harmoniously coordinated into a management strategy, based on cost/benefit analyses that take into account the interests of and impacts on producers, society, and the environment.	[15]
IPM is a dynamic and constantly evolving approach to crop protection in which all the suitable management tactics and available surveillance and forecasting information are utilized to develop a holistic management programme as part of a sustainable crop production technology.	[43]
IPM is a systemic approach in which interacting components (mainly control measures) act together to maximize the advantages (mainly producing a profitable crop yield) and minimize the disadvantages (mainly causing risk to human and environment) of pest control programmes.	[The authors of this chapter]

Table 1. Some of the proposed definitions for IPM

3. Systems in agriculture and the situation of IPM as a sub-system

Spedding [44] defined a system as a group of interacting components, operating together for a common purpose, capable of reaching as a whole to external stimuli. A system is unaffected by its own output and has a specified boundary based on the inclusion of all significant feedbacks. However, four types of systems are generally acknowledged in agriculture including ecosystem, agroecosystem, farming systems and cropping systems (Figure 1). In this hierarchy, a system may consist of several sub-systems. IPM is a subsystem of cropping system and considered as the operating system used by farmers to manage population of crop pests. This sub-system has a degree of independence and can be studied in isolation of the cropping system. It has its own inputs and has the same output as the main system (i. e. yield) but relates to only some of the components and therefore, to only some of the inputs [45].

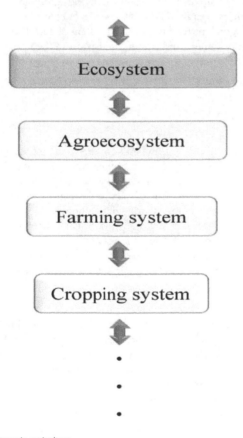

Figure 1. A hierarchy of systems in agriculture

IPM systems have a goal of providing the farmer with an economic and appropriate means of controlling crop pest. The aim should be to devise an IPM system which is sufficiently robust to maintain control over a prolonged period of time [46]. However, to achieve an IPM system a number of attributes will require including: (a) provide effective control of pest; (b) be economically viable; (c) simplicity and flexibility; (d) utilize compatible control measure; (e) sustainability and (f) minimum harmful effect on the environment, producer and consumer.

First and foremost the IPM system must be effective. For the farmer this means that this system should be at least as good as the conventional control methods. The system should be economic. No farmer will adopt and sustain use of uneconomic pest management practices. On the other hand, an IPM system must be designed to be as simple as possible, utilizing the minimum number of control measures compatible with maintaining pest populations at appropriate levels. The individual control measures should of course be compatible and optimize natural mortality factors. It is important during design of an IPM system to consider the level of control which is required and the best mix of control measures that will achieve this with minimal antagonism. Finally, the IPM system should be sustainable, have minimum impact on the environment and present no hazard to the farmer, their families or the consumers of the crop products [45].

4. Decision making in IPM

Following widespread concerns about the adverse effects of insecticides it became clear that calendar spraying was not the appropriate approach to pest control. In fact, determining whether an insect control measure (usually an insecticide) is "needed" is one of the basic principles of any IPM programme. "Need" can be defined in a number of ways, but most growers associate the need for an insecticide with economics. In other words, most growers ask some form of these questions: "How many insects cause how much damage?", "Are the damage levels all significant?" and "Will the value of yield protection with an insecticide offset the cost of control?" Therefore, researchers from different agricultural disciplines came to realize that a decision rule or threshold should answer such questions and that pest control must be viewed as a decision making process (Figure 2).

Pest management is a combination of processes that include obtaining the information, decision making and taking action [41]. In assessing, evaluating and choosing a particular pest control option, farmer's perception of the problem and of potential solutions is the most important factor (Figure 2). Decision making in pest management, like other economic problems in agriculture, involves allocating scarce resources to meet food demand of a growing population. In this process, agricultural producers have to make choices regarding the use of several inputs including labor, insecticides, herbicides, fungicides, and consulting expenses related to the level and intensity of pest infestation and the timing of treatment. However, decision making process for pest control takes place in many levels at the fields. These various layers of decision making affect the whole strategy of pest control in a given cropping system, region or country as well as the set of approaches and measures that are chosen to implement pest control programmes.

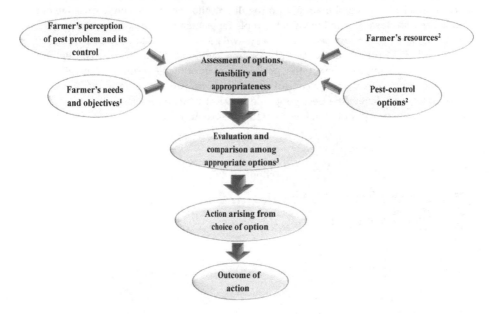

Figure 2. The process of decision making in IPM. (after Reichelderfer *et al.* [47]) 1. The way in which control options are assessed will depend on the farmer's objectives. Subsistence farmers may select for a guaranteed food supply, while commercial farmers are more concerned with profit. 2. The number of options that a farmer can feasibly use will depend on the constraints set by the resources available. 3. Compare the cost-effectiveness of alternative practices.

5. Crucial role of economic thresholds for implementation of IPM programmes

In most situations it is not necessary, desirable, or even possible to eradicate a pest from an area. On the other hand, the presence of an acceptable level of pests in a field can help to slow or prevent development of pesticide resistance and maintain populations of natural enemies that slow or prevent pest population build-up. Therefore, the concepts of economic injury level (EIL) and economic threshold (ET [sometimes called an action threshold]) were developed (Figure 3). EIL and ET constitute two basic elements of the IPM [48]. Economic injury level was defined as the lowest population density that will cause economic damage [49]. The EIL is the most essential of the decision rules in IPM. In addition, the economic injury level provides an objective basis for decision making in pest management and the backbone for the management of pests in an agricultural system is the concept of EIL [48]. Ideally, an EIL is a scientifically determined ratio based on results of replicated research trials over a range of environments. In practice, economic injury levels tend to be less rigorously defined, but instead are nominal or empirical thresholds based on grower experience or generalized pest-crop response data from research trials. Although not truly comprehensive, such informal EILs in combination with regular monitoring efforts and knowledge of pest biology and life history provide valuable tools for planning and implementing an effective IPM programme. However, because growers will generally want to act before a population reaches EIL, IPM programmes use the economic threshold (Figure 3). The concept of economic threshold implies that if the pest population and the resulting damage are low enough, it does not pay to take control measures. In practice, the term economic threshold has been used to denote the pest population level at which economic loss begins to occur and indicate the pest population level at which pest control should be initiated [50].

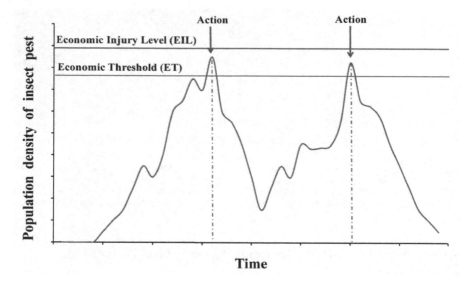

Figure 3. Graph showing the relationship between the economic threshold (ET) and economic injury level (EIL). The arrows indicate when a pest control action is taken.

5.1. EIL and ET for *H. armigera* on different crops

Economic injury level and economic threshold of *H. armigera* on some crops was estimated by several researchers (Table 2). In the case of *H. armigera* on soybean, these thresholds are poorly defined and a little information in this regard is available. However, economic thresholds; especially economic injury level; are dynamic and can be varied from year to year or even from field to field within a year depending on crop variety, market conditions, development stages of plant, available management options, crop value and management costs (Table 2).

Crop	Economic threshold (ET)	Economic Injury Level (EIL)	References
Chickpea	-	> 4 larvae / m²	[51]
Chickpea	-	1.0 larva / m row	[52]
Chickpea	1.0 larva / m row	-	[53]
Chickpea	-	1 larva / 10 plants	[54]
Chickpea	-	0.6 larva / plant	[55]
Chickpea	1.77 - 2.00 larvae / m row	-	[56]
Chickpea	-	1.0 larva / m row	[57]
Chickpea	0.81 larva / m row	1.1 larva / m row	[58]
Pigeon pea	-	0.78-0.80 larvae / plant	[59]
Tomato	1.0 larva / plant	-	[60]
Cotton	-	19.86 larvae / 100 plants	[61]
Mung bean	1-3 larvae / m²	-	[62]
Peanuts	4 larvae / m²	-	[62]
Soybean	-	8 larvae / m²	[63]

Table 2. Economic threshold (ET) and Economic injury level (EIL) of *Helicoverpa armigera* on different crops.

6. Monitoring activity in integrated management of *H. armigera*

In an IPM programme, pest managers use regular inspections, called monitoring, to collect the information they need to make appropriate decisions. A central idea in IPM is that a treatment is only used when pest numbers justify it, not as a routine measure. Keeping this in view, in IPM programmes, chemical control is applied only after visual inspection or monitoring devices indicate the presence of pests in that specific area, the pest numbers have exceeded the economic threshold (ET) and adequate control cannot be achieved with non-chemical methods within a reasonable time and cost. Therefore, it was considered that monitoring could reduce spraying costs by withholding a spray until a given threshold is reached [64].

For many years, light traps have been used to monitor *Helicoverpa* moth populations. Hartstack *et al.* [65] developed a model for estimating the number of moths per hectare from these light-trap catches, to evaluate the possible use of light traps for controlling *Helicoverpa* spp. Walden [66] presented the first comprehensive report on seasonal occurrence and abundance of the *Helicoverpa zea* (Boddie), based on light trap collections. Beckham [67] used light traps to index the populations of *Helicoverpa* spp. and reported that a significantly lower percentage of the *Helicoverpa virescens* (Fabricius) populations responded to black-light lamps in traps than did *H. zea*.

In the last few years, pheromone traps (containing virgin females or synthetic pheromones) have replaced light traps as moth-monitoring devices. These traps provide the pest manager with a convenient and effective tool for monitoring adult moth [68]. However, pheromone traps are highly efficient, simple to construct, inexpensive, and portable (requiring no power). Furthermore, only the single species for which the trap is baited is attracted and caught, making identification and counting quick and easy. As an added bonus, pheromone traps also detect spring emergence of moths 2 or 3 weeks earlier than light traps, which should give more precision to forecasts.

Several group of researchers made the comparison of indexing populations of *Helicoverpa* spp. in light traps versus pheromone traps. There results revealed that light trap catches may index seasonal fluctuation of populations more accurately than pheromone traps, however, pheromone traps are more sensitive to low populations early in the season and decline in efficiency with high populations late in the season [69].

There has been a considerable improvement into synthesis of the pheromones of *Helicoverpa* spp. in recent years. However, preliminary studies have already revealed that the catches in pheromone traps do not correlate very well with light-trap catches and field counts of the pest in all circumstances. In fact, trap catch data do not provide a quantitative threshold for intervention because a relationship between catch number and subsequent crop damage has proved to be lacking in most cases [70].

Egg count provides a better quantitative threshold for monitoring activity of *H. armigera* but egg desiccation, egg infertility or egg parasitism (biasing data) together with skill needed for field scouting, too often promote weak correlations between egg number and larval damage [71]. On the other hand, fruit inspection in the field has proved to be a valuable tool when develop against a number of fruit damaging pest species including *H. armigera* [72]. The major advantage of thresholds based on fruit inspection is that the short time between plant scouting for larval injury and fruit damage greatly increases correlation between both of these variables. Moreover, damaged fruit-count-based decision making may also be easily learned and carried out by growers [70].

Finally, we must now determine whether these pheromone traps are going to be of practical value in *Helicoverpa* management. For this, there is first a need to standardize trap design, pheromone dosage and release rates from the chosen substrate, and siting of the traps. As the next step, catches in these traps should be compared with other measures of *Helicoverpa* populations (light traps and actual counts of *Helicoverpa* eggs/larvae on the host plants in the same area). However, data from pheromone traps have already been shown to be valuable in some studies in the USA, where the data have been used in prediction models and have given useful information on the timing of infestations [64].

7. Importance of thermal modeling in successful implementation of integrated management of *H. armigera*

For decades, models have been an integral part of IPM. For instance, the use of models has helped pest managers decide how the agroecosystem should be changed to favor economy and conservation and not to favor pests. Moreover, models have allowed scientists to conduct simulated experiments when the conduct of those experiments would not have been possible. Furthermore, models have been used whenever scientists wanted to explore as well as understand the complexities of agroecosystems [26, 73, 74]. However, among the different types of models developed for implication of IPM programmes, forecasting models (especially thermal models) have a highlighted situation. Understanding the factors governing the pest development and implementing this knowledge into forecast models enable effective timing of interventions and increases efficacy and success of control measures [74]. For a pest manager, being able to predict abundance and distribution of a pest species, and its timing and level, is crucial to both strategic planning and tactical decision making. Thermal models, based on insect physiological timescales, have been relatively successful at predicting the timing of population peaks and are useful for timing sampling and control measures.

Temperature is a critical abiotic factor influencing the dynamics of insect pests and their natural enemies [75-77]. Temperature has a direct influence on the key life processes of survivorship, development, reproduction, and movement of poikilotherms and hence their population dynamics [78]. The importance of predicting the seasonal occurrence of insects has led to the formulation of many mathematical models that describe developmental rates as a function of temperature [79]. Thermal models have been developed for insect pests to predict emergence of adults from the overwintering generation, eclosion of eggs, larval and pupal development, and generation time. These models, all based on a linear relationship between temperature and developmental rate, have been used with varying degrees of success to time pesticide application for pest control [80]. However, linear approximation enables the estimation of lower temperature thresholds (T_{min}) and thermal constants (K) within a limited temperature range and to describe the developmental rate more realistically and over a wider temperature range, several nonlinear models have been applied [74, 76]. These nonlinear models provide value estimates of lower and upper temperature thresholds and optimal temperature for development of a given stage.

Several studies have been conducted on the effects of temperature on developmental time of *H. armigera* reared on host plant materials or artificial diets [81, 82]. In a recent study by Mironidis and Savopoulou-Soultani [83], a comprehensive analysis of survivorship and development rates at all life stages of *H. armigera* reared under constant and corresponding alternating temperatures regimes was performed (some of the most important results are listed in Tables 3 and 4).

Stage	Temperature	Lower temperature threshold (T_{min}°C)	Thermal constant [a] (K DD)
Egg	Constant	11.95	39.68
	Alternating	5.53	57.47
Larva	Constant	10.52	238.09
	Alternating	2.17	416.16
Pupa	Constant	10.17	192.30
	Alternating	1.06	285.71
Total immature stages	Constant	9.57	476.19
	Alternating	2.23	769.23

[a] Cumulative degree-day (DD) required for stage development

Table 3. Lower temperature threshold and thermal constant of different life stages of *Helicoverpa armigera* (after Mironidis and Savopoulou-Soultani [83]).

The results obtained by Mironidis and Savopoulou-Soultani [83] revealed that over a wide constant thermal range (15-27.5°C) total survivorship is stable and apparently not affected by temperature. Below 15°C, survivorship decreased rapidly, reached zero at 12.5°C. At higher temperatures, survivorship also decreased very quickly above 28°C and fell to zero at 40°C. Furthermore, their results showed that *H. armigera*, when reared at constant temperatures, could not develop from egg to adult stage (capable of egg production) out of the temperature range of 17.5-32.5°C. Nevertheless, alternating temperatures allowed *H. armigera* to complete its life cycle over a much wider range, 10-35°C, compared with constant temperatures.

Stage	Temperature	Lower temperature threshold (T_{min}°C)	Optimal temperature (T_{opt}°C)	Upper temperature threshold (T_{max}°C)
Egg	Constant	10.58	34.84	39.99
	Alternating	2.33	39.26	40.56
Larva	Constant	11.17	34.22	39.11
	Alternating	1.55	39.35	40.95
Pupa	Constant	12.31	35.37	40.00
	Alternating	1.01	41.92	43.54
Total immature stages	Constant	9.42	34.61	39.81
	Alternating	1.85	42.35	42.92

Table 4. Lower temperature threshold, optimal temperature and upper temperature threshold of different life stages of *Helicoverpa armigera* obtained by nonlinear Lactin model (after Mironidis and Savopoulou-Soultani [83]).

In another study, temperature-dependent development of *H. armigera* was studied in the laboratory conditions at eight constant temperatures (15, 17.5, 20, 22.5, 25, 30, 32.5 and 35°C) [Adigozali and Fathipour, unpublished data]. In this study, two linear (Ordinary linear and Ikemoto and Takai) and 9 nonlinear (Briere-1, Briere-2, Lactin-1, Lactin-2, Polynomial, Kontodimas-16, Analytis-1, Analytis-2 and Analytis-3) models were fitted to describe development rate of *H. armigera* as a function of temperature. The lower temperature threshold and thermal constant of different life stages of *H. armigera* estimated by linear models are listed in Table 5. The obtained results revealed that both models have acceptable accuracy in prediction of T_{min} and K for different life stages of *H. armigera* [Adigozali and Fathipour, unpublished data].

Stage	Model	Lower temperature threshold (T_{min} °C)	Thermal constant (K DD)
Egg	Ordinary linear	8.61	47.85
	Ikemoto and Takai	9.52	44.60
Larva	Ordinary linear	6.07	367.65
	Ikemoto and Takai	7.18	343.00
Pre-pupa	Ordinary linear	11.70	42.55
	Ikemoto and Takai	10.80	46.70
Pupa	Ordinary linear	14.29	132.28
	Ikemoto and Takai	13.20	150.00
Total immature stages	Ordinary linear	10.39	561.78
	Ikemoto and Takai	10.30	566.00

Table 5. Lower temperature threshold and thermal constant of different life stages of *Helicoverpa armigera* estimated by Ordinary linear and Ikemoto and Takai models.

According to results obtained by Adigozali and Fathipour [unpublished data], of the nonlinear models fitted, the Lactin-2, Lactin-2, Polynomial, Polynomial and Briere-2 models were found to be the best for modeling development rate of egg, larva, pre-pupa, pupa and total immature stages of *H. armigera*, respectively (Table 6). However, estimated values for crucial temperatures of different life stages of *H. armigera* by Adigozali and Fathipour [unpublished data] conflict with those reported by Mironidis and Savopoulou-Soultani [83] (Tables 3-6). Some possible reasons for these disagreements are: physiological difference depending on the food quality, genetic difference as a result of laboratory rearing and techniques/equipment of the experiments. In general, the results obtained from constant temperature experiments are often not applicable directly to the field where pests are subjected to diurnal variation of temperature and such information need to be validated under fluctuating temperatures before using for predictive purpose in the field. Finally, such data provide fundamental information describing development of *H. armigera*, when this information to be used in association with other ecological data may be valuable in integrated management of this noctuid pest in soybean cropping systems.

Stage	Model	Optimal temperature ($T_{opt}°C$)	Upper temperature threshold ($T_{max}°C$)
Egg	Lactin-2	33.00	41.98
Larvae	Lactin-2	34.50	35.38
Pre-pupa	Polynimial	29.00	-
Pupa	Polynimial	32.50	-
Total immature stages	Briere-2	34.00	35.00

Table 6. Lower temperature threshold, optimal temperature and upper temperature threshold of different life stages of *Helicoverpa armigera* obtained by nonlinear models.

8. Strategies for integrated management of *H. armigera*

8.1. Chemical control

Historically pest management on many crops has relied largely on synthetic pesticides and in intensive cropping systems, pesticides are main components of pest management programmes that represents a significant part of production costs [84]. However, chemical control is still the most reliable and economic way of protecting crops from pests. Beside, over reliance on chemical pesticides without regarding to complexities of the agroecosystem is not sustainable and has resulted in many problems like environment pollution, secondary pest outbreak, pest resurgence, pest resistance to pesticides and hazardous to human health. Furthermore, over dependence on chemical pesticides has also resulted in increased plant protection, thus leading to high cost of production.

Insecticide treatments, whether or not included in IPM programmes, are currently indispensable for the control of *H. armigera* in almost all cropping systems around the world [85], so, this pest species has been subjected to heavy selection pressure. Some of the synthetic insecticides currently used for controlling this pest are indoxacarb, methoxyfenozide, emamectin benzoate, novaluron, chlorfenapyr, imidacloprid, fluvalinate, endosulfan, spinosad, abamectin, deltamethrin, cypermethrin, lambda-cyhalothrin, carbaryl, methomyl, profenofos, thiodicarb and chlorpyrifos [21, 85-87]. Because of indiscriminate use of these chemicals to minimize the damage caused by *H. armigera*, however, it has developed high levels of resistance to conventional insecticides such as synthetic pyrethroids, organophosphates and carbamates [88].

8.1.1. Sustainable use of insecticides and obtaining maximum benefits from their application

However, selection for resistance to pesticides will occur whenever they are used [87]. After a pest species develops resistance to a particular pesticide, how do you control it? One method is to use a different pesticide, especially one in a different chemical class or family of pesticides that has a different mode of action against the pest. Of course, the ability to use other pesticides in order to avoid or delay the development of resistance in pest populations depends on the availability of an adequate supply of pesticides with differing modes of action. This method is perhaps not the best solution, but it allows a pest to be controlled until other management strategies can be developed and brought to bear against the pest [21]. However, suggestions will now be made as to how the maximum benefit can be obtained from the unique properties of the insecticides.

a.   Given the decreasing susceptibility of older caterpillars than early ones, it is important to use the insecticides early. Not only young larvae of *H. armigera* are more susceptible, but first and second instars are also more exposed than later instars [89].

b.   To decide whether the infestation by a pest has reached the economic threshold and an insecticide is required, more attention should be devoted to monitoring programmes. On crops where *Helicoverpa* is the main target, synthetic insecticides should not be used until these pests have reached the economic threshold [90].

c.   It is best to be used selective insecticides, a practice that will help to conserve beneficial insects. They can assist in delaying the onset and reducing the intensity of mid-season *Helicoverpa* attack [91].

d.   If infestation is high and the growth of the plants rapid, spray applications should be made at short intervals to protect the new growth, which may from otherwise be attacked by larvae repelled treated older foliage. Furthermore, short interval strategy will give better spray distribution and increase the chance of obtaining direct spray impingement on adults, larvae, and eggs [90].

e.   For crops in which higher economic thresholds are acceptable, integration of synthetic insecticide and beneficial insects becomes a practical possibility. The integration of chemical and biological control is often critical to the success of an IPM programme for arthropod pests [92, 93]. To combine the use of natural enemies with insecticides application, the chemical residues must be minimally toxic to the natural enemies to prevent its population being killed and the target pests increasing again [94]. Toxicological studies that only evaluate the lethal effects may underestimate the negative effects of insecticides on natural enemies and hence, sublethal effects should be assessed to estimate the total effect of insecticides on biological performance of natural enemies [95]. However, even though several studies showed that sublethal effects of insecticides can affect efficiency of natural enemies [96, 97], such effects on these organisms are rarely taken into account when IPM programmes are established and only mortality tests are considered when a choice between several insecticides must be made. Accordingly, to achieve maximum benefit from insecticides application and to reduce the selective pressure and development of insecticide resistance, insecticide at low concentrations may be used in combination with biological control.

8.2. Biological control

The most interesting component of IPM for many people is biological control. It is also the most complicated as there is a diverse range of species and types of predators, parasitoids and pathogens. The value of biological control agents in integrated pest management is becoming more apparent as researches are conducted. Natural enemies clearly play an important role in integrated management of *Helicoverpa* spp., particularly in low value crops where they may remove the need for any chemical intervention. Likewise in high value crops (such as cotton and tomato) beneficial species provide considerable benefit but are unable to provide adequate control alone, especially in situations where migratory influxes of *Helicoverpa* result in significant infestations [14]. However, although parasitoids and predators cannot be relied upon for complete control of *H. armigera* in unsprayed area, knowledge about their role in cropping systems where *H. armigera* is an important pest is an essential component in the development of integrated management.

Before using a natural enemy in a biological control programme, it is essential to know about its efficiency. However, study of demographic parameters and foraging behaviors of natural enemies is the reliable criteria for assessment of their efficiency. Among the demographic parameters, intrinsic rate of increase (r_m) is a key parameter in the prediction of population growth potential and has been widely used to evaluate efficiency of natural enemies [22, 76, 98, 99]. In addition to demographic parameters, another important aspect for assessing the efficiency of natural enemies is the study of their foraging behaviors including functional, numerical and aggregation responses, mutual interference, preference and switching [100-110]. Such information is essential to interpret how the natural enemies live, how they influence the population dynamics of their hosts/preys, and how they influence the structure of the insect communities in which they exist [111].

8.2.1. Parasitoids

The most common parasitoids that contribute to mortality of *Helicoverpa* spp. are shown in Table 7. Studies on the effects of parasitoids in biological control of *H. armigera* focused on monitoring parasitism of eggs and larvae. In Botswana, parasitism of larvae collected from different crops averaged up to 50% on sorghum, 28% on sunflower, 49% on cowpeas and 76% on cotton [112]. These results showed that parasitoids had a crucial role in management of *H. armigera*. However, this level of parasitism is higher when compared to the results from East Africa where the level of parasitism was generally low (<5%) or absent [113]. Surveys made of the parasitoid of *Helicoverpa* spp. in cotton fields of Texas by Shepard and Sterling [114] showed that larval parasitoids accounted for approximately 7% regulation of *Helicoverpa* spp. Such investigations highlight importance of parasitoids in integrated management of *H. armigera* in different cropping systems around the world.

Order	Family	Parasitoid species	References
Hymenoptera	Trichogrammatidae	*Trichogramma pretiosum* Riley	[115]
		Trichogramma exiguum Pinto and Platner	[114]
		Trichogramma australicum Girault	[116]
		Trichogramma pretiosum Riley	[117]
	Braconidae	*Microplitis croceipes* (Cresson)	[115]
		Habrobracon brevicornis Wesm.	[118]
		Habrobracon hebetor Say	[17]
		Cardiochiles nigriceps Vierick	[114]
		Chelonus insularis Cresson	[119]
		Apanteles marginiventris (Cresson)	[114]
		Meteorus sp.	[120]
		Apanteles ruficrus Hal.	[120]
		Apanteles kazak Telenga	[121]
		Microplitis demolitor Wilkinson	[120]
		Microplitis rufiventris Kok.,	[118]
		Chelonus inanitus (L.),	[118]
		Chelonus versalis Wilkn	[112]
	Ichneumonidae	*Campoletis sonorensis* Cameron	[123]
		Netelia sp.	[120]
		Hyposoter didymator (Thunb.)	[123]
		Heteropelma scaposum (Morley)	[120]
		Barylypa humeralis Brauns	[118]
		Campoletis chlorideae Uchida	[124, 125]
		Pristomerus spp.	[112]
		Charops spp.	[112]
	Scelionidae	*Telonomus* spp.	[112]
Diptera	Tachinidae	*Archytas marmoratus* (Townsend)	[114]
		Eucelatoria bryani Sabrosky	[114]
		Lespesia archippivora (Riley)	[126]
		Winthemia sp.	[120]
		Chaetophthalmus dorsalis (Malloch)	[127]
		Palexorista laxa (Curran)	[124]
		Exorista fallax Mg.,	[124]
		Goriophthalmus halli Mesnil	[124]
		Palexorista sp.	[112]
		Paradrino halli Curran	[112]

Table 7. The most common parasitoids of *Helicoverpa* spp.

8.2.2. *Predators*

The most important predators of *Helicoverpa* spp. are listed in Table 8. In some cropping systems these predators have considerable impact on population of *Helicoverpa* spp. These biological control agents have been reported as major factors in mortalities of *H. armigera* in cotton agroecosystems in South Africa and in smallholder crops in Kenya. In South Africa the average daily predation rates of 37% and 30% of *H. armigera* eggs and larvae, respectively were found in absence of insecticides [128]. Regarding this considerable potential, some of these predators could be candidated for implementation of biological control programmes. Accordingly, the species of *Sycanus indagator* (Stal) was imported from India to the USA. In another programme, *Pristhesancus papuensis* Stal was introduced from Australia to the USA and its efficiency was evaluated in laboratory [120].

Order	Family	Predator species	References
Coleoptera	Coccinelliadae	*Scymnus moreletti* Sic	[118]
		Exochomus flavipes (Thunberg)	[128]
		Cheilomenes propinqua (Mulsant)	[128]
		Hippodamia varigata Goeze	[128]
		Coccinella sp.	[118]
	Carabidae	*Calosoma* spp.	[129]
	Staphilinidae	-	[128]
Hymenoptera	Formicidae	*Pheidole* spp.	[128]
		Myrmicaria spp.	[128]
		Dorylus spp.	[128]
Hemiptera	Miridae	*Campylomma* sp.	[128]
	Anthocoridae	*Orius thripoborus* (Hesse)	[113]
		Cardiastethus exiguous (Poppius)	[113]
		Orius albidipenrzis (Reuter),	[113]
		Orius tantillus (Motschulsky)	[113]
		Blaptostethus sp.	[113]
		Cardiastethus sp.	[113]
	Reduviidae	*Sycanus indagator* (Stal)	[130]
	Reduviidae	*Pristhesancus papuensis* Stal	[120]
	Pentatomidae	*Podisus maculiventris* (Say)	[131]
	Nabidae	*Nabis* spp.	[129]
	Lygaeidae	*Geocoris punctipes (Say)*	[131]
Neuroptera	Chrysopidae	*Chrysoperla carnea* (Stephens)	[132]

Table 8. Important predators of *Helicoverpa* spp.

8.2.3. Pathogens

Naturally occurring entomopathogens are important regulatory factors in insect populations. The application of microorganisms for control of insect pests was proposed by notable
early pioneers in invertebrate pathology such as Agostino Bassi, Louis Pasteur and Elie
Metchnikoff [133]. However, it was not until the development of the bacterium *Bacillus thuringiensis* Berliner (*Bt*) that the use of microbes for the control of insects became widespread.
Today a variety of entomopathogens (bacteria, viruses, fungi, protozoa, and nematodes) are
used for the control of insect pests [134]. However, when environmental benefits of these
pathogens including safety for humans and other nontarget organisms, reduction of pesticide residues in food and environment, increased activity of most other natural enemies and
increased biodiversity in managed ecosystems are taken into account, their advantages are
numerous. There are also some disadvantages, mostly linked with their persistence, speed
of kill, specificity (too broad or too narrow host range) and cost relative to conventional
chemical insecticides. However, their increased utilization will require (*a*) increased pathogen virulence and speed of kill; (*b*) improved pathogen performance under challenging environmental conditions; (*c*) greater efficiency in their production; (*d*) improvements in
formulation that enable ease of application, increased environmental persistence, and longer
shelf life; (*e*) better understanding of how they will fit into integrated systems and their interaction with the environment and other IPM components and (*f*) acceptance by growers
and the general public [134].

The critical need for safe and effective alternatives to chemical insecticides in integrated
management of *H. armigera* has stimulated considerable interest in using pathogens as biological control agents. A list of some isolated microorganisms from *Helicoverpa* spp. is presented in Table 9. Among these microorganisms, nuclear polyhedrosis virus (NPV) and *B.
thuringiensis* have a considerable effect on population of *H. armigera*. Potential of these
pathogens in management programmes of *H. armigera* were evaluated by several researchers. Roome [135] tested a commercial preparation of NPV against *H. armigera*. The results
showed that NPV was as effective as a standard insecticide in reducing yield loss on sorghum due to damage by *H. armigera*. In addition, the long survival of NPV on sorghum (80
days) indicated that a single application of NPV was adequate to protect the crop for a
growing season. In another study, Moore *et al.* [136] showed that NPV has potential in management of *H. armigera* on citrus trees. Recent work by Jeyarani *et al.* [137] revealed that NPV
has an acceptable efficiency in control of *H. armigera* on cotton and chickpea.

Pathogenicity of *B. thuringiensis* for management of *H. armigera* population was investigated
by several researchers [19, 144]. They showed that larvae ingest enough quantities of *B. thuringiensis* toxins to die, or at least to reduce its weight and development, depending on the
toxin and conditions of the experiment. In a recent study, sublethal effects of *B. thuringiensis*
on biological performance of *H. armigera* were investigated [Sedaratian and Fathipour, unpublished data]. According to results obtained, values recorded for duration of total immature stages increased from 31.87 days in control to 37.17 days in LC_{25}. Furthermore, female
longevity decreased from 13.14 days to 7.23 days. Fecundity was also negatively affected in
female moths developed from treated neonates, with the rate of egg hatchability reaching

zero. The results obtained also revealed that the sublethal effects of *B. thuringiensis* could carry over to the next generation. The intrinsic and finite rates of increase (r_m and λ, respectively) were significantly lower in insects treated with sublethal concentrations compared to control. Consequent with the reduce rate of development observed for *H. armigera* treated with *B. thuringiensis*, the doubling time (*DT*) were significantly higher in insects exposed to any concentration tested compared to control (Table 10). However, according to results obtained, *B. thuringiensis* could play a critical role in integrated management of *H. armigera*.

Group	Family	Pathogen species	References
Bacteria	Enterobacteriaceae	*Pantoea agglomerans* (Ewing and Fife)	[138]
	Bacillaceae	*Bacillus thuringiensis* Berliner	[19]
Fungi	Cordycipitaceae	*Beauveria bassiana* (Balsamo)	[139]
	Clavicipitaceae	*Metarhizium anisopliae* (Metsch.)	[140]
	Moniliaceae	*Nomuraea rileyi* (Farlow) Samson	[141]
Viruses	Baculoviridae	Nuclear Polyhedrosis Virus	[142]
Nematoda	Heterorhabditidae	*Heterorhabditis bacteriophora* Poinar	[143]
	Steinernematidae	*Steinernema carpocapsae* (Weiser)	[143]
		Steinernema feltiae (Filipjev)	[143]

Table 9. Some of the isolated microorganisms from *Helicoverpa* spp.

Generation	Parameter	Treatments					
		Control	LC$_5$	LC$_{10}$	LC$_{15}$	LC$_{20}$	LC$_{25}$
Parental	Total immature	31.87±0.38	34.81±0.24	35.63±0.52	34.74±0.31	36.17±0.42	37.17±0.43
	stages	c	b	b	b	ab	a
	Female	13.14±0.40	11.69±0.48	10.28±0.86	10.06±0.67	9.27±0.44	7.23±0.44
	longevity	a	ab	bc	bc	c	d
	Total fecundity	789.52±42.68	665.13±52.46	601.00±45.72	532.53±33.70	376.00±21.95	98.46±12.33
		a	b	b	b	c	d
Offspring	r_m (day^{-1})	0.19±0.00	0.18±0.00	0.16±0.00	0.14±0.00	0.13±0.00	-*
		a	a	b	c	d	
	λ (day^{-1})	1.21±0.00	1.20±0.00	1.1±0.00	1.16±0.01	1.14±0.00	-
		a	a	b	c	d	
	DT (day)	3.59±0.05	3.75±0.05	4.29±0.08	4.78±0.08	5.33±0.16	-
		d	d	c	b	a	

Means in a row followed by the same letters are not significantly different (P <0.05) (S.N.K.)

* In this treatment hatch rate reaching zero.

Table 10. Sublethal effects of *Bacillus thuringiensis* on biological performance of *Helicoverpa armigera* in two subsequent generations.

The reliance on the entomopathogens for management of *H. armigera*, however, is risky since the different factors that govern epizootics. Accordingly, in most cases no single microbial control agent will provide sustainable control of this pest. Nevertheless, as components of an integrated management programme, entomopathogens can provide significant and selective control [134]. In the not too distant future we envision a broader appreciation for the attributes of entomopathogens and expect to see synergistic combinations of microbial control agents with other technologies that will enhance the effectiveness and sustainability of integrated management of *H. armigera*.

8.3. Cultural control

Cultural control is the deliberate manipulation of the cropping or soil system environment to make it less favorable for pests or making it more favorable for their natural enemies. Many procedures such as tillage, host plant resistance, planting, irrigation, fertilizer applications, destruction of crop residues, use of trap crops, crop rotation, etc. can be employed to achieve cultural control. Early workers used cultural practices as the mainstay of their insect control efforts. Newsom [145] pointed out that the rediscovery of the importance of cultural control tactics has provided highly effective components of pest management systems. Although some cultural practices have a noticeable potential in integrated management, use of some cultural controls is not universally beneficial. For example, providing nectar sources for beneficial insects may also provide nectar sources for pests.

8.3.1. Uncultivated marginal areas and abundance of natural enemies

Monoculture in modern agriculture, especially in annual crops, often discriminates against natural enemies and favors development of explosive pest populations. According to Fye [146], management of naturally occurring populations of insect predators may depend on knowledge of the succession of winter weeds and crops that provide natural hosts for food and shelter. The results obtained by Whitcomb and Bell [147] revealed that very few predators move directly from overwintering sites to field and pass one or two generations on weeds in the uncultivated marginal areas. In a 2-year study on the abundance of predators of *Helicoverpa* spp. in the various habitats in the Delta of Mississippi, predator populations in all the marginal areas were observed to be much higher than in the more homogeneous areas such as soybean fields.

8.3.2. Intercropping and its effect on natural enemies

Dispersal from target area often reduces the effectiveness of natural enemies especially in augmentation programmes. To minimize this shortcoming, provision of supplemental resources such as food to maintain, arrest or stimulate the released natural enemy could provide mechanisms for managing parasitoids and predators [148]. Accordingly, some environmental manipulation could affect efficiency of a natural enemy during biological control programmes of *Helicoverpa* spp. Roome [149] suggested that increasing plant diversity in cropping systems by intercropping crops carrying nectars could enhance effectiveness of natural enemies. When different host plants of *H. armigera* are interplanted, population of

H. armigera and its natural enemies on a crop are influenced by neighboring crops, both directly and indirectly. Direct influences include preference for one crop over the other by ovipositing moths and the movement of larvae and natural enemies between interplanted crops. Indirect influences arise when *H. armigera* infestation on one crop is influenced by the population build-up or mortality level on neighboring crops [113].

8.3.3. Ploughing and early planting effects on Helicoverpa populations

An alternative and often complementary strategy for management of *H. armigera* is the control of overwintering pupae through the practice of pupae busting which has been used in several cropping areas. Ploughing in late maturing crops in winter increase the mortality of any pupae formed in cropland by exposing them to heat and predation. The other cultural control method is early planting which avoids the seasonal peaks of population thereby avoiding very heavy larval infestations and reducing the overwintering population [150].

8.3.4. Trap crops and management of H. armigera

The recent resurgence of interest in trap cropping as an IPM tool is the result of concerns about potential negative effects of pesticides. Prior to the introduction of modern synthetic insecticides, trap cropping was a common method of pest control for several cropping systems [150]. Trap crops have been defined as "plant stands that are, per se or via manipulation, deployed to attract, divert, intercept and retain targeted insects or the pathogens they vector in order to reduce damage to the main crop" [151, 152]. Trap cropping is essentially a method of concentrating a pest population into a manageable area by providing the pest with an area of a preferred host crop and when strategically planned and managed, can be utilized at different times throughout the year to help manage a range of pests. For example, spring trap crops are designed to attract *H. armigera* as they emerge from overwintering pupae. A trap crop, strategically timed to flower in the spring, can help to reduce the early season buildup of *H. armigera* in a district. Spring trap cropping, in conjunction with good *Helicoverpa* control in crops and pupae busting in autumn, is designed to reduce the size of the local *Helicoverpa* population. On the other hand, summer trap cropping has quite a different aim from that of spring trap cropping. A summer trap crop aims to draw *Helicoverpa* away from a main crop and concentrate them in another crop. Once concentrated into the trap crop, the *Helicoverpa* larvae can be controlled. Finally, in addition to diverting insect pests away from the main crop, trap crops can also reduce insect pest populations by enhancing populations of natural enemies within the field. For example, a sorghum trap crop used to manage *H. armigera*, also increases rates of parasitism by *Trichogramma chilonis* Ishii [153]. However, to avoid creating a nursery for *H. armigera*, the trap crop must be destroyed prior to the pupation of the first large *H. armigera* larvae. Furthermore, to protect the trap crop from large infestations of *Helicoverpa* spp. spraying may be required.

8.3.4.1. Trap crops and push-pull strategy in integrated management of H. armigera

The term push-pull was first applied as a strategy for IPM by Pyke *et al.* in Australia in 1987 [154]. They investigated the use of repellent and attractive stimuli, to manipulate the distribution of *Helicoverpa* spp. in cotton fields. Push-pull strategies involve the "be-havioral manipulation of insect pests and their natural enemies via the integration of stimuli that act to make the protected resource unattractive or unsuitable to the pests (push) while luring them toward an attractive source (pull) from where the pests are subsequently removed". The strategy is a useful tool for integrated pest management programmes reducing pesticide input [155].

In plant-based systems, naturally generated plant stimuli can be exploited using vegetation diversification, including trap cropping and these crops have a crucial role as one of the most important stimuli for pull components. The host plant stimuli responsible for making a particular plant growth stage, cultivar, or species naturally more attractive to pests than the plants to be protected can be delivered as pull components by trap crops [155]. However, the relative attractiveness of the trap crop compared with the main crop, the ratio of the main crop given to the trap crop, its spatial arrangement (i.e., planted as a perimeter or in-tercropped trap crop), and the colonization habits of the pest are crucial to success and re-quire a thorough understanding of the behavior of the pest [156].

In the case of *Helicoverpa* spp. on cotton in Australia, the potential of combining the applica-tion of neem seed extracts to the main crop (push) with an attractive trap crop, either pi-geonpea or maize (pull), to protect cotton crops from *H. armigera* and *H. punctigera* has been investigated [153]. Trap crops, particularly pigeonpea, reduced the number of eggs on cot-ton plants in target areas. In trials, the push-pull strategy was significantly more effective than the individual components alone. The potential of this strategy was supported by a re-cent study in India. Neem, combined with a pigeonpea or okra trap crop, was an effective strategy against *H. armigera* [157].

8.3.5. Host plant resistance

Plants that are inherently less damaged or infested by insect pests in comparable environ-ments are considered resistant [158]. Host plant resistance (HPR) is recognized as the most effective component of IPM and has been considered to replace broad spectrum insecticides. A resistant host plant provides the basic foundation on which structures of IPM for different pests can be built [159]. The advantage that farmers gain in using cultural control with sus-ceptible cultivars would certainly be enhanced when combined with the resistant cultivars. Adkisson and Dyck [160] stated that resistant cultivars are highly desirable in a cultural con-trol systems designed to maintain pest numbers below the economic injury level (EIL) while preserving the natural enemies. Besides, even low level of resistance is important because of reduction of the need for other control measures in the crop production systems. Further-more, with low value crops, where chemical control is not economical, the use of HPR may be the only economic solution to a pest problem [46]. However, the most advantageous fea-tures of HPR are the following: (*a*) cheapest technology; (*b*) easiest to introduce; (*c*) is specify to one or several pests; (*d*) cumulative effectiveness makes high level of resistance unneces-sary; (*e*) is persistence; (*f*) can easily be adopted into normal farm operations; (*g*) is compati-ble with other control tactics in IPM such as chemical, biological and cultural control; (*H*) reducing the costs to the growers and (*I*) it is not detrimental to the environment [160, 161].

Generally, the phenomena of resistance are based on heritable traits. However, some traits fluctuate widely in different environmental conditions. Accordingly, plant resistance may be classified as genetic, implying the traits that are under the primary control of genetic factors; or ecological, implying the traits that are under the primary control of environmental factors. Host plants with genetic resistance to insect pest are very pleasure in IPM programmes [159]. This type of resistance is subdivided into two categories including induced and constitutive resistance. If biotic and abiotic environmental factors reduces insect fitness or negatively affects host selection processes, the effect is called induced resistance. On the other hand, constitutive resistance involves inherited characters whose expression, although influenced by the environment, is not triggered by environmental factors [41]. However, genetic resistance to insect pest could be results of three distinct mechanisms including antixenosis, antibiosis and tolerance. Antixenosis is the resistance mechanism employed by plant to deter or reduce colonization by insect. Antibiosis is the resistance mechanism that operates after the insect have colonized and started utilizing the plant. This mechanism could affect growth, development, reproduction and survivorship of insect pests and therefore, is the most important mechanism for IPM purposes. Tolerance is a characteristic of some plants that enable them to withstand or recover from insect damage [159].

Plant resistance to insect pests can be inherited in two distinct ways including vertical (monogenic) and horizontal (polygenic) resistance. Vertical resistance is generally controlled by a single gene, referred as R-gene. These R-genes can be remarkably effective in suppression of pest populations and can confer complete resistance. However, each R-gene confers resistance to only one insect pest and thus, depending on the pest species in specific area a cultivar may appear strongly resistant or completely susceptible. Horizontal resistance is also known as polygenic resistance due to this type of resistance is controlled by many genes. Unlike vertical resistance, horizontal resistance generally does not completely prevent a plant from becoming damaged. For insect pests, this type of resistance may slow the infection process so much that the pest does not grow well or spread to other plants. However, because of the large number of genes involved, it is much more difficult to breed cultivars with horizontal resistance to insect pests [162].

8.3.5.1. Plant resistance to H. armigera

To evaluate plant resistance to *H. armigera* several researchers evaluate population growth parameters of this pest on different host plants. Table 11 presents the main finding of several studies regarding to population growth parameters of this noctuid pest on different crop plants. However, information about population growth parameters of *H. armigera* on different host plants could reveal the suitability of one crop for this noctuid pest than other host plants.

In the case of *H. armigera* on different soybean cultivars, resistance of some cultivars to this noctuid pest was evaluated under laboratory conditions [6]. Results obtained by these researchers showed that various soybean cultivars differed greatly in suitability as diets for *H. armigera* when measured in terms of development, survivorship, life table parameters and nutritional indices. Fathipour and Naseri [11] presented detailed information regarding

evaluation of soybean resistance to *H. armigera* in a book chapter entitled " Soybean cultivars affecting performance of *Helicoverpa armigera* (Lepidoptera: Noctuidae) ".
However, a review of literature showed that a little information regarding resistance evaluation in field conditions is available and hence, for sustainable management of *H. armigera* in soybean cropping systems more attention should be devoted to fill this gap.

Crop	Experimental conditions		R_0 (female offspring)	r_m (day⁻¹)	T (day)	DT (day)	References
	Temperature °C	Diet type					
Canola (10) [a]	25	artificial [b]	157.4 - 331.5	0.153 - 0.179	31.10 - 36.10	3.80 - 4.50	[12]
Chickpea	25	artificial	359.67	0.161	33.28	4.27	[c]
Chickpea (4)	25	artificial	59.49 - 195.00	0.140 - 0.205	24.11 - 30.36	3.40 - 4.88	[d]
Common bean	27	leaf and fruit	19.50	-	-	-	[163]
Corn	25	artificial	203.14	0.130	40.56	5.29	[84]
Corn	25	artificial	147.40	0.126	37.90	5.62	[c]
Corn	27	leaf and fruit	44.50	-	-	-	[163]
Corn		cob	50.1	0.0853	46.6	-	[164]
Cotton	27	leaf and fruit	117.60	-	-	-	[163]
Cowpea	25	artificial	228.5	0.131	34.88	5.28	[e]
Cowpea	25	artificial	365.66	0.180	31.62	3.92	[c]
Cowpea	25	artificial	250.60	0.178	30.38	3.85	[d]
Hot pepper	27	leaf and fruit	5.10	-	-	-	[163]
Navy bean	25	artificial	294.28	0.164	32.31	4.14	[c]
Pearl millet	-	-	374.01	0.142	-	-	[165]
Soybean	25	artificial	239.69	0.161	33.28	4.23	[c]
Soybean (10)	25	artificial	16.00 - 270.00	0.084 - 0.114	36.72 - 45.28	6.08 - 8.10	[6]
Soybean (13)	25	leaf and pod	89.35 - 354.92	0.132 - 0.185	28.85 - 36.61	3.75 - 5.23	[166]
Sunflower	-	-	143.77	0.113	-	6.11	[167]
Tobacco	27	leaf	11.70	-	-	-	[163]
Tomato	27	leaf and fruit	9.5	-	-	-	[163]
Tomato (10)	25	leaf and fruit	1.36 - 62.32	0.008 - 0.137	30.26 - 37.34	5.06 - 27.41	[f]

[a] Digits in parentheses show number of tested cultivars.

[b] Artificial diet based on the seed of host plant.

[c] Baghery and Fathipour, unpublished data; [d] Fallahnejad-Mojarrad and Fathipour, unpublished data; [e] Adigozali and Fathipour, unpublished data; [f] Safuraie and Fathipour, unpublished data.

Table 11. Effects of different host plants on some population growth parameters of *Helicoverpa armigera*

8.3.5.2. Integration of HPR with other control measures and possible interactions

Several studies have been performed to investigate the possible interactions of host plant resistance to insect with other control measures. Results obtained revealed both incompatibility and compatibility of HPR in an integrated programme. However, in IPM programmes there can be three types of interactions between different control measures including additive, synergistic, and antagonistic. Additive interaction means the combined effect of two control measures is equal to the sum of the effect of the two measures taken separately. In synergistic interaction, the effect of two control measures taken together is greater than the sum of their separate effect. Finally, antagonistic interaction means that the effect of two control measures is actually less than the sum of their effects taken independently of each other. However, despite importance of such information in IPM, a little knowledge in this field is available.

8.3.5.2.1. HPR and biological control

Plant resistance and biological control are the key components of IPM for field crops and generally considered to be compatible. Insects feeding on HPR commonly experience retarded growth and an extended developmental period. Under field conditions, such poorly developed insect herbivores are more vulnerable to natural enemies for a longer period and the probability of their mortality is higher. Insect herbivores that develop slowly on resistant cultivars are more effectively regulated by the predators than those developed robustly on the susceptible cultivars. This is because the predator has to consume more small-sized prey to become satiated [168]. Wiseman *et al.* [169] found that populations of *Orius insidiosus* (Say), a predator on *H. zea* larvae, were higher on the resistant corn hybrids than on the susceptible ones, an indication of the compatibility of HPR and the predator.

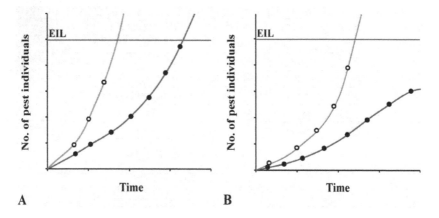

Figure 4. Influence of a low level of plant resistance to pest attack on the effectiveness of natural enemies. o, without predator; ●, with predator. Predator activity fails to exert economic control of insect pests on susceptible plants (A), whereas the same degree of predator activity exert economic control of the insect pest on plants with some degree of resistance (B) (after van Emden and Wearing [170]).

van Emden and Wearing [170] developed a simple model on the interaction of HPR with natural enemies. On the basis of this model, the reduced rate of multiplication of aphids on moderator resistant cultivars should magnify the plant resistance in the presence of natural enemies (Figure 4). Danks *et al.* [123] stated that a number of predators and parasitoids attack early instars of *Helicoverpa* sp. on soybean and tobacco but generally do not attack bigger larvae. But because of moderately resistance of host plants, the larvae remain in early instars for longer period and are more likely to be parasitized. However, such interactions are valuable phenomenon in the development of practical IPM.

However, there are instances of deleterious interactions between HPR and biological control which could be more important in the IPM. Sometimes, plant morphological traits and plant defense chemicals had adverse effects on the natural enemies. For example, certain genotypes of tobacco with glandular trichomes have been shown to severely limit the parasitization of the eggs of *Manduca sexta* (Linnaeus) by *Trichogramma minutum* Riley [171]. Resistant eggplant cultivars to *Tetranychus urticae* Koch adversely affected biological performance of *Typhlodromus bagdasarjani* Wainstein and Arutunjan [22]. These researchers stated that antibiotic compounds in resistant cultivars are also toxic for *T. bagdasarjani* and concentrated compounds in the *T. urticae* reduced effectiveness of this predator. Barbour *et al.* [172] found that methylketone adversely affected the egg predators of *H. zea* that fed on the foliage of wild tomato. However, plant breeders can sometimes manipulate plant traits to promote the effectiveness of natural enemies [159]. For example, a reduction in trichomes density of cucumber leaves significantly increase effectiveness of *Encarsia formosa* Gahan on the greenhouse whitefly [173].

8.3.5.2.2. HPR and insect pathogens

Schultz [174] hypothesized that the effectiveness of insect pathogens may be reduced or improved, depending upon plant chemistry and variability of plant resistance. Interactions among HPR, herbivores and their pathogens can alter pathogenicity of *B. thuringiensis* on *M. sexta* [175]. Furthermore, insect susceptibility to the entomopathogenic fungus can also be affected by HPR. Felton and Duffey [176] reported the possible incompatibility of resistant cultivars of tomato with NPV control of *H. zea*. These researchers revealed that chlorogenic acid in resistant cultivars of tomato is oxidized by foliar phenol oxidases and generated components binds to the occlusion bodies of NPV, thereby decreasing its pathogenicity against *H. zea*.

8.3.5.2.3. HPR and chemical control

There is usually a beneficial interaction between HPR and chemical control. Because the toxicity of an insecticide is a function of insect bodyweight, it is expected that a lower concentration is needed to control insect feeding on a resistant cultivar than those feeding on a susceptible ones [177]. van Emden [178] pointed out that there is a potentially useful interaction in the possibility of using reduced dosses of insecticide on resistant cultivar, when spray is needed. This theory relies on the selectivity of the insecticide in favor of natural enemies as dose rate is reduced (Figure 5). However, it appears that even in the presence of small levels of plant resistance, insecticide concentration can be reduced to one-third of that required on a susceptible cultivar [178]. This reduced use of pesticide not only benefits the agroecosystems and natural enemies but also results in lower pesticide residues in the human food chain. Accordingly, Wiseman *et al.* [179] showed that even one low-dose application of insecticide to the resistant hybrid of corn gave an *H. zea* control equal to that achieved with seven applications to the susceptible hybrid. Fathipour *et al.* [25] compared the chemical control of *Eurygaster integriceps* Put. on resistant and susceptible cultivars of wheat. The results obtained by these researchers revealed that the sensitivity of 4th and 5th instar nymphs and new adults of *E. integriceps* to insecticide Fenitrothion was enhanced on resistant cultivar compared with those on susceptible cultivar. Accordingly, the LC_{50} of insecticide on susceptible and resistant cultivars for 4th instar nymphs was 42.16 and 33.48, for 5th instar nymphs was 147.03 and 114.01 and for new adults was 303.35 and 227.88 ppm, respectively.

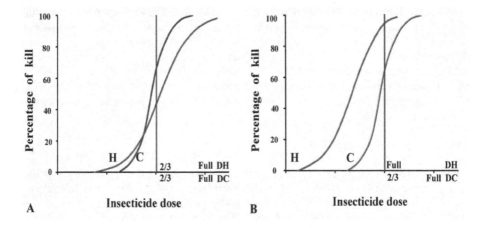

Figure 5. Effect of plant resistance on the selectivity of an insecticide. A: susceptible cultivar; B: resistant cultivar where dose for the herbivore can be reduced by one third; C: dose mortality curve for carnivore; H: dose mortality curve for herbivore; DC: dose scale for carnivore; DH: dose scale for herbivore (after van Emden, [178]).

However, in addition to negative effects on insect bodyweight, repellent chemicals or morphological traits of resistant cultivars may be effective in reduction of insecticide spray. Repellency is to be effective in limiting pest damage to treated crops and it may also keep the pests away from their suitable resources and therefore cause indirect mortality or lower fecundity. Furthermore, repellency is equivalent to using low doses of insecticides along with the repellent properties of the host plant [159].

8.4. Semiochemicals and their possible use in suppression of *H. armigera* populations

Many insects and other arthropods rely on chemical messages to communicate with each other or to find suitable hosts. Chemical messages that trigger various behavioral response are collectively referred to as semiochemicals. Generally, semiochemicals is subdivided into two distinct groups including pheromones and allelochemicals (Table 12). The term pheromone is used to describe compounds that operate intraspecifically, while allelochemical is the general term for an interspecific effector [26]. However, the realization that behaviors critical to insect survival were strongly influenced by semiochemicals rapidly led to proposals for using these agents as practical tools for pest suppression [180].

	Pheromones [a]	Sex ph.	A volatile chemical substance produced by one sex of an insect which produces some specific reaction in the opposite sex.
		Aggregation ph.	Also known as arrestants. These are chemicals that cause insects to aggregate or congregate.
		Alarm ph.	A substance produced by an insect to repel and disperse other insects in the area.
		Trail ph.	A substance laid down in the form of a trail by one insect and followed by another member of the same species.
		Host-marking ph.	A substance placed inside/outside of the host body at the time of oviposition to distinguish unparasitized from parasitized hosts.
		Caste-regulating ph.	A substance used by social insects to control the development of individuals in a colony.
Semiochemicals	Allelochemicals [b]	Allomone	A substance produced by a living organism that evokes in receiver a behavioral or physiological reaction that is adaptively favorable to the sender.
		Kairomone	A substance produced by a living organism that evokes in receiver a behavioral or physiological reaction that is adaptively favorable to the receiver.
		Synomone	A substance produced by an organism that evokes in the receiver a behavioral or physiological reaction that is adaptively favorable to both sender and receiver.
		Antimone	A substance produced by an organism that evokes in the receiver a behavioral or physiological reaction that activates a repellent response to the sender and is unfavorable to both sender and receiver.
		Apneumone	A substance emitted by a nonliving material that evokes a behavioral or physiological reaction that is adaptively favorable to a receiving organism but detrimental to an organism of another species that may be found in or on the nonliving material.

[a] classified according to function

[b] classified according to the advantage to receiver or sender

Table 12. Classification of behavior-modifying chemicals (semiochemicals)

As discussed in previous section (see section 6) and in addition to using the pheromones of *Helicoverpa* spp. for essential monitoring of infested areas, these compounds have been shown to be useful for suppression of *Helicoverpa* infestation. Attractant-baited lures form the basis for three direct control measures: (1) mass trapping of male, (2) attract-and-kill strategy and (3) mating disruption via permeating the atmosphere of crop environments with sex pheromones. Potential of synthesized pheromones for mass trapping of *H. armigera* was investigated by several researchers. According to Pawar *et al.* [181], *H. armigera* will readily respond to synthesized pheromones and traps are capable of capturing hundreds of male moths per trap per night. The same results were reported by Reddy and Manjunath [18]. Attract-and-kill is a promising new strategy that involves an attractant such as a pheromone and a toxicant. Unlike mating disruption, which functions by confusing the insect, this strategy attracts the insect to a pesticide laden gel matrix and kills them. This strategy has been successfully used on several lepidopteran species [18] but no information is available in the case of *Helicoverpa* spp. However, the most developed tactic is mating disruption. This approach entails releasing large amounts of synthetic sex pheromone into the atmosphere of a crop to interfere with mate-finding, thereby controlling the pest by curtailing the reproductive phase of its life cycle. Mating disruption through the use of some synthesized pheromone such as (Z)-9-tetradecen-1-ol for air permeation is a potentially valuable development in integrated management of *Helicoverpa* spp. It has been shown to be very effective with *H. zea* and *H. virescens* and should certainly be pursued for the same purpose with *H. armigera* [182].

9. Biotechnology in IPM

Recent advantage in biotechnology, particularly cellular and molecular biology have opened new avenues for developing resistant cultivars. From this diagnostic perspective, molecular techniques are likely to play an important role in identification, quantification and genetic monitoring of pest populations [183]. The diagnostic information is a necessary prerequisite for implementing rational control strategy. Appropriate molecular techniques can be employed to study the species composition of the pest population and to identify strains, races or biotypes of the same species.

Another important application of molecular diagnostic techniques is for monitoring both the presence and frequency of genes of particular interest. For example, genes for resistance to a specific class of pesticides and their frequency in particular region can be assessed. Such information is very useful for designing and implementing rational pest management strategies [159].

The most important application of biotechnology in IPM is the introduction of novel genes for resistance into crop cultivars through genetic engineering. HPR is a highly effective management option, but cultivated germplasm has only low to moderate resistance levels to some key pests. Furthermore, some sources of resistance have poor agronomic characteristics. On the other hand, development of cultivars with enhanced resistance will strengthen the control of *H. armigera* in different cropping systems. Therefore, we need to make a concerted effort to transfer pest resistance into genotypes with desirable agronomic and grain characteristics. Recent achievements of genetics and molecular biology have been widely implemented into breeding new crop cultivars and brought in many various traits absent from parent species and cultivars. Furthermore, new progress in biotechnology makes it feasible to transfer genes from totally unrelated organisms, breaking species barriers not possible by conventional genetic enhancement. Today, transgenic plants expressing insecticidal proteins from the bacterium *B. thuringiensis*, are revolutionizing agriculture. *Bacillus thuringiensis* has become a major insecticide because genes that produce *B. thuringiensis* toxins have been engineered into major crops grown on 11.4 million ha worldwide (including soybean, cotton, peanut, tomato, tobacco, corn and canola). These crops have shown positive economic benefits to growers and reduced the use of other insecticides. Genetically engineered cottons expressing delta-endotoxin genes from *B. thuringiensis* offer great potential to dramatically reduce pesticide dependence for control of *Helicoverpa* spp. and consequently offer real opportunities as a component of sustainable and environmentally acceptable IPM systems [16]. Certainly, for sustainable management of *H. armigera* in soybean cropping systems, such soybean resistant cultivars could play pivotal role. Therefore, to achieve this goal, much works should be conducted in breeding new soybean cultivars expressing *Bt* toxins against *H. armigera*.

The potential ecological and human health consequences of *Bt* crops, including effects on nontarget organisms, food safety, and the development of resistant insect populations, are being compared for *Bt* plants and alternative insect management strategies. However, *Bt* plants were deployed with the expectation that the risks would be lower than current or alternative technologies and that the benefits would be greater. Based on the data to date, these expectations seem valid [16]. The major challenge to sustainable use of transgenic *Bt* crops is the risk that target pests may evolve resistance to the *B. thuringiensis* toxins. *Helicoverpa armigera* is a particular resistance risk having consistently developed resistance to synthetic pesticides in the past [21]. For this reason a pre-emptive resistance management strategy was implemented to accompany the commercial release of transgenic cultivars. The strategy, based on the use of structured refuges to maintain susceptible individuals in the population, seeks to take advantage of the polyphagy and local mobility of *H. armigera* to achieve resistance management by utilizing gene flow to counter selection in transgenic crops. However, refuge crops cannot be treated with Bt sprays, and must be in close proximity to the transgenic crops (within 2 km) to maximize the chance of random mating among sub-populations [184].

10. Tritrophic interactions and its manipulation for IPM

Plant quality can affect herbivore fitness directly (as food of herbivores) and indirectly (by affecting foraging cues for natural enemies) [12, 23]. Until recently, there has been a tendency by those involved in IPM to be principally concerned with effects on herbivores or interactions between just two trophic levels [185]. However, interest in the importance of interactions among the three or four trophic levels (Figure 6) that characterize most natural systems and agroecosystems has been increased rapidly during the last two decades [26].

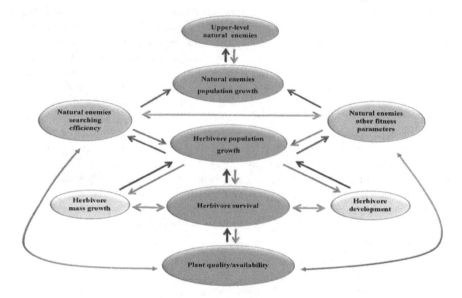

Figure 6. Simple diagram of multitrophic interactions representing some important causal relationships among the trophic levels mediated by some important insect fitness parameters

It is interesting that many traditional cultural practices exert their effects through complex multitrophic interactions, but it is exactly this complexity that makes such systems difficult to assess experimentally or validate conclusively across a broad range of environments. For example, it has been demonstrated that toxic secondary compounds in an herbivore diet may affect development, survivorship, morphology and size of its natural enemies. This effect of poor-quality plants can thus indirectly lead to poor-quality natural enemies [186].

As knowledge of interactions across multitrophic systems both in nature and in agroecosystems expands, researchers and pest management practitioners are beginning to find ways of manipulating interactions across different trophic levels in order to develop more sustainable approaches to pest management. Accordingly, population ecologists are actively debating the relative importance of bottom-up (resource-driven) and top-down (natural enemy-driven) processes in the regulation of herbivores populations [22, 187, 188]. However, there are a number of key areas where manipulation of host plant-pest-natural enemy interactions could provide substantial benefits in pest management systems (manipulation of host plant quality, allelochemicals and crop diversification and genetic manipulation of insect) [26].

For many years, there was a widely held view that HPR should be seen as an integral component of IPM programmes, but it has been demonstrated that HPR is by no means always compatible with biological control [178]. The significant and growing evidence from fundamental research in allelochemically mediated interactions hold substantial promise with regard to the development of novel IPM techniques. Allelochemicals mediated interactions in insect-host plant relationship have been recognized as the most important factors in the successful establishment of an insect species on a crop [189]. Furthermore, allelochemicals produced by plants also have considerable influence on the prey/host selection behavior of natural enemies, so that plants, herbivores, and natural enemies are interconnected through the well-knit array of chemicals. The host plant volatiles play a key role in attracting/repelling or retaining the natural enemies, thereby causing considerable changes in pest populations on different plant cultivars [190]. Hare [191] cited 16 studies where interactions between resistant cultivars and natural enemies (parasitoids) were studied and the outcomes show a spectrum of interactions, ranging from synergistic, to additive, to none apparent through to disruptive or antagonistic. Negative interactions can occur due to the presence of secondary chemicals that are ingested or sequestered by natural enemies feeding on hosts present on resistant or partially resistant plants [192]. For example, specific toxic components in partially resistant soybean plants can be particularly problematic in this regard [193]. In addition to allelochemicals, morphological traits of host plants such as trichome density and color complexion can affect insect fitness and effectiveness of its natural enemies. It was observed that plant cultivars were sufficiently differing in their trichome density and color complexion which were considered as main resource of variations in rate of parasitism on different plant cultivars. Cotton cultivars with low density of hairs on the upper leaf surface and high hair density on the lower leaf surface help in reduction of pest incidence [194]. The rates of parasitism were negatively associated with trichome density as revealed by Mohite and Uthamasamy [195]. In another study, Asifulla *et al.* [196] noticed higher parasitism by *T. chilonis* on *H. armigera* eggs in glabrous cotton species compared to hairy types.

In conclusion, as a novel strategy for IPM programmes, well understanding of multitrophic interactions is critical to develop the sustainable, less pesticide-dependent or pesticide-free pest management programmes [197]. In the interest of agricultural sustainability, tritrophic manipulation, as a distinct approach to biological or cultural control, is probably to be prioritized increasingly by both researchers and those responsible for the development and practical implementation of pest management programmes. This process will be facilitated if improvements in the understanding of crop-pest-natural enemy evolution and their interactions are achieved [26, 197]. Information in this regard are essential in finding out what role the plant play in supporting the action of natural enemies and how this role could be manipulated reserving the natural enemies.

11. Conclusion

Helicoverpa armigera represents a significant challenge to soybean cropping systems in many parts of the world and remain the target for concentrated management with synthetic insecticides. However, the extensive use of insecticides for combating *H. armigera* populations is of environmental concern and has repeatedly led to the development of resistance in this pest as well as the deleterious effects on nontarget organisms and environment. The common trend towards reducing reliance on chemicals for control of insect pests in agriculture renewed worldwide interest in integrated pest management (IPM) programmes and it seems that in most areas the aim must be integrated management, particularly on crops such as soybean where *H. armigera* is part of a diverse pest complex. Accordingly, in this chapter we attempt to introduce basic elements for implementation of sustainable management of *H. armigera*. For this, we reviewed the main findings of different researchers and in some cases present our data. However, our findings revealed that for successful management of *H. armigera*, more attention should be devoted to some basic information such as monitoring efforts, forecasting activities and economic thresholds. In addition, more studies are needed to evaluate potential of novel control measures including selective insecticides and sublethal doses, HPR and genetically modified soybean cultivars and microbial pathogens (especially commercial formulations of *B. thuringiensis* and NPV) for control of this noctuid pest. However, for future outlook of integrated management of *H. armigera* in soybean cropping systems, the development and use of resistant cultivars will play a crucial role. In other words, more works should be conducted to evaluate resistance of soybean cultivars to *H. armigera* in field conditions. Moreover, a further need is to evaluate tritrophic interactions among the soybean cultivars, *H. armigera* and its natural enemies and new studies should be included to evaluate such interactions. However, the information gathered in the current chapter could be valuable for integrated management of *H. armigera* in soybean cropping systems.

Author details

Yaghoub Fathipour* and Amin Sedaratian

*Address all correspondence to: fathi@modares.ac.ir

Department of Entomology, Faculty of Agriculture, Tarbiat Modares University, Tehran, Iran

References

[1] Sharma HC. Integrated Pest Management Research at ICRISAT: Present Status and Future Priorities. Andhra Pradesh: International Crops Research Institute for the Semi-Arid Tropics; 2006.

[2] Naseri B, Fathipour Y, Moharramipour S, Hosseininaveh V. Comparative life history and fecundity of *Helicoverpa armigera* (Lepidoptera: Noctuidae) on different soybean varieties. Entomological Science 2009; 12 147-154.

[3] Sedaratian A, Fathipour Y, Moharramipour S. Evaluation of resistance in 14 soybean genotypes to *Tetranychus urticae* (Acari: Tetranychidae). Journal of Pest Science 2009; 82 163-170.

[4] Naseri B, Fathipour Y, Moharramipour S, Hosseininaveh V, Gatehouse AM. Digestive proteolytic and amylolytic activities of *Helicoverpa armigera*in response to feeding on different soybean cultivars. Pest Manageent Science 2010; 66 1316-1323.

[5] Sedaratian A, Fathipour Y, Talebi AA, Farahani S. Population density and spatial distribution pattern of *Thrips tabaci* (Thysanoptera: Thripidae) on different soybean varieties. Journal of Agricultural Science and Technology 2010; 12 275-288.

[6] Soleimannejad S, Fathipour Y, Moharramipour S, Zalucki MP. Evaluation of potential resistance in seeds of different soybean cultivars to *Helicoverpa armigera* (Lepidoptera: Noctuidae) using demographic parameters and nutritional indices. Journal of Economic Entomology 2010; 103 1420-1430.

[7] Sedaratian A, Fathipour Y, Moharramipour S. Comparative life table analysis of *Tetranychus urticae* (Acari: Tetranychidae) on 14 soybean genotypes. Insect Science 2011; 18 541-553.

[8] Mehrkhou F, Talebi AA, Moharramipour S, Hosseininaveh V. Demographic parameters of *Spodoptera exigua* (Lepidoptera: Noctuidae) on different soybean cultivars. Environmental Entomology 2012; 41 326-332.

[9] Taghizadeh R, Talebi AA, Fathipour Y, Khalghani J. Effect of ten soybean cultivars on development and reproduction of lima bean pod borer, *Etiella zinckenella* (Lepidoptera: Pyralidae) under laboratory conditions. Applied Entomology and Phytopathology 2012; 79 15-28.

[10] Zalucki MP, Murray DAH, Gregg PC, Fitt GP, Twine PH, Jones C. Ecology of *Helicoverpa armigera* (Hübner) and *H. punctigera* (Wallengren) in the inland of Australia: larval sampling and host plant relationships during winter and spring. Australian Journal of Zoology 1994; 42 329-346.

[11] Fathipour Y, Naseri B. Soybean Cultivars Affecting Performance of *Helicoverpa armigera* (Lepidoptera: Noctuidae). In: Ng TB. (ed.) Soybean - Biochemistry, Chemistry and Physiology. Rijeka: InTech; 2011. p599-630.

[12] Karimi S, Fathipour Y, Talebi AA, Naseri B. Evaluation of canola cultivars for resistance to *Helicoverpa armigera* (Lepidoptera: Noctuidae) using demographic parameters. Journal of Economic Entomology 2012 [in press].

[13] Reed W, Pawar CS. *Heliothis*: A Global Problem. In: Reed W, Kumble V. (eds.) Proceedings of the International Workshop on *Heliothis* Management, 15-20 November 1981, Patancheru, India. International Crops Research Institute for the Semi-Arid Tropics; 1982. p9-14.

[14] Fitt GP. The ecology of *Heliothis* in relation to agroecosystems. Annual Review of Entomology 1989; 34 17-52.

[15] Kogan M. Integrated pest managements: historical perspectives and contemporary developments. Annual Review of Entomology 1998; 43 243-270.

[16] Shelton AM, Zhao JZ, Roush RT. Economic, ecological, food safety, and social consequences of the deployment of Bt transgenic plants. Annual Review of Entomology 2002; 47 845-881.

[17] Abdi-Bastami F, Fathipour Y, Talebi AA. Comparison of life table parameters of three populations of braconid wasp, *Habrobracon hebetor* Say (Hym.: Braconidae) on *Ephestia kuehniella* Zell (Lep.: Pyralidae) in laboratory conditions. Applied Entomology and Phytopathology 2011; 78: 131-152.

[18] Reddy GVP, Manjunatha M. Laboratory and field studies on the integrated pest management of *Helicoverpa armigera* (Hübner) in cotton, based on pheromone trap catch threshold level. Journal of Applied Entomology 2000; 124 213-221.

[19] Liao C, Heckel DG, Akhursta R. Toxicity of *Bacillus thuringiensis* insecticidal proteins for *Helicoverpa armigera* and *Helicoverpa punctigera* (Lepidoptera: Noctuidae), major pests of cotton. Journal of Invertebrate Pathology 2002; 80 55-63.

[20] Jallow MFA, Cunningham JP, Zalucki MP. Intra-specific variation for host plant use in *Helicoverpa armigera* (Hübner) (Lepidoptera: Noctuidae): implications for management. Crop Protection 2004; 23 955-964.

[21] Rafiee-Dastjerdi H, Hejazi MJ, Nouri-Ganbalani G, Saber M. Toxicity of some biorational and conventional insecticides to cotton bollworm, *Helicoverpa armigera* (Lepidoptera: Noctuidae) and its ectoparasitoid, *Habrobracon hebetor* (Hymenoptera: Braconidae). Journal of Entomological Society of Iran 2008; 28 27-37.

[22] Khanamani M, Fathipour Y, Hajiqanbar H, Sedaratian A. Antibiotic resistance of eggplant to two-spotted spider mite affecting consumption and life table parameters of *Typhlodromus bagdasarjani* (Acari: Phytoseiidae). Experimental and Applied Acarology 2012 [in press].

[23] Soufbaf M, Fathipour Y, Hui C, Karimzadeh J. Effects of plant availability and habitat size on the coexistence of two competing parasitoids in a tri-trophic food web of canola, diamondback moth and parasitic wasps. Ecological Modelling 2012; 244 49-56.

[24] Ranjbar-Aghdam H, Fathipour Y. Physiological time model for predicting the codling moth (Lepidoptera: Tortricidae) phenological events by using hourly recorded environmental temperature. Journal of Economic Entomology 2012 [in press].

[25] Fathipour Y, Kamali K, Abdollahi G, Talebi AA, Moharramipour S. Integrating wheat cultivar and fenitrothion for control of sunn pest (*Eurygaster integriceps* Put.). Seed and Plant 2003; 19 245-261.

[26] Koul O, Dhaliwal GS, Cuperus GW. Integrated Pest Management: Potential, Constraints and Challenges. Wallingford: CAB International; 2004.

[27] Gaines JC. Cotton insects and their control in the United States. Annual Review of Entomology 1957; 2 319-338.

[28] Hoskins WM, Borden AD, Michelbacher AE. Recommendations for a more discriminating use of insecticides. Proceedings of the Sixth Pacific Science Congress of the Pacific Science Association 1939; 5 119-123.

[29] Smith RF, Allen WW. Insect control and the balance of nature. Scientific American 1954; 190 38-92.

[30] Rabb RL, Guthrie FE. Concepts of Pest Management. Raleigh, North Carolina: North Carolina State University Press; 1970.

[31] Waterhouse DL. Some aspects of Australian entomological research. Proceeding of 12th International Congress of Entomology. London 1965.

[32] Geier PW. Management of insect pests. Annual Review of Entomology 1966; 11 471-490.

[33] Zadoks JC. On the conceptual basis of crop loss assessment: the threshold theory. Annual Review of Phytopathology 1985; 23 455-473.

[34] CEQ. Integrated Pest Management. Washington DC: Council on Environmental Quality;1972.

[35] FAO. Intercountry Programme for the Development and Application of Integrated Pest Control in Rice in South and South-East Asia. Rome: Food and Agriculture Organization; 1995.

[36] NAS. Principles of Plant and Animal Pest Control: Insect Management and Control. Washington DC: National Academy of Sciences; 1969.

[37] Smith RF. History and Complexity of Integrated Pest Management. In: Smith EH, Pimetel D. (eds.) Pest Control Strategies. New York: Academic Press; 1978. p41-53.

[38] Frisbie RE, Adkisson PC. Integrated Pest Management on Major Agricultural Systems. Texas: Texas A & M University; 1985.

[39] Allen WA, Rajotte EG. The changing role of extension entomology in the IPM era. Annual Review of Entomology 1990; 25 379-397.

[40] Pedigo LP. Entomology and Pest Management. New York: Macmillan; 1991.

[41] Metcalf RL, Luckmann WH. Introduction to Insect Pest Management. New York: John Wiley and Sons; 1994.

[42] USDA. Managing Cover Crops Profitably. Washington DC: United State Department of Agriculture Division of Entomology Bulletin; 1998.

[43] Dhaliwal GS, Arora R. Integrated Pest Management: Concepts and Approaches. New Delhi: Kalyani Publishers; 2001.

[44] Spedding CRW. An Introduction to Agricultural Systems. London: Elsevier Applied Science; 1988.

[45] Dent D. Integrated Pest Management. London: Chapman and Hall; 1995.

[46] Matthews GA. Pest Management. New York: Longman; 1984.

[47] Reichelderfer KH, Carlson GA, Norton GA. Economic Guidelines for Crop Pest Control. Rome: Food and Agriculture Organization; 1984.

[48] Pedigo LP, Hutchins SH, Higley LG. Economic injury levels in theory and practice. Annual Review of Entomology 1996; 31 341-358.

[49] Stem VM, Smith RF, van den Bosch R, Hagen KS. The integrated control concept. Hilgardia 1959; 29 81-101.

[50] Davidson A, Norgaard RB. Economic aspects of pest control. European Plant Protection Organization Bulletin 1973; 3:63-75.

[51] Odak SK, Thakur BS. Preliminary Studies on the Economic Threshold of Gram Pod Borer *Heltothis armigera* (Hübner) on Gram. All India Workshops on Rabi Pulses. Hydrabad, India; 1975.

[52] Singh BR, Reddy AR. Studies on Minimum Population Level of Gram Pod Borer which Caused Economic Damage to Gangal Gram Crop. Report on All India Rabi Pulses Workshops. Varanasi, India; 1976.

[53] Patel AJ. Estimation of economic injury level and economic threshold on level for *Helicoverpa armigera* on gram crop. Gujarat Agricultural University Research Journal 1994; 20 88-92.

[54] Sekhar PR, Rao NV, Venkataiah M, Rajasri M. Sequential sampling plan of gram pod borer *Helicoverpa armigera* in chickpea. Indian Journal of Pulses Research 1994; 7 153-157.

[55] Venkataiah M, Sekhar PR, Rao NV, Singh TVK, Rajastri M. Distribution pattern and sequential sampling of pod borer, *Heliothis armigera* in pigeonpea. Indian Journal of Pulses Research 1994; 7 158-161.

[56] Nath P, Rai R. Study of the bioecology and economic injury levels of *Helicoverpa armigera* infesting gram crop. Proceeding of national seminar on international pest management (IPM) in agriculture, 19-30 December 1995, Nagpur, India; 1995.

[57] Whitman JA, Anders MM, Row VR, Reddy LM. Management of *Helicoverpa armigera* (Lepidoptera: Noctuidae) on chickpea in South India: thresholds and economics of host plant resistance and insecticide application. Crop Protection 1995; 437- 446.

[58] Zahid MA, Islam MM, Reza MH, Prodhan MHZ, Begum MR. Determination of economic injury levels of *Helicoverpa armigera* (Hübner) chickpea. Bangladesh Journal of Agricultural Research 2008; 33 555-563.

[59] Reddy CN, Singh Y, Singh VS. Economic injury level of gram pod borer (*Helicoverpa armigera*) on pegionpea. Indian Journal of Entomology 2001; 63 381-387.

[60] Cameron PJ, Walker GP, Herman TJ, Wallace AR. Development of economic thresholds and monitoring systems for *Helicoverpa armigera* (Lepidoptera: Noctuidae) in tomatoes. Journal of Economic Entomology 2001; 94 1104-1012.

[61] Alavi1 J, Gholizadeh M. Estimation of economic injury level (EIL) of cotton bollworm *Helicoverpa armigera* Hb. (Lep., Noctuidae) on cotton. Journal of Entomological Research 2010; 2 203-212.

[62] Brier H, Quade A, Wessels J. Economic thresholds for *Helicoverpa* and other pests in summer pulses-challenging our perceptions of pest damage. Proceedings of the 1st Australian summer grains conference, 21- 24 June 2010, Australia, Gold Coast; 2010.

[63] Rogers DJ, Brier HB. Pest-damage relationships for *Helicoverpa armigera* (Hübner) (Lepidoptera: Noctuidae) on vegetative soybean. Crop Protection 2010; 29 39-46.

[64] Nyambo BT. Problems and Progress in *Heliothis* Management in Tanzania, with Special Reference to Cotton. In: Reed W, Kumble V. (eds.) Proceedings of the International Workshop on *Heliothis* Management, 15-20 November 1981, Patancheru, India. International Crops Research Institute for the Semi-Arid Tropics; 1982. p355-362.

[65] Hartstack AW, Hollingsworth JP, Ridgway RL, Coppedge JR. A population dynamics study of the bollworm and the tobacco budworm with light traps. Environmental Entomology 1973; 2 244-252.

[66] Walden HH. Owlet Moths (Phalaenidae) Taken at Light Traps in Kansas and Nebraska. Washington: United State Department of Agriculture Division of Entomology Bulletin; 1942.

[67] Beckham CM. Seasonal abundance of *Heliothis* spp. in the Georgia Piedmont. Journal of the Georgia Entomological Society 1970; 5 138-142.

[68] Klun JA, Plimmer JR, Bierl-Leonhardt BA, Sparks AN, Chapman OL. Trace chemicals: the essence of sexual communication systems in *Heliothis* species. Science 1979; 204 1328-1330.

[69] Roach SH. *Heliothis zea* and *H. virescens* moth activity as measured by black light and pheromone traps. Journal of Economic Entomology 1975; 68 17-21.

[70] Torres-Vilaa LM, Rodrıguez-Molinab MC, Lacasa-Plasenciac A. Testing IPM protocols for *Helicoverpa armigera* in processing tomato: egg-count- vs. fruit-count-based damage thresholds using Bt or chemical insecticides. Crop Protection 2003; 22 1045-1052.

[71] van Hamburg H. The inadequacy of egg counts as indicators of threshold levels for control of *Heliothis armigera* on cotton. Journal of the Entomological Society of South Africa 1981; 44 289-295.

[72] Zalom FG, Wilson LT, Hoffmann MP. Impact of feeding by tomato fruitworm, *Heliothis zea* (Boddie) (Lepidoptera: Noctuidae), and beet armyworm, *Spodoptera exigua* (Hübner) (Lepidoptera: Noctuidae), on processing tomato fruit quality. Journal of Economic Entomology 1986; 79 822-826.

[73] Soufbaf M, Fathipour Y, Zalucki MP, Hui C. Importance of primary metabolites in canola in mediating interactions between a specialist leaf-feeding insect and its specialist solitary endoparasitoid. Arthropod-Plant Interactions 2012; 6 241-250.

[74] Ranjbar-Aghdam H, Fathipour Y, Radjabi G, Rezapanah M. Temperature-dependent development and temperature thresholds of codling moth (Lepidoptera: Tortricidae) in Iran. Environmental Entomology 2009; 38 885-895.

[75] Huffaker C, Berryman A, Turchin P. Dynamics and Regulation of Insect Populations. In: Huffaker CB, Gutierrez AP. (eds.) Ecological Entomology. New York: Wiley; 1999. p269-305.

[76] Taghizadeh R, Fathipour Y, Kamali K. Influence of temperature on life-table parameters of *Stethorus gilvifrons* (Mulsant) (Coleoptera: Coccinellidae) fed on *Tetranychus urticae* Koch. Journal of Applied Entomology 2008; 132 638-645.

[77] Zahiri B, Fathipour Y, Khanjani M, Moharramipour S, Zalucki MP. Preimaginal development response to constant temperatures in *Hypera postica* (Coleoptera: Curculionidae): picking the best model. Environmental Entomology 2010; 39 177-189.

[78] Price PW. Insect Ecology. New York: Wiley; 1997.

[79] Wagner TL, Wu HI, Sharpe PJH, Schoolfield RM, Coulson RN. Modeling insect developments rates: a literature review and application of a biophysical model. Annals of the Entomological Society of America 1984; 77 208-225.

[80] Howell JF, Neven LG. Physiological development time and zero development temperature of the codling moth (Lepidoptera: Tortricidae). Environmental Entomology 2000; 29 766-772.

[81] Qureshi MH, Murai T, Yoshida H, Shiraga T, Tsumuki H. Effects of photoperiod and temperature on development and diapause induction in the Okayama population of *Helicoverpa armigera* (Hb.) (Lepidoptera: Noctuidae). Applied Entomology and Zoology 1999; 34 327-331.

[82] Jallow MFA, Matsumura M. Influence of temperature on the rate of development of *Helicoverpa armigera* (Hübner) (Lepidoptera: Noctuidae). Applied Entomology and Zoology 2001; 36 427-430.

[83] Mironidis GK, Savopoulou-Soultani M. Development, survivorship and reproduction of *Helicoverpa armigera* (Lepidoptera: Noctuidae) under constant and alternating temperatures. Environmental Entomology 2008; 37 16-28.

[84] Fitt GP. An Australian approach to IPM in cotton: integrating new technologies to minimize insecticide dependence. Crop Protection 2000; 19 793- 800.

[85] Mahdavi V, Saber M, Rafiee-Dastjerdi H, Mehrvar A. Comparative study of the population level effects of carbaryl and abamectin on larval ectoparasitoid *Habrobracon hebetor* Say (Hymenoptera: Braconidae) BioControl 2011; 56 823-830.

[86] Avilla C, Gonzalez-Zamora JE. Monitoring resistance of *Helicoverpa armigera* to different insecticides used in cotton in Spain. Crop Protection 2010; 29 100-103.

[87] Babariya PM, Kabaria BB, Patel VN, Joshi MD. Chemical control of gram pod bore *Helicoverpa armigera* Hübner infesting pigeonpea. Legume Research 2010; 33 224-226.

[88] Daly JC, Hokkanen HMT, Deacon J. Ecology and resistance management for *Bacillus thuringiensis* transgenic plants. Biocontrol Science and Technology 1994; 4 563-571.

[89] Mabett TH, Dareepat P, Nachapong M. Behavior studies on *Heliothis armigera* and their application to scouting techniques for cotton in Thailand. Tropical Pest Management 1980; 26 268-273.

[90] Kohli A. The Likely Impact of Synthetic Prethroids on *Heliothis* Management. In: Reed W, Kumble V. (eds.) Proceedings of the International Workshop on *Heliothis* Management, 15-20 November 1981, Patancheru, India. International Crops Research Institute for the Semi-Arid Tropics; 1982. p197-204.

[91] Gentz MC, Murdoch G, King GF. Tandem use of selective insecticides and natural enemies for effective, reduced-risk pest management. Biological Control 2010; 52 208-215.

[92] Saber M, Hejazi MJ, Kamali K, Moharramipour S. Lethal and sublethal effects of fenitrothion and deltamethrin residues on the egg parasitoid *Trissolcus grandis* (Hymenoptera: Scelionidae). Journal of Economic Entomology 2005; 98 35-40.

[93] Hamedi N, Fathipour Y, Saber M, Sheikhi-Garjan A. Sublethal effects of two common acaricides on the consumption of *Tetranychus urticae* (Prostigmata: Tetranychidae) by *Phytoseius plumifer* (Mesostigmata: Phytoseiidae). Systematic and Applied Acarology 2009; 14 197-205.

[94] Issa GI, Elbanhawy EM, Rasmy AH. Successive release of the predatory mite *Phytoseius plumifer* for combating *Tetranychus arabicus* (Acarina) on fig seedlings. Zeitschrift für Angewandte Entomologie 1974; 76 442-444.

[95] Hamedi N, Fathipour Y, Saber M. Sublethal effects of fenpyroximate on life table parameters of the predatory mite *Phytoseius plumifer*. BioControl 2010; 55 271-278.

[96] Desneux N, Decourtye A, Delpuech J. The sublethal effects of pesticides on beneficial arthropods. Annual Review of Entomology 2007; 52 81-106.

[97] Hamedi N, Fathipour Y, Saber M. Sublethal effects of abamectin on the biological performance of the predatory mite, *Phytoseius plumifer* (Acari: Phytoseiidae). Experimental and Applied Acarology 2011; 53 29-40.

[98] Haghani M, Fathipour Y. The effect of the type of laboratory host on the population growth parameters of *Trichogramma embryophagum* Hartig (Hym., Trichogrammatidae). Journal of Agricultural Science and Natural Resources 2003; 10 117-124.

[99] Ganjisaffar F, Fathipour Y, Kamali K. Temperature-dependent development and life table parameters of *Typhlodromus bagdasarjani* (Phytoseiidae) fed on two-spotted spider mite. Experimental and Applied Acarology 2011; 55 259-272.

[100] Fathipour Y, Kamali K, Khalghani J, Abdollahi G. Functional response of *Trissolcus grandis* (Hym., Scelionidae) to different egg densities of *Eurygaster integriceps* (Het., Scutelleridae) and effects of different wheat genotypes on it. Applied Entomology and Phytopathology 2001; 68 123-136.

[101] Fathipour Y, Dadpour-Moghanloo H, Attaran M. The effect of the type of laboratory host on the functional response of *Trichogramma pintoi* Voegele (Hym., Trichogrammatidae). Journal of Agricultural Science and Natural Resources 2002; 9 109-118.

[102] Fathipour Y, Jafari A. Functional response of predators *Nabis capsiformis* and *Chrysoperla carnea* to different densities of *Creontiades pallidus* nymphs. Journal of Agricultural Science and Natural Resources 2003; 10 125-133.

[103] Fathipour Y, Hosseini A, Talebi AA, Moharramipour S. Functional response and mutual interference of *Diaeretiella rapae* (Hymenoptera: Aphidiidae) on *Brevicoryne brassicae* (Homoptera: Aphididae). Entomologica Fennica 2006; 17 90-97.

[104] Kouhjani-Gorji M, Fathipour Y, Kamali K. The effect of temperature on the functional response and prey consumption of *Phytoseius plumifer* (Acari: Phytoseiidae) on the two-spotted spider mite. Acarina 2009; 17 231-237.

[105] Pakyari H, Fathipour Y, Rezapanah M, Kamali K. Temperature-dependent functional response of *Scolothrips longicornis* (Thysanoptera: Thripidae) preying on *Tetranychus urticae*. Journal of Asia-Pacific Entomology 2009; 12 23-26.

[106] Jalilian F, Fathipour Y, Talebi AA, Sedaratian A. Functional response and mutual interference of *Episyrphus balteatus* and *Scaeva albomaculata* (Dip.: Syrphidae) fed on *Myzus persicae* (Hom.: Aphididae). Applied Entomology and Phytopathology 2010; 78 257-274.

[107] Asadi R, Talebi AA, Khalghani J, Fathipour Y, Moharramipour S, Askari-Siahooei M. Age-specific functional response of *Psyllaephagus zdeneki* (Hymenoptera: Encyrtidae), parasitoid of *Euphyllura pakistanica* (Hemiptera: Psyllidae). Journal of Crop Protection 2012; 1 1-15.

[108] Farazmand A, Fathipour Y, Kamali K. Functional response and mutual interference of *Neoseiulus californicus* and *Typhlodromus bagdasarjani* (Acari: Phytoseiidae) on *Tetranychus urticae* (Acari: Tetranychidae). International Journal of Acarology 2012; 38 369-376.

[109] Heidarian M, Fathipour Y, Kamali K. Functional response, switching, and prey-stage preference of *Scolothrips longicornis* (Thysanoptera: Thripidae) on *Schizotetranychus smirnovi* (Acari: Tetranychidae). Journal of Asia-Pacific Entomology 2012; 15 89-93.

[110] Jafari S, Fathipour Y, Faraji F. The influence of temperature on the functional response and prey consumption of *Neoseiulus barkeri* (Phytoseiidae) on two-spotted spider mite. Journal of Entomological Society of Iran 2012 [in press].

[111] Jervis M, Kidd N. Insect Natural Enemies: Practical Approaches to Their Study and Evaluation. London: Chapman and Hall; 1996.

[112] Obopile M, Mosinkie KT. Integrated pest management for African bollworm *Helicoverpa armigera* (Hübner) in Botswana: review of past research and future perspectives. Journal of Agricultural, Food, and Environmental Science 2007; 1 1-9.

[113] van den Berg H, Cock MJW, Onsongo GIEK. Incidence of *Helicoverpa armigera* (Lepidoptera: Noctuidae) and its natural enemies on smallholder crops in Kenya. Bulletin of Entomological Research 1993; 83 321-328.

[114] Shepard M, Sterling W. Incidence of parasitism of *Heliothis* spp. (Lepidoptera: Noctuidae) in some cotton fields of Texas. Annals of the Entomological Society of America 1972; 65 759-760.

[115] Quaintance AL, Brues CT. The Cotton Bollworm. Washington DC: United State Department of Agriculture Division of Entomology Bulletin; 1905.

[116] Huang K, Gordh G. Does *Trichogramma australicum* Girault (Hymenoptera: Trichogrammatidae) use kairomones to recognize eggs of *Helicoverpa armigera* (Hübner) (Lepidoptera: Noctuidae)? Australian Journal of Entomology 1998; 37 269-274.

[117] van Den Bosch R, Hagen KS. Predaceous and Parasitic Arthropods in California Cotton Fields. California: California Agricultural Experiment Station Bulletin; 1966.

[118] Ibrahim AE. Biotic Factors Affecting Different Species in the Genera *Heliothis* and *Spodoptera* in Egypt. Egypt: Institute of Plant Protection; 1981

[119] Butler GDJ. Braconid wasps reared from lepidopterous larvae in Arizona. Pan-Pacific Entomologist 1958; 34 222-223.

[120] King EG, Powell JE, Smith JW. Prospects for Utilization of Parasites and Predators for Management of *Heliothis* Spp. In: Reed W, Kumble V. (eds.) Proceedings of the International Workshop on *Heliothis* Management, 15-20 November 1981, Patancheru, India. International Crops Research Institute for the Semi-Arid Tropics; 1982. p103-122.

[121] Carkl KP. *Heliothis armigera*; Parasite Survey and Introduction of *Apanteles kazak* to New Zealand. Delemont: Commonwealth Institute of Biological Control; 1978.

[122] Danks HV, Rabb RL, Southern PS. Biology of insect parasites of *Heliothis* larvae in North Carolina. Journal of the Georgia Entomological Society 1979; 14 36-64.

[123] Mironidis GK, Savopoulou-Soultani M. Development, survival and growth rate of the *Hyposoter didymator-Helicoverpa armigera* parasitoid-host system: effect of host in-star at parasitism. Biological Control 2009; 49 58-67.

[124] Rao VP. Biology and Breeding Techniques for Parasites and Predators of *Ostrinia* spp. and *Heliothis* spp. India: CIBC Final Technical Report; 1974.

[125] Gupta RK, Rai D, Devil N. Biological and impact assessment studies on *Campoletis chlorideae* Uchida: a promising solitary larval endoparasitiod of *Helicoverpa armigera* (Hübner). Journal of Asia-Pacific Entomology 2004; 7 239-247.

[126] Young JH, Price RG. Incidence, parasitism, and distribution patterns of *Heliothis zea* on sorghum, cotton, and alfalfa for southwestern Oklahoma. Environmental Entomology 1975; 4 777-779.

[127] Walker PW. Biology and development of *Chaetophthalmus dorsalis* (Malloch) (Diptera: Tachinidae) parasitising *Helicoverpa armigera* (Hübner) and *H. punctigera* Wallengren (Lepidoptera: Noctuidae) larvae in the laboratory. Australian Journal of Entomology 2011; 50 309-318.

[128] van Hamburg H, Guest PJ. The impact of insecticides on beneficial arthropods in cotton agroecosystems in South Africa. Archives of Environmental Contamination and Toxicology 1997; 32 63-68.

[129] Lincoln C, Phillips JR, Whitcomb WH, Powell GC, Boyer WP, Bell K, Dean J, Matthews GL, Atthews EJ, Graves JB, Newsom LD, Clowre DF, Braley JR, Bagent JL. The Bollworm-Tobacco Budworm Problem in Arkansas and Louisiana. Arkansas Agricultural Experiment Station Bulletin; 1967.

[130] Greene GT, Shepard M. Biological studies of a predator, Sycanus indagator: Field survival and predation potential. Florida Entomologist 1974; 57 33-38.

[131] Lopez JD, Ridgway RL, Pinnell RE. Comparative efficacy of four insect predators of the bollworm and tobacco budworm. Environmental Entomology 1976; 5 1160-1164.

[132] Hassanpour M, Mohaghegh J, Iranipour S, Nouri-Ganbalani G, Enkegaard A. Functional response of *Chrysoperla carnea* (Neuroptera: Chrysopidae) to *Helicoverpa armigera* (Lepidoptera: Noctuidae): effect of prey and predator stages. Insect Science 2011; 18 217-224.

[133] Steinhaus EA. Microbial control: the emergence of an idea. Hilgardia 1956; 26 107-160.

[134] Lacey LA, Frutos R, Kaya HK, Vailss P. Insect pathogens as biological control agents: do they have a future? Biological Control 2001; 21 230-248.

[135] Roome RE. Field trials with a nuclear polyhedrosis virus and *Bacillus thuringiensis* against larvae of *Heliothis armigera* (Hb.) (Lepidoptera, Noctuidae) on sorghum and cotton in Botswana. Bulletin of Entomological Research 1975; 65 507-514.

[136] Moore SD, Pittaway T, Bouwer G, Fourie JG. Evaluation of *Helicoverpa armigera* nucleo polyhedron virus (HearNPV) for control of *Helicoverpa armigera* (Lepidoptera: Noctuidae) on citrus in South Africa. Biocontrol Science and Technology 2004; 14 239-250.

[137] Jeyarani S, Sathiah N, Karuppuchamy P. Field efficacy of *Helicoverpa armigera* nucleo polyhedron virus isolates against *H. armigera* (Hübner) (Lepidoptera: Noctuidae) on cotton and chickpea. Plant Protection Science 2010; 46 116-122.

[138] Yaman M, Aslan I, Calmasur O, Sahin F. Two bacterial pathogens of *Helicoverpa armigera* (Hübner) (Lepidoptera: Noctuidae). Proceeding of Entomological Society of Washington 2005; 107 623-626.

[139] Prasad A, Syed N. Evaluating prospects of fungal biopesticide *Beauveria bassiana* (Balsamo) against *Helicoverpa armigera* (Hübner): an ecosafe strategy for pesticidal pollution. Asian Journal of Experimental Biological Science 2010; 1 596-601.

[140] Rijal JP, Dhoj GCY, Thapa RB, Kafle L. Virulence of native isolates of *Metarhizium anisopliae* and *Beauveria bassiana* against *Helicoverpa armigera* in Nepal. Formosan Entomology 2008; 28 21-29.

[141] Tang LC, Hou RF. Effects of environmental factors on virulence of the entomopathogenic fungus, *Nomuraea rileyi*, against the corn earworm, *Helicoverpa armigera* (Lep., Noctuidae). Journal of Applied Entomology 2001; 125 243-248.

[142] Yearian WC, Hamm JJ, Carner GR. Efficacy of *Heliothis* Pathogens. In: Johnson SJ, King EG, Bradley JR. (eds.) Theory and Tactics of *Heliothis* Population Management. Southern Coop Series Bulletin; 1986.

[143] Kary NE, Golizadeh A, Rafiee-Dastjerdi H, Mohammadi D, Afghahi S, Omrani M, Morshedloo MR, Shirzad A. A laboratory study of susceptibility of *Helicoverpa armigera* (Hübner) to three species of entomopathogenic nematodes. Munis Entomology and Zoology 2012; 7 372-379.

[144] Avilla C, Vargas-Osuna E, Gonzalez-Cabrera J, Ferre J, Gonzalez-Zamora JE. Toxicity of several δ-endotoxin of *Bacillus thuringiensis* against *Helicoverpa armigera* (Lepidoptera: Noctuidae) from Spain. Journal of Invertebrate Pathology 2005; 90 51-54.

[145] Newsom LD. Pest Management: Concept to Practice. In: Pimentel D. (ed.) Insects, Science and Society. New York: Academic Press; 1975. p257-277.

[146] Fye RE. The interchange of insect parasites and predators between crops. Pest Articles and News Summaries 1972; 18 143-146.

[147] Whitcomb WH, Bell K. Predaceous Insects, Spiders, and Mites of Arkansas Cotton Fields. Arkansas Agricultural Experiment Station Bulletin; 1964.

[148] Nordlund DA, Jones RL, Lewis WJ. Semiochemicals: Their Role in Pest Control. New York: Wiley; 1981.

[149] Roome RE. A note on the use of biological insecticide against *H. armigera* (Hb.) in Botswana. Proceedings, cotton insect control conference, Malawi; 1971.

[150] Duffield SJ. Evaluation of the risk of overwintering *Helicoverpa* spp. pupae under irrigated summer crops in south-eastern Australia and the potential for area-wide management. Annals of Applied Biology 2004; 144 17-26.

[151] Talekar NS, Shelton AM. Biology, ecology, and management of the diamondback moth. Annual Review of Entomology 1993; 38 275-301.

[152] Shelton AM, Badenes-Perez FR. Concepts and application of trap cropping in pest management. Annual Review of Entomology 2006; 51 285-308.

[153] Virk JS, Brar KS, Sohi AS. Role of trap crops in increasing parasitation efficiency of *Trichogramma chilonis* Ishii in cotton. Journal of Biological Control 2004; 18 61-64.

[154] Pyke B, Rice M, Sabine B, Zalucki MP. The push-pull strategy-behavioral control of *Heliothis*. The Australian Cotton Grower 1987; 7-9.

[155] Cook SM, Khan ZR, Pickett JA. The use of push-pull strategies in integrated pest management. Annual Review of Entomology 2007; 52 375-400.

[156] Cook SM, Smart LE, Martin JL, Murray DA, Watts NP, Williams IH. Exploitation of host plant preferences in pest management strategies for oilseed rape (*Brassica napus*). Entomologia Experimentalis et Applicata 2006; 119 221-229.

[157] Regupathy DP. Push-pull strategy with trap crops, neem and nuclear polyhedrosis virus for insecticide resistance management in *Helicoverpa armigera* (Hübner) in cotton. American Journal of Applied Science 2005; 2 1042-1048.

[158] Painter RH. Insect Resistance in Crop Plants. New York: Macmillan; 1951.

[159] Panda N, Khush GS. Host Plant Resistance to Insect. Wallingford: CAB International; 1995.

[160] Adkisson PL, Dyck VA. Resistance Varieties in Pest Management Systems. In: Maxwell FG, Jennings PR. (eds.) Breeding Plant Resistance to Insect. New York: John Wiley and Sons; 1980. p233-251.

[161] Maxwell FG. Utilization of host plant resistance in pest management. Insect Science and its Application 1985; 6 437-442.

[162] Smith CM. Plant Resistance to Insects: A Fundamental Approach. New York: John Wiley and Sons; 1989.

[163] Liu Z, Li D, Gong P, Wu K. Life table studies of the cotton bollworm, *Helicoverpa armigera* (Hübner) (Lepidoptera: Noctuidae), on different host plants. Environmental Entomology 2004; 33 1570-1576.

[164] Jha RK, Chi H, Tang LC. Comparison of artificial diet and hybrid sweet corn for the rearing of *Helicoverpa armigera* (Hübner) (Lepidoptera: Noctuidae) based on life table characteristics. Environmental Entomology 2012; 41 30-39.

[165] Patal CC, Koshyia DJ. Life table and innate capacity of *Helicoverpa armigera* (Hübner) on pearl millet. Indian Journal of Entomology 1997; 59 389-395.

[166] Naseri B, Fathipour Y, Moharramipour S, Hosseininaveh V. Life table parameters of the cotton bollworm, *Helicoverpa armigera* (Lepidoptera: Noctuidae) on different soybean cultivars. Journal of Entomological Society of Iran 2009; 29 25-40.

[167] Reddy KS, Rao GR, Rao PA, Rajasekhar P. Life table studies of the capitulum borer, *Helicoverpa armigera* (Hübner) infesting sunflower. Journal of Entomological Research 2004; 28 13-18.

[168] Price PW, Bouton CE, Gross P, McPheron BA, Thopmson JN, Weis AE. Interactions among three trophic levels: influence of plants on interactions between insect herbivores and natural enemies. Annual Review of Ecology, Evaluation and Systematics 1980; 11 41-65.

[169] Wiseman BR, Widstrom NW, McMillian WW, Waiss JAC. Relationship between maysin concentration in corn silk and corn earworm (Lepidoptera: Noctuidae) growth. Journal of Economic Entomology 1985; 78 423-427.

[170] van Emden HF, Wearing CH. The role of the aphid host plant in delaying economic damage levels in crops. Annals of Applied Biology 1965; 56 323-324.

[171] Rabb RL, Bradely JR. The influence of host plants on parasitism of eggs of the tobacco hornworm. Journal of Economic Entomology 1968; 61 1249-1252.

[172] Barbour JD, Farra JRR, Kennedy GG. Influence of *Manduca sexta* resistance in tomato with insect predator of *Helicoverpa zea*. Entomologia Experimentalis et Applicata 1993; 68 143-155.

[173] van Lenteren JC. Bilogical Control in Tritrophic System Approach. In: Peters DC, Webster JA. (eds.) Aphid-Plant Interactions: Populations to Molecules. Stillwater, OK: Oklahoma State University Press; 1991. p3-28.

[174] Scultz JC. Impact of Variable Plant Chemistry on Susceptibility of Insect to Natural Enemies. In: Hedin PA. (ed.) Plant Resistance to Insect. Washington DC: American Chemistry Society; 1983. p37-54.

[175] Krischik VA, Barbosa P, Reichelderfer CF. Three trophic level interactions: allelochemicals, *Manduca sexta*, and *Bacillus thuringiensis* var. *kurstaki*. Environmental Entomology 1983; 17 476-482.

[176] Felton GW, Duffey SS. Inactivation of Baculovirus by quinones formed in insect-damaged plant tissues. Journal of Chemical Ecology 1990; 16 1221-1236.

[177] van Emden HF. Principals of Implementation of IPM. In: Cameron PJ, Wearing CH, Kain WM. (eds.) Proceedings of Australian Workshop on Development and Implementation of IPM. Auckland: Government printer; 1982. p9-17.

[178] van Emden HF. The interaction of host plant resistance to insect with other control measures. Proceedings of brighton crop protection conference-pest and disease. Suffolk: The Lavenham Press Limited; 1990.

[179] Wiseman BR, Harrell EA, McMillian WW. Continuation of tests resistant sweet corn hybrid plus insecticides to reduce losses from corn earworm. Environmental Entomology 1974; 2 919-920.

[180] Shorey HH, McKelvey JJ. Chemical Control of Insect Behavior. New York: John Wiley and Sons; 1977.

[181] Pawar CS, Sithanantham S, Bhatnagar VS, Srivastava CP, Reed W. The development of sex pheromone trapping of *H. armigera* at ICRISAT, India. Tropical Pest Management 1988; 34 39-43.

[182] Jacobson M. The Potential Role of Natural Product Chemistry Research in *Heliothis* Management. In: Reed W, Kumble V. (eds.) Proceedings of the International Workshop on *Heliothis* Management, 15-20 November 1981, Patancheru, India. International Crops Research Institute for the Semi-Arid Tropics; 1982. p233-239.

[183] Whitten MJ. Pest Management in 2000: What We Might Learn from the Twenty Century. In: Kadi AASA, Barlow HS. (eds.) Pest Management and Environment in 2000. Wallingford: CAB International; 1992. p9-44.

[184] Roush RT. Two-toxin strategies for management of insect resistant transgenic crops: can pyramiding succeed where pesticide mixtures have not? Philosophical Transactions of the Royal Society of London 1998; 353 1777-1786.

[185] Denyer R. Integrated crop management: introduction. Pest Management Science 2000; 56 945-946.

[186] Harvey JA, van Dam N, Witjes LA, Soler R, Gols R. Effects of dietary nicotine on the development of an insect herbivore, its parasitoid and secondary hyperparasitoid over four trophic levels. Ecological Entomology 2007; 32 15-23.

[187] Denno RF, Gratton C, Peterson MA, Langellotto GA, Finke DL, Huberty AF. Bottom-up forces mediate natural-enemy impact in a phytophagous insect community. Ecology 2002; 83 1443-1458.

[188] Soufbaf M, Fathipour Y, Karmizadeh J, Zalucki MP. Bottom-up effect of different host plants on *Plutella xylostella* (Lepidoptera: Plutellidae): a life-table study on canola. Journal of Economic Entomology 2010; 103 2019-2027.

[189] Detheir VG. Chemical Interaction between Plants and Insects. In: Sondheimer E, Simeone JB. (eds.) Chemical Ecology. New York: Academic Press; 1970. p33-102.

[190] Vinson SB. Biochemical Coevolution between Parasitoids and their Hosts. In: Price PW. (ed.) Evolutionary Strategies of Parasitic Insects and Mites. New York: Plenum Press; 1975. p14-48.

[191] Hare DJ. Effects of Plant Variation on Herbivore-Enemy Interactions. In: Fritz RS, Simms EL. (eds.) Plant Resistance to Herbivores and Pathogens. Chicago: University of Chicago Press; 1992. p278-298.

[192] Godfray HJH. Parasitoids-Behavioral and Evolutionary Ecology. New Jersey: Princeton University Press; 1994.

[193] Orr DB, Boethel DJ. Comparative development of *Copidosoma truncatellum* (Hymenoptera: Encyrtidae) and its host, *Pseudoplusia includens* (Lepidoptera: Noctuidae), on resistant and susceptible soybean genotypes. Environmental Entomology 1985; 14 612-616.

[194] Khadi BM, Kulkarni VN, Narjji SS. Achieving multiple pest tolerance through manipulation of morphological features in cotton. World proceedings cotton research conference; 1998.

[195] Mohite PB, Uthamasamy S. Host plant resistance and natural enemies interaction in the management of *Helicoverpa armigera* (Hübner) on cotton. Indian Journal of Agricultural Research 1998; 32 28-30.

[196] Asifulla HR, Awaknavar JS, Rajasekhar DW, Lingappa S. Parasitisation of *Trichogramma chilonis* Ishii on bollworm eggs in different cotton genotypes. Advances in Agricultural Research in India 1998; 9 143-146.

[197] Bottrell DG, Barbosa P. Gould F. Manipulating natural enemies by plant variety selection and modification: a realistic strategy? Annual Review of Entomology 1998; 43 347-367.

Chapter 9

Mycotoxins in Cereal and Soybean-Based Food and Feed

Małgorzata Piotrowska, Katarzyna Śliżewska and
Joanna Biernasiak

Additional information is available at the end of the chapter

1. Introduction

1.1. Toxigenic fungi and mycotoxins in cereal and soybean products

Cereals and soybean are plants used extensively in food and feed manufacturing as a source of proteins, carbohydrates and oils. These materials, due to their chemical composition, are particularly susceptible to microbial contamination, especially by filamentous fungi. Cereals, soybean, and other raw materials can be contaminated with fungi, either during vegetation in the field or during storage, as well as during the processing.

Fungi contaminating grains have been conventionally divided into two groups – field fungi and storage fungi. Field fungi are those that infect the crops throughout the vegetation phase of plants and they include plant pathogens such as *Alternaria, Fusarium, Cladosporium,* and *Botrytis* species. Their numbers gradually decrease during storage. They are replaced by storage fungi of *Aspergillus, Penicillium, Rhizopus* and *Mucor* genera that infect grains after harvesting, during storage [1]. Both groups of fungi include toxigenic species. Currently, this division is not so strict.

Therefore, according to [2], four types of toxigenic fungi can be distinguished:

- Plant pathogens as *Fusarium graminearum* and *Alternaria alternata*;

- Fungi that grow and produce mycotoxins on senescent or stressed plants, e.g. *F. moniliforme* and *Aspergillus flavus*;

- Fungi that initially colonize the plant and increase the feedstock's susceptibility to contamination after harvesting, e.g. *Aspegillus flavus*.

- Fungi that are found on the soil or decaying plant material that occur on the developing kernels in the field and later proliferate in storage if conditions permit, e.g. *Penicillium verrucosum and Aspergillus ochraceus.*

Fungal growth is influenced by complex interaction of different environmental factors such as temperature, pH, humidity, water activity, aeration, availability of nutrients, mechanical damage, microbial interaction or the presence of antimicrobial compounds. Poor hygiene, inappropriate temperature and moisture during harvesting, storage, processing and handling may contribute to increased contamination extent.

Fungal contamination can cause damage in cereal grains and oilseeds, including low germination, low baking quality, discoloration, off-flavours, softening and rotting, and formation of pathogenic or allergenic propagules.

It may also decrease the kernel size and thus affect the flour yield. Moulds growing on stored cereals produce a range of volatile odour compounds, including 3-octanone, 1-octen-3-ol, geosmin, 2-methoxy-3-isopropylpyrazine, and 2-methyl-1-propanol which are responsible for an earthy-musty off-odour and affect the quality of raw materials even when present in very small amounts [3]. Moulds produce a vast number of enzymes: lipases, proteases, amylases, which are able to break down food into components leading to its spoilage. Fungi growing on stored grains can reduce the germination rate and decrease the content of carbohydrate, protein and oils. During storage of soybean seed lasting 12 months, the moisture content was at the level of 10-11%. It was observed that the germination rate decreased from initial 75% to 4% prior to the lapse of a 9-month period. In prolonged storage under natural conditions, the total carbohydrate content decreased from 21% to 16.8%, and protein and the total oil contents became slightly reduced [4]. Moulds as food and feed spoilage microorganisms have been characterized in several review articles [2, 5].

The largest producers of soybean in the World are the United States of America, Brazil, Argentina, China, and India. The climatic conditions in soybean-growing regions (moderate mean temperature and relative humidity between 50 and 80%) provide optimal conditions for fungal growth. Soybean (*Glyccine max* L.Merr.) is often attacked by fungi during cultivation, which significantly decreases its productivity and quality in most production areas. Fungi associated with cereal grains and oilseeds are important in assessing the potential risk of mycotoxin contamination. Mycotoxins are fungal secondary metabolites which are toxic to vertebrate animals even in small amounts when introduced orally or by inhalation.

Table 1 summarises the occurrence of contamination of different raw materials in various countries. Some of them are of mycotoxicological interest.

Soybean matrix has been rarely studied compared to cereals in relation to fungal and mycotoxin contamination. The fungi associated with soybean seeds, pods and flowers in North America were reviewed by [20]. The most common species belong to *Aspergillus, Fusarium, Chaetomium, Penicillium, Alternaria* and *Colletotrichum* genera. Most of these fungi were recorded in mature seeds prior to storage. About 10% of them are commonly referred to as storage moulds. Most of the isolated fungi are facultative parasites or saprophytes.

Commodities	Country	Fungal species	Ref
Soybean	Ecuador	*Aspergillus flavus, A.niger, A.ochraceus, A.parasiticus, Fusarium verticillioides, F.semitectum, Penicillium janthinellum, P.simplicissimum, Nigrospora oryzae, Cladosporium cladosporioides, Arthrinium phaeospermum*	[6]
	Romania	*Aspergillus flavus, A.parasiticus, A.candidus, A.niger, Penicillium griseofulvum, P.variabile, Fusarium culmorum, F.graminearum, F.oxysporum*	[7]
	India	*Aspergillus flavus, A.candidus, A.versicolor, Eurotium repens, A.sulphureus, Fusarium sp., Alternaria sp., Curvularia sp.*	[4]
	USA	*Diaporthe phaseolorum var. sojae, Fusarium sp., Alternaria alternata, Alternaria sp., Fusarium sp., Curvularia sp., Cladosporium sp., Fusarium equiseti, F.oxysporum, F.solani*	[8-10]
	Croatia	*Fusarium sporotrichides, F.verticillioides, F.equiseti, F.semitecium, F.pseudograminearum, F.chlamydosporum, F.sambucinum*	[11]
	Argentina	*Aspergillus flavus, A.niger, A.candidus, A.fumigatus, Fusarium verticillioides, F.equiseti, F.semitecium, F.graminearum, Penicillium funiculosum, P.griseofulvum, P.canenscens, Erotium sp. Cladosporium sp., Alternaria alternata, A.infectoria, A.oregonensis*	[12, 13]
Rice	Ecuador	*Aspergillus flavus, A.ochraceus, Fusarium verticillioides, F.oxysporum, F.proliferatum, F.semitectum, F.solani, Penicillium janthinellum, Epicoccum nigrum, Curvularia lunata, Nigrospora oryzae, Rhizopus stolonifer, Bipolaris oryzae*	[6]
Wheat	Argentina	*Aspergillus flavus, A.niger, A.oryzae, Fusarium verticillioides, Penicillium funiculosum, P.oxalicum*	[12]
	Germany	*Aspergillus candidus, A.flavus, A.versicolor, Eurotium sp., Penicillium auriantogriseum, P.verrucosum, P.viridicatum, Alternaria sp.*	[14]
	Poland	*Alternaria tenuis, Aspergillus aculeatus, A.parasiticus, Fusarium moniliforme, F.verticillioides, Penicillium verrucosum, P.viridicatum P.crustosum*	[15]
	Croatia	*Fusarium graminearum, F.poae, F.avenaceum, F.verticillioides*	[11]
Maize	Ecuador	*Aspergillus flavus, A.parasiticus, Fusarium graminearum, F.verticillioides, Mucor racemosus Rhizopus stolonifer, Acremonium strictum, Alternaria alternata, Cladosporium sp.*	[6]
	Poland	*Aspergillus aculeatus, Aspergillus parasiticus, Fusarium moniliforme, F.verticillioides*	[15]
	Argentina	*Fusarium verticillioides, F.proliferatum, F.subglutinans, F.dlamini, F.nygamai, Alternaria alternata, Penicillium funiculosum, P.citrinum, Aspergillus flavus*	[16, 17]

Commodities	Country	Fungal species	Ref
	Croatia	*Fusarium verticillioides, F.graminearum*	[11]
Oats	Poland	*Cladosporium sp., Aspergillus sp., Penicillium sp.*	[18]
Breakfast cereals	Poland	*Aspergillus versicolor, A.flavus, A.sydowi, A.niger, A.ochraceus, Fusarium graminearum, Penicillium chrysogenum, Eurotium repens*	[19]
Wheat flour	Germany	*Aspergillus candidus, A.flavus, A.niger, Eurotium sp. Penicillium auriantogriseum, P.brevicompactum, P.citrinum, P.griseofulvum, P.verrucosum, Cladosporium cladosporioides*	[14]

Table 1. Fungal species dominated in cereals and cereal products

Fusarium graminearum is associated with cereals and soybean growing in warmer areas such as South and North America or China, and *F.culmorum* in cooler areas such as Finland, France, Poland or Germany. Mechanical damage of kernels by birds or insects, e.g. European corn borer and sap beetles, predisposes corn to infections caused by *Fusarium* and other "field fungi". *Fusarium moniliforme* and *F.proliferatum* are the most common fungi associated with maize. It was found that the levels of contamination with *Fusarium* sp. were significantly greater on the conventional than the transgenic cultivars in 2000, but in 1999 the difference between the cultivars was not statistically significant. In case of *Alternaria*, a greater frequency of contamination in transgenic varieties was observed. The authors concluded that the isolation frequency can vary by years and is more dependent on the environmental and cultural practices than on varieties [9]. The isolation frequencies of fungi from seeds and pods of soybean cultivars varied annually, in part due to some differences in environmental conditions (rainfall) [8].

Fusarium species occur worldwide in a variety of climates and on many plant species as epiphytes, parasites, or pathogens. *Fusarium*-induced diseases of soybean have been attributed to different species: *Fusarium oxysporum* (fusarium blight, wilt and root rot), *Fusarium semitectum* (pod and collar rot), *F.solani* (sudden death syndrome) [21, 22]. *Fusarium* infections are spread by air-borne conidia on the heads or by a systemic infection. The species belonging to *Fusarium* genera are of particular interest due to the formation of a wide range of secondary metabolites, many of which are toxic to humans or animals. Infections by *Fusarium* spp. were determined by [11] in different crops. The contamination expressed as the percentage of seeds with *Fusarium* colonies ranged from 5% to 69% for wheat, from 25% to 100% for maize, from 4% to 17% for soybean. The dominant species were *F.graminearum* on wheat (27% of isolates), *F.verticillioides* on maize (83 % of isolates), and *F.sporotrichioides* on soybean (34 % of isolates) [11]. This study suggested that the risk of contamination with *Fusarium* toxins is higher for maize and wheat than for soybean.

The mycological state of grain can be considered as good when the number of CFU is within the range 10^3-10^5 per gram [23]. In our research, the contamination of feed components such as barley, maize and wheat was in the range from 10^2 to 10^4 CFU/g, depending on the crop, region and mills [15]. It was found that wheat from organic farms was contaminated with fungi by 70.5% more and barley by 24.8% less as compared to the crops from conventional farms [24]. Similarly, the total number of fungi in Polish ecological oat products was about a hundred times higher than in conventional ones. In samples of ecological origin, the mean value of fungi was 1.1×10^4 CFU/g, whereas for conventional grains it was 5.0×10^2 CFU/g [18].

The results obtained by [14] showed that the most common moulds isolated from whole wheat and wheat flour belong to the *Aspergillus* and *Penicillium* genera. From the whole wheat flour, 83.7% of *Aspergillus* followed by *Penicillium* (7.6%), *Eurotium* (2.9%) and *Alternaria* (2.5%) species were isolated. The white flour contained 77.3% of *Aspergillus*, 15% of *Penicillium* and 4.1% of *Cladosporium* genera. *Aspergillus candidus* was the dominant species. Among all the isolated fungal species, 93.2% belonged to the group of toxigenic fungi. Several toxin-producing *Aspergillus* species were reported to dominate on cereals, especially *A.flavus*, *A.candidus*, *A.niger*, *A.versicolor*, *A.penicillioides*, and *Eurotium* sp. at lower water activity [25]. Among *Aspergillus* species isolated from Ecuadorian soybean seeds, *Aspergillus flavus* and *A.ochraceus* were the most prevalent ones. The most frequent *Fusarium* species were *F.verticillioides* and *F.semitecium*. All the examined samples were contaminated with these species [6]. The presence of mycobiota in raw materials and finished fattening pig feed was determined in eastern Argentina. All samples of soybean seeds were contaminated with fungi in the range from 10 to 9.0×10^2 CFU/g, depending on the sampling period. The most prevalent species in soybean and wheat bran were *Aspergillus flavus* and *Fusarium verticillioides* [12].

The fungal microflora changes during post-harvest drying and storage. The field fungi are adapted to growth at high water activity and they die during drying and storage, to be replaced by storage fungi that are capable of growing at lower a_w. For most grains, moisture content in the range from 10% to 14% is recommended, depending on the grain type and desired storage life [1].

A wide range of microorganisms have been isolated from storage grains, including psychrotolerant, mesophilic, thermophilic, xerophilic and hydrophilic species. The extremely xerophilic species are *Eurotium* spp. and *Aspergillus restrictus*, the moderate xerophilic ones include *A.candidus* and *A.flavus*, and the slighty xerophilic one is *A.fumigatus*. An example of psychrotolerant species belonging to *Penicillium* genera is *P.aurantiogriseum* and *P.verrucosum*, mesophilic species can be represented by *P.corylophilum*, and thermophilic species by *Talaromyces thermophilus*. Among the hydrophiles, the most common are *Fusarium* and *Acremonium* species [25]. The minimum a_w for conidial formation is influenced by temperature, for instance, *P.aurianogriseum* produces conidia to a minimum of 0.86 a_w at 30°C, but to 0.83 a_w at 23°C. Many species belonging to *Aspergillus* and *Penicillium* genera are highly adapted to the rapid colonisation of substrates of reduced water activity. Modifying several factors in grain storage may facilitate safe storage. Stores should be monitored for relative humidity, temperature and airflow efficiency. Moisture migration may occur during storage and create damp pockets. In addition to this, insect infestations may cause heating and the generation of moisture. Aeration with cool air may help to protect the stored commodities against fungal development.

2. Conditions affecting mycotoxin production

Cereals in the field are exposed to fungi from the soil, birds, animals, insects, organic fertilizers, and from other plants in the field. The mechanical damage of raw material or food due to insects and pests is a disturbing problem mainly in tropical regions, particularly as food contaminants are present in the field more abundantly than in the storage. Many different insects, e.g. European corn borer and sap beetles have the capability of promoting infections of various crops with mycotoxigenic fungi [25].

Mycotoxin production is determined by genetic capability related to strain and environmental factors including the substrate and its nutritious content. Toxin production is dependent on physical (temperature, moisture, light), chemical (pH value, nutrients, oxygen content, preservatives), and biological factors (competitive microbiota). Each fungus requires special conditions for its growth and other conditions for its toxin production.

2.1. Physical factors

The most important factor governing colonisation of grains and mycotoxin production is the availability of water which on the field comes mainly with rainfall. The second important factor is temperature. The moisture and temperature effects on mycotoxin production often differ from those on germination and growth. Table 2 presents the moisture and temperature requirements of most common toxigenic fungi for their growth and mycotoxin production.

It was found that optimal temperature for *F.graminearum* growth on soybean contained in the range 15-20°C (in isothermal temperature) and 15/25°C (in cycling temperature). The optimal temperature for mycotoxin production on soybean was 20°C for deoxynivalenol (DON) and 15°C for zearalenone (ZEA). After 15 days of incubation, the maximum levels 39 ppm and 1040 ppm for ZEA and DON, respectively, were detected. Fumonisins were produced by *Fusarium graminearum* only the on culture medium at 30°C; on soybean no fumonisins were detected [31].

Most fungi need at least 1-2% of O_2 for their growth. The influence of high carbon dioxide and low oxygen concentrations on the growth and mycotoxin production by the foodborne fungal species was investigated by [32]. Three groups of species were distinguished: first, which did not grow in 20% CO_2 <0.5% O_2 (*Penicillium commune*, *Eurotium chevalieri* and *Xeromyces bisporus*); second, which grew in 20% CO_2 <0.5% O_2, but not 40% CO_2 <0.5% O_2 (*Penicillium roqueforti* and *Aspergillus flavus*); and third, which grew in 20%, 40% and 60% CO_2 <0.5% O_2 (*Mucor plumbeus*, *Fusarium oxysporum*, *F.moniliforme*, *Byssochlamys fulva* and *B.nivea*). The production of aflatoxin, patulin, and roquefortine C was greatly reduced under all of the atmospheres tested. For example, aflatoxin was not produced by *A. flavus* during growth under 20% CO_2 for 30 days. Patulin was produced by *B.nivea* in the atmospheres of 20% and 40% CO_2, but only at low levels [32].

2.2. Chemical factors

Nutritional factors such as carbonohydrate and nitrogen sources and microelements (copper, zinc, cobalt) affect mycotoxin production, but the mechanisms of this impact are still

Species	For growth			For mycotoxin production			Ref.
	Temperature [°C]		Minimal a$_w$	Temperature [°C]		Minimal a$_w$	
	Range	Optimum	Range	Range	Optimum	Range	
Alternaria alternata	0 – 35	20 – 25	0.88	5-30	20-25	0.95-1.0 AOH 0.90 TeA	[25, 26, 28]
Fusarium culmorum	<0 – 31	21	0.89	11-30	25-26	Nd	[25]
Fusarium graminearum	Nd	24 – 26	0.89	Nd	24-26	Nd	[25]
Fusarium sporotrichoides	-2 – 35	22 – 28	0.88	6-20	Nd	Nd	[25]
Penicillium verrucosum	0-31	20	0.81-0.83	4-31	20	0.86	[28, 29]
Penicillium expansum	-6 – 35	25 – 26	0.82 – 0.85	0-31	25	0.95	[25]
Aspergillus ochraceus	8-37	24-30	0.76-0.83	12-37	25-31	0.85 OTA 0.88 PA	[28, 29]
Aspergillus parasiticus	10-43	32-33	0.84	12-40	25-30	0.87	[28, 29]
Aspergillus flavus	6 – 45	35 – 37	0.78	12-40	30	0.82	[25]
Aspergillus versicolor	4 – 39	25 – 30	0.75	15-30	23 – 29	"/>0.76	[25, 30]

OTA – ochratoxin A; PA – penicillic acid AOH – alternariol, TeA – tenuazonic acid, ND – no data

Table 2. Environmental requirements for growth and mycotoxin production

unclear. A relationship between mycotoxin production and sporulation has been documented in several toxigenic fungi. For example, chemical substances that inhibit sporulation of *Aspergillus parasiticus* have also been shown to inhibit the production of aflatoxin [33]. Chemical preservatives such as organic acids (sorbic, propionic, acetic, benzoic) or fungicides have been used to restrict the growth of mycotoxigenic fungi. It was found that propionic acid at the concentration of up to 0.05% inhibited the growth and ochratoxin production by *Penicillium auriantogriseum*. A more effective result in higher temperature was observed [34]. Inhibiting fungal growth and toxigenic properties by organic acids is connected with lowering the pH value. It was found that ammonium and sodium bicarbonate at the concentration of 2% fully inhibited the development of the cultures of *Aspergillus ochraceus*, *Fusarium graminearum* and *Penicillium griseofulvum* inoculated into corn. The production of ochratoxin A by *Aspergillus ochraceus* was reduced from 26 ppm in untreated corn to 0.26 ppm in bicarbonate-treated corn samples [35].

2.3. Biological factors

The simultaneous presence of different microorganisms, such as bacteria or other fungi, could disturb fungal growth and the production of mycotoxins. For instance, *Alternaria* and *Fusarium* are antagonistic, and *Alternaria* was less abundant in grain with a high incidence rate of *F.culmorum*. *Epicoccum* is a strong antagonist too [25].

At 30°C, the ochratoxin production by *Aspergillus ochraceus* was inhibited by *A.candidus*, *A.flavus*, and *A.niger* in 0.995 a_w. At 18°C and 0.995 a_w, the interaction between *Aspergillus ochraceus* and *Alternaria alternata* resulted in a significant stimulation of ochratoxin A production [36]. Therefore, several microorganisms were reported as effective biocontrol agents against several fungal plant pathogens [37]. It was determined that *Trichoderma harzianum* produces a lytic enzyme, chitinase, which manifests antifungal activity against a wide range of fungal strains. It was found that non-toxigenic *T.harzianum* isolates significantly reduce the production of six types of A trichothecenes in cereals [38].

According to [39], soybean is not a favourable medium for ZEA production since it possesses some features that limit the production of this toxin by *Fusarium* isolates. Similarly, the production of aflatoxin B_1 by *Aspergillus flavus* was suppressed by soybean phytoalexin – glyceollin [40].

3. Main mycotoxins

The worldwide contamination of foods and feeds with mycotoxins is a significant problem. It was estimated that 25% of the world's crops may be contaminated with these metabolites. Mycotoxigenic fungi involved with the human food chain belong mainly to three genera *Aspergillus*, *Penicillium* and *Fusarium*. The toxins produced by *Alternaria* have recently been of particular interest. The biochemistry, physiology and genetics of mycotoxigenic fungi have been discussed in several review articles [28, 41, 42].

Mycotoxins diffuse into grain and can be found in all grind fractions and, due to their thermo-resistant properties, also in products subjected to thermal processing [43].

The characteristics of major toxins that contaminate foods and feeds in the EU, described from the economic and toxicological point of view, are presented below.

3.1. Aflatoxins (AFs)

Aflatoxins are difuranocumarin derivatives. The main naturally produced aflatoxins based on their natural fluorescence (blue or green) are called B_1, B_2, G_1, and G_2. Aflatoxin M_1 is a monohydroxylated derivative of AFB_1 which is formed and excreted in the milk of lactating animals. AF_s are very slightly soluble in water (10–30 µg/mL); insoluble in non-polar solvents; freely soluble in moderately polar organic solvents (e.g. chloroform and methanol) and extremely soluble in dimethyl sulfoxide. They are unstable under the influence of ultraviolet light in the presence of oxygen, to extremes of pH ($< 3, > 10$) and to oxidizing agents [44].

Aflatoxins are produced only by a closely related group of aspergilli: *Aspergillus flavus*, *A.parasiticus*, and *A.nomius* strains [45]. These species are very widespread in the tropical and subtropical regions of the world. Other species such as *A.bombycis*, *A.ochraceoroseus*, and *A.pseudotamari* are also aflatoxin-producing species, but they are found less frequently [46, 47]. Aflatoxins constitute a problem concerning many commodities (nuts, spices), however, in terms of grain they are primarily problematic in case of maize. This is because only maize can be colonised by *A.flavus* and related species in the field. Out of the other grains, rice is an important dietary source of aflatoxins in tropical and subtropical areas. In regions with moderate climate, the problem is connected with imported commodities or the local crops that are wet or stored in improper conditions [45]. The carcinogenicity, mutagenicity and acute toxicology of AFB_1 have been well documented. The IARC determined it to be a human carcinogen (group 1A).

3.2. Ochratoxin A (OTA)

Ochratoxin A is a chlorinated isocumarin derivative, which contains a chlorinated isocoumarin moiety linked through a carboxyl group to L-phenylalanine via an amide bond. It is colourless, crystalline, and soluble in polar organic solvents compounds. This toxin is more stable in the environment than AFs. The studies of [45] reported that thermal destruction of OTA occurs after exceeding 250ºC. OTA is produced by *Penicillium* species such as *P.verrucosum*, *P.auriantiogriseum*, *P.nordicum*, *P.palitans*, *P. commune*, *P.variabile* and by *Aspergillus* species e.g. *A.ochraceus*, *A.melleus*, *A.ostanius*, as well as the aspergilli species of section *Nigri*. In moderate climates, the main producers of OTA are *Penicillium* species, while *Aspergillus* species dominate in tropical and subtropical climates. Ochratoxin A is often found with citrinin produced by *Penicillium aurantiogriseum*, *P.citrinum*, and *P.expansum* [48]. Significant human exposure comes from the consumption of grape juice, wine, coffee, spices, dried fruits and cereal-based products, e.g. whole-grain breads, and in addition to this from products of animal origin, e.g. pork and pig blood-based products. The Scientific Panel on Contaminants in the Food Chain of the European Food Safety Authority (EFSA) has derived an OTA tolerable weekly intake (TWI) on the level of 120ng/kg b.w. The IARC [49] determined it to be a possible human carcinogen (group 2B). Ochratoxins are the cause of urinary tract cancers and kidney damage. In ruminants, ochratoxin A is divided to non-toxic ochratoxin alfa and phenylalanine [44].

3.3. Citrinin

Citrinin is a polyketide nephrotoxin produced by several species of the genera *Aspergillus*, *Penicillium* and *Monascus*. Some of the citrinin-producing fungi are also able to produce ochratoxin A or patulin. Citrinin is insoluble in cold water, but soluble in aqueous sodium hydroxide, sodium carbonate, or sodium acetate; in methanol, acetonitrile, ethanol, and most other polar organic solvents. Thermal decomposition of citrinin occurs at >175 °C under dry conditions, and at > 100 °C in the presence of water. The known decomposition products include citrinin H_2 which did not show significant cytotoxicity, whereas the decomposition product citrinin H_1 showed an increase in cytotoxicity as compared to the parent compound [50].The most commonly contaminated commodities are barley, oats, and corn, but contamination can also occur in case of other products of plant origin e.g. beans, fruits, fruit and vegetable juices, herbs and spices, and also in spoiled dairy products [50].

3.4. Fumonisins (F_s)

Fumonisins are a group of diester compounds with different tricarboxylic acids and polyhydric alcohols and primary amine moiety. There are several fumonisins, but only fumonisins

B_1 (FB$_1$) and B_2 (FB$_2$) have been found in significant amounts. Some technological processes hydrolyze the tricarboxylic acid chain in fumonisin B_1. The product of this reaction is more toxic than fumonisin [51].

FB$_1$ is produced by fungi from *Fusarium* genera, especially by *F.moniliforme* and *F.prolifera-tum*. The study of [11] suggests that the risk of contamination with *Fusarium* toxins is higher for maize and wheat than for soybean and pea. High concentrations of fumonisins are associated with hot and dry weather, followed by the periods of high humidity. Studies on fumonisin residues in milk, meat and eggs are incomplete [52, 53]. Human exposure assessments on fumonisin B_1 have rarely been reported. The mean daily intake in Switzerland is estimated to be 0.03 µg/kg bw/day. In the Netherlands the exposure estimates ranged from 0.006 to 7.1 µg/kg bw/day. In South Africa, the estimates ranged from 14 to 440 µg/kg bw/day, showing that the exposure to FB$_1$ is considerably higher than in the other countries in which exposure assessments were performed [54]. It was concluded that for F_s there was inadequate evidence in humans for carcinogenicity. Therefore, the IARC classified *Fusarium monilliforme* toxins, including fumonisins, as potential carcinogens to humans (group 2B).

3.5. Zearalenone

Zearalenone is a macrocyclic lactone with high binding affinity to oestrogen receptors. ZEA is produced mainly by *Fusarium graminearum* and *F.sporotrichoides* in the field and during storage of commodities such as maize, barley, sorghum, and soybean. The IARC has evaluated the carcinogenicity of zearalenone and found it to be a possible human carcinogen (group 2B). Residues of zearalenone in meat, milk and eggs do not appear to be a practical problem [53, 54].

3.6. Trichotecenes

Trichothecenes constitute a group of 50 mycotoxins produced by *Fusarium, Cephalosporium* and *Stachybotrys* genera in different commodities. There are including T-2 toxins, deoxyniva-lenol, nivalenol, and diacetoxyscirpenol. Beside trochothecenes, deoxynivalenol (DON, wo-mitoxin) is probably the most widely distributed in cereal and soybean foods and feeds. In contaminated cereals, DON derivatives such as 3-acetyl DON and 15-acetyl DON can occur in significant amounts (10 – 20%) with DON. DON is produced by closely related *Fusarium graminearum, F.culmorum* and *F.crokwellense* species [55].

T-2 toxin produced mainly by *F.sporotrichoides* and *F.poae* is primarily associated with mould millet, wheat, rye, oats, and buckwheat. This toxin can be transmitted from dairy cattle feed to milk [56].

3.7. Alternaria toxins

Alternaria species, besides *Fusarium,* is the most isolated fungi from soybean and other cereals. Several species are known producers of toxic metabolites called *Alternaria* mycotoxins. The most important *Alternaria* mycotoxins include alternariol (AOH), alternariol monometh-yl ether (AME), altertoxins I, II, and III (ATX-I, -II, III), tenuazonic acid (TeA), and altenuene

(ALT). They belong to three structural classes: dibenzopyrone derivatives, perylene derivatives, and tetramic acid derivatives. Alternariol and related metabolites (AME and ALT) are produced by *Alternaria alternate, A.brassicae, A.citri, A.cucumerina, A.dauci, A.kikuchiana, A.solani, A.tenuissima,* and *A.tomato.* These strains are known as plant, especially fruit and vegetable pathogens. In cereals, soybean and oilseeds, AOH, AME and ALT are produced mainly by *Alternaria alternata, A.tennuisima,* and *A.infectoria.* AOH has been reported to possess cytotoxic, genotoxic, mutagenic, carcinogenic, and oestrogenic properties [27]. Tenuazonic acid (TeA) is a mycotoxin and phytotoxin produced primarily by *Alternaria alternata* and other phytopatogenic *Alternaria* species. The overview of the chemical characterisation, producers, toxicity, analysis and occurrence in foodstuffs was summarised by [27].

3.8. Sterigmatocystin

Sterigmatocystin (STC) is a precursor of the aflatoxins produced mainly by many *Aspergillus* species such as *A.versicolor, A.chevalieri, A.ruber, A.aureolatus, A.quadrilineatus, A.sydowi, Eurotium amstelodami,* and less often by *Penicillium, Bipolaris, Chaetomium,* and *Emericella* genera [30]. Sterigmatocystin was reported as a fungal metabolite in mouldy wheat, rice, barley, rapeseed, peanut, corn, and cheeses or salami. The STC producers, occurrence and toxic properties were reviewed by [30, 57].

4. Contamination level in cereal and soybean-based food and feed products

Food security strategy in the European Union (EU) includes the Rapid Alert System for Food and Feed. The RASFF was established by the European Parliament and Council Regulation No. 178/2002 laying down the general principles and requirements of food law, establishing the European Food Safety Authority and specifying the procedures in matters concerning food safety [58].

In 2002 – 2011, the number of notifications to the RASFF system due to mycotoxin contamination of food was respectively: 302, 803, 880, 996, 878, 760, 933, 669, 688, 631 notifications identifying the presence of aflatoxin B_1 (AFB_1) and the amount of AFB_1, B_2, G_1, G_2, AFM_1, ochratoxin A (OTA), fumonisins B_1 and B_2 (FB_1, FB_2), patulin, deoxynivalenol (DON) and zearalenone (ZEA) in such groups of foods, as nuts and milk, oilseeds, cereal, dried fruit, fruit, cocoa, coffee, herbs and spices, wine, milk, products for children. Approximately 95% of the notifications concerned foodstuffs contaminated with aflatoxins. During this period, the number of notifications regarding mycotoxin contamination of grains did not exceed 15% of the total number of notifications. The data in Figure 1 show that in 2002-2011 aflatoxins, ochratoxin A and fumonisins were the main contaminants isolated from cereals [59].

In the research of [60], ninety-fife cereal samples from retail shops and local markets of different locations in Pakistan were examined in terms of the presence of aflatoxins. The results showed the percentage of aflatoxin contamination samples in the commodities such as in:

Mycotoxin	Produced species	Commodities
Aflatoxins	Aspergillus flavus, A.parasiticus, A.nomius, A.bombycis, A.ochraceoroseus, A.pseudotamari	Nuts, spices, Cereals, maize, soybean, rice
Ochratoxin A	Penicillium verrucosum, P.auriantiogriseum, P.nordicum, P.palitans, P.commune, P.variabile, Aspergillus ochraceus, A.melleus, A.niger, A.carbonarius, A.sclerotiorum, A.sulphureus	Cereals, fruits, spices, coffee, Food of animal origin
Citrinin	Penicillium citrinum, P.verrucosum, P.viridicatum, Monascus purpureus	Oats, rice, corn, beans, fruits, fruit and vegetable juices, herbs and spices
Sterigmatocystin	Aspergillus versicolor, A.nidulans, A.chevalieri, A.ruber, A.aureolatus, A.quadrilineatus, Eurotium amstelodami	Cereals, cheese
Zearalenone	Fusarium graminearum, F.sporotrichoides, F.culmorum, F.cerealis, F.equiseti, F.incarnatum	Maize, soybean, cereals
Deoksynivalenol	Fusarium graminearum, F.culmorum, F.crokwellense	Maize, soybean, cereals
Fumonisins	Fusarium proliferatum, F.verticillioides,	Maize, soybean, cereals
Alternariol, alternariol monomethyl ether	Alternaria alternata, A.brassicae, A.capsici-anui, A.citri, A.cucumerina, A.dauci, A.kikuchiana, A.solani, A.tenuissima, A.tomato, A.longipes, A.infectoria, A.oregonensis	Vegetables, fruit, cereals, soybean
Tenuazonic acid	Alternaria alternata, A.capsici-anui, A.citri, A.japonica, A.kikuchiana, A.mali, A.solani, A.oryzae, A.porri, A.radicina, A.tenuissima, A.tomato, A.longipes	Vegetables, fruit, cereals, soybean

Table 3. Mycotoxigenic fungi and mycotoxins

rice (25%), broken rice (15%), wheat (20%), maize (40%), barley (20%) and sorghum (30%), while in soybean (15%). The highest contamination levels of aflatoxins were found in one wheat sample (15.5 ppb), one maize sample (13.0 ppb) and one barley sample (12.6 µg/kg). In the research of [61], seventeen samples of wheat grain from Morocco were tested for OTA and DON contamination. The results show that only two samples (11.76%) out of 17 were contaminated with OTA, at the mean concentration of 29.4 ppb. However, seven samples (41.17%) were contaminated with DON at the mean concentration of 65.9 ppb.

The aim of our own research [15] was mycotoxic analysis of grains included in the standard mixtures used in feed formulations. Eighteen samples were tested containing seeds evenly divided into three types: barley, wheat and corn. The tested seeds were from randomly selected Polish mills: the central, western, eastern and south ones (Figure 2). The aflatoxins content in 51% of the screened barley samples and in 34% of the screened wheat and maize samples did not exceed the limit set in the European Union Regulation, i.e. 4 ppb [62]. In reference to the grain origin, it was established that grains from the central and western parts of Poland exhibited the highest extent of AF_s contamination. To compare, the AFs level in wheat grains from various regions of Turkey was very low, ranging from 10.4 to 634.5

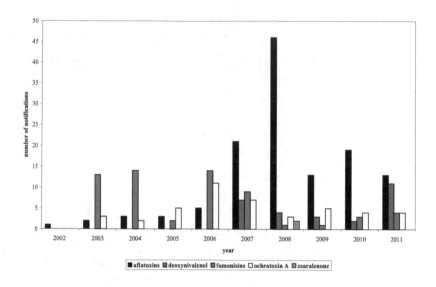

Figure 1. The number of notifications received by RASFF on mycotoxins in cereals in 2002-2011

ng/kg [63], whereas in the samples of barley, wheat, and oat grains from Sweden it was contained between 50 and 400 ppb [64].

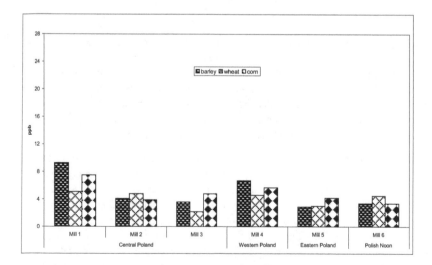

Figure 2. Level of contamination with aflatoxins in grains coming from different regions of Poland

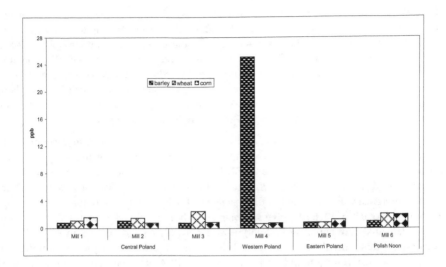

Figure 3. Level of contamination with ochratoxina A in grains coming from different regions of Poland

The OTA level in the examined grains collected from mills in central, eastern and southern Poland was low and ranged from 0.5 to 2.5 ppb (Figure 3). Therefore, it did not exceed the permissible limit set by the European Union (Commission Regulation No. 105/2010), i.e. 5 ppb [65]. Only in barley coming from a mill located in western Poland, the OTA level exceeded the limits fivefold. The extent of OTA contamination of barley, wheat, and maize grain from various regions of Mexico was also low and recorded 0.17 ppb, 0.42 ppb, and 1.08 ppb, respectively. Only 1 out of 20 examined maize grains showed the OTA level of 7.22 [66]. To compare, the OTA concentration in barley and wheat grain from the UK equalled from 1 to 33 ppb [67]. In the research of [68], among others, the levels of AFs and OTA in 532 grain and feed samples from Poland from 2002 and 2003 were determined. The average mycotoxin concentration levels were similar and quite low, i.e. AFs - 0.3 ppb and OTA - 1.1 ppb in grains and feeds from 2002, and respectively, AF$_s$ 3.1 and 1.0 ppb and OTA 0.5 and 0.7 OTA in samples from 2003. The authors of the study stressed that in 2002 and 2003 the harvesting seasons were hot and dry, which might have resulted in the low extent of fungi contamination of the examined grain. Although the extent of mycotoxin contamination of grain in the quoted studies varies, their authors concur that it is a serious issue whose scale depends on the microclimate during arable farming and the subsequent phases, i.e. grain storage. It was reported that no mycotoxins were found in barley samples stored for 20 weeks at 15% seed humidity, whereas the samples of wheat stored for the same period of time at 19% humidity recorded relatively high concentration levels: OTA - 24 ppb, citrinin - 38 ppb, and sterigmatocystin even up to 411 ppb [69].

The aim of our research was the assessment of cereal products available in trade and meant for direct consumption as for contamination with selected mycotoxins. The research included corn flakes, corn flakes with nuts and honey, various kinds of breakfast cereal products and muesli containing dried fruit, nuts as well as cereal and coconut flakes (15 samples). None of the products was contaminated with AB_1 on the level exceeding the acceptable limits (2 ppb). The presence of ochratoxin A exceeding the amount of 3 ppb was discovered in four samples (two kinds of corn flakes, exotic muesli and traditional muesli). The contamination with that toxin equalled 4.5 ppb on average. According to the current regulation, contamination of breakfast flakes with deoxynivalenol DON should not exceed 500 ppb. Four samples (containing corn) exceeded this limit by 50%. In case of one sample, DON contamination was very high, almost three times higher than the acceptable level [19].

Mycotoxin contamination of soybean is not considered a significant problem as compared to commodities such as corn, cottonseed, peanuts, barley and other grains. In the early surveys conducted by the U.S. Department of Agriculture (USDA), 1046 soybean samples collected from different regions of the United States were examined for aflatoxins contamination. Aflatoxin presence was confirmed at low levels (7-14 ppb) in only two of the tested samples [70]. In the research of [71], fifty-five samples of soybean meals were analysed for the content of aflatoxins, deoxynivalenol (DON), zearalenone (ZEA) and ochratoxin A (OTA). Regarding aflatoxins, only AFB_1 was detected in 32 out of the 51 non-suspicious samples, but the maximal concentration found was only 0.41 ppb. ZEA was detected in 23 out of the 51 samples with a maximum concentration of 18 ppb. DON could be detected only in one suspicious sample in a low concentration of 104 ppb. OTA was found in 5 samples, with the greatest concentration being only 1 ppb.

The research of [72] tested 122 soybean samples that came from Asia and the Pacific region. Aflatoxin was found in only in 2% (maximum of 13 ppb, median 9 ppb), zearalenone in 17% (maximum 1078 ppb, median 57 ppb), ochratoxin in 13% (maximum 11 ppb, median 7 ppb), and DON and fumonisins each in 7% of the analyzed samples (DON: maximum 1347ppb, median 264 ppb; fumonisins: maximum 331 ppb, median 154 ppb). In maize and maize products, the levels of fumonisins varied from 0.07 to 38.5 ppm in Latin America, from 0.004 to 330 ppm in North America, from 0.02 to 8.85 ppm in Africa, and from 0.01 to 153 ppm in Asia. The data available for Europe varied from 0.007 to 250 ppm in maize, and from 0.008 to 16 ppm in maize products. [54].

5. Influence of mycotoxins on human and animal organisms

Effects of mycotoxins on human and animal health are now increasingly recognised. Mycotoxins enter human and animal dietary systems mainly through ingestion, but increasing evidence also points to inhalation as another entry route. Mycotoxins exhibit a wide array of biological effects and individual mycotoxins can be [73]:

- carcinogenic - aflatoxins, ochratoxins, fumonisins, and possibly patulin;

- mutagenic - aflatoxins and sterigmatocystin;

- hematopoietic - aflatoxins and trichothecenes. Hemotopoiesis refers to the production of all types of blood cells from the primitive cells stem cells in the bone marrow. The dysfunction of hematopoiesis leads firstly to the decrease in the number of neutrophils, thus perturbing the animal's immune system and subsequently to the decrease in red blood cells, which leads to anemia;

- hepatotoxic - aflatoxins, ochratoxins, fumonisins. All of them induce significant liver damage when given to animals;

- nephrotoxigenic - ochratoxins, citrinin, trichothecenes, and fumonisins;

- teratogenic - aflatoxin B_1, ochratoxin A, T-2 toxin, sterigmatocystin, and zearalenone;

- oestrogenic - zearalenone;

- neurotoxic - ergot alkaloids, fumonisins, deoksynivalenol. The effects of mycotoxins are best evidenced by vomiting and taste aversion produced by DON, seizures, focal malata and liquefaction of the brain tissue, possibly mediated by sphingolipid synthesis under the influence of fumonisins, staggering and trembling produced by many tremorgenic penitrem mycotoxins seizures and other neural effects of ergot alkaloids and parasympathomimetic activity resulting from the effects of the metabolite slaframine for selected receptors in the nervous system

- immunosupresive - several mycotoxins. The predominant mycotoxins in this regard are aflatoxins, trichothecenes, and ochratoxin A. However, several other mycotoxins such as fumonisins, zearalenone, patulin, citrinin, and fescue and ergot alkaloids have been shown to produce some effects on the immune system.

Table 4 presents the groups of mycotoxins which are most harmful to human and animal organisms, together with the chosen disease symptoms they cause.

5.1. Negative effects of mycotoxins on humans

Mycotoxicoses can be divided into acute and chronic. Acute toxicity usually has a rapid onset and obvious toxic response, chronic exposure is characterized by chronic doses over a long period of time and may lead to cancer and other effects that are generally irreversible. The symptoms of mycotoxicosis depend on the type, amount and duration of exposure, age, health and sex of the exposed individual, and many poorly understood synergistic effects involving genetics, dietary status, and interaction with other toxic contaminants. Thus, the severity of mycotoxin poisoning can be compounded by factors such as vitamin deficiency, caloric deprivation, alcohol abuse, and infectious disease status. Mycotoxicosis is difficult to diagnose because doctors do not have experience with this disease and its symptoms are so wide that it mimics many other conditions [74, 75].

Aflatoxicosis is toxic hepatitis leading to jaundice and, in severe cases, death. AFB_1 has been extensively linked to human primary liver cancer and was classified by the International Agency for Research on Cancer (IARC) as a human carcinogen (Group 1A - carcinogens) [49]. Although acute aflatoxicosis in humans is rare, several outbreaks have been reported. In 2004, one of the largest aflatoxicosis outbreaks in Kenya, resulting in 317 cases and 125 deaths was observed. Contaminated corn was responsible for the outbreak, and officials

Mycotoxin	Toxicity class according to International Agency for Research on Cancer (IARC)	Symptoms and diseases
Aflatoxins	I *	aflatoxicosis, primary liver cancer, lung neoplasm, lung cancer, failure of the immune system, vomiting, depression, hepatitis, anorexia, jaundice, vascular coagulation
Ochratoxins	II B **	renal diseases, nephropathy, anorexia, vomiting, intestinal haemorrhage, tonsillitis, dehydration
Fumonisins	II B **	diseases of the nervous system, cerebral softening, pulmonary oedema, liver cancers, kidney diseases, oesophagus cancers, anorexia, depression, ataxia, blindness, hysteria, vomiting, hypotension
Zearalenone	-	reproduction disruptions, abortions, pathological changes in the reproductive system
Trichothecenes	-	nausea, vomiting, haemorrhages, anorexia, alimentary toxic aleukia, failure of the immune system, infants' lung bleeding, increased thirst, skin rash

*The agent (mixture) is carcinogenic to humans. The exposure circumstance entails exposures that are carcinogenic to humans

**The agent (mixture) is possibly carcinogenic to humans. The exposure circumstance entails exposures that are possibly carcinogenic to humans.

Table 4. The list of adverse effects of the chosen mycotoxins.

found the level of aflatoxin B_1 as high as 4400 ppb [76]. Research in Gambian children and adults reported a strong association between aflatoxin exposure and impaired immunocompetence suggesting that the consumption of aflatoxin reduces resistance to infections in human populations [77, 78]. In 1974, an epidemic of hepatitis in India affected 400 people resulting in 100 deaths. The death was due to consumption of corn that was contaminated with *A. flavus* containing up to 15000 ppb of aflatoxins [79].

Ochratoxin A was the cause of epithelial tumours of the upper urinary tract in the Balkans [80, 81]. The condition is known as Balkan endemic nephropathy. Despite the seriousness of the problem, the study did not explain the mechanism of action and the size of OTA carcinogenicity in humans [82]. Ochratoxin has been detected in blood in 6-18% of the human population in some areas where Balkan endemic nephropathy is prevalent. Ochratoxin A has also been found in human blood samples from outside the Balkan Peninsula. In some survey, over 50% of the tested samples were contaminated. A highly significant correlation was observed between Balkan nephropathy and urinary tract cancers, particularly tumours of the renal pelvis and ureter. However, no data have been published that establishes a direct causal role of ochratoxin A in the etiology of these tumours [81].

Fumonisin B_1 was classified by the IARC as a group 2B carcinogen (possibly carcinogenic for humans) [44]. Fumonisins, which inhibit the absorption of folic acid through the foliate receptor, have also been implicated in the high incidence of neural tube defects in the rural population known to consume contaminated corn, such as the former Transkei region of South Africa and some areas of Northern China [75, 83].

Trichothecenes have been proposed as potential biological warfare agents. In the years 1975-1981, T-2 toxin was implicated as a chemical agent "yellow rain" used against the Lao Peoples Democratic Republic. A study conducted from 1978 to 1981 in Cambodia revealed the presence of T-2 toxin, DON, ZEA, and nivalenol in water and leaf samples taken from the affected areas [75, 84]. Clinical symptoms proceeding to death included vomiting, diarrhoea, bleeding, and difficulty with breathing, pain, blisters, headache, fatigue and dizziness. There also occurred necrosis of the mucosa of the stomach as well as the small intestine, lungs and liver [85]. One disease outbreak was recorded in China and was associated with the consumption of scabby wheat containing 1000-40000 ppb of DON. The disease is characterized by gastrointestinal symptoms. Also, in India there took place a reported infection associated with the consumption of bread made from contaminated wheat (DON 350-8300 ppb, acetyldeoxynivalenol 640-2490 ppb, NIV 30-100 ppb and T-2 toxin 500-800 ppb). The disease is characterized by gastrointestinal symptoms and throat irritation, which developed within 15 minutes to one hour after ingestion of the contaminated bread [81].

5.2. Negative effects of mycotoxins on animal

Animals may show varied symptoms upon contact with mycotoxins, depending on the genetic factors (species, breed, and strain), physiological factors (age, nutrition) and environmental factors (climatic conditions, rearing and management). The natural contamination with mycotoxins in animal feed usually does not occur at the levels that may cause acute or overt mycotoxicosis, such as hepatitis, bleeding, nephritis and necrosis of the oral and enteric epithelium, and even death. It is often difficult to observe and diagnose the symptoms of the disease, but it certainly is the most common form of mycotoxicosis in farm animals, affecting such parameters as productivity, growth and reproductive performance, feed efficiency, milk and egg production.

The negative effects of mycotoxins on the performance of poultry have been shown in numerous studies. For example, feeding the broilers with feed containing an AF_s mixture (79% AFB_1, 16% AFG_1, AFB_2 4% and 1% AFG_2) in the concentration of 3.5 ppm decreased their body weight and increased their liver and kidney weight [75, 86]. Feeding OTA (0.3-1 ppm) to broilers reduced glycogenolysis and dose-dependent accumulation of glycogen in the liver. These negative metabolic reactions were attributed to inhibition of cyclic adenosine 3',5'-monophosphate-dependent protein kinase, and were reflected in reduced efficiency of feed utilization and teratogenic malformations [75].

Fusarium mycotoxins proved to be harmful to poultry. In addition to reduced feed intake and weight gain, sore mouth, cheeks and plaque formation was observed after 7-day-old chicks were exposed to T-2 toxin (4 or 16 ppm) [75, 87]. Pigs are among the most sensitive species to mycotoxins. In the study by [88], pigs in response to AF_s (2 ppm), OTA (2 ppm), or both were evaluated. Compared to the control group, the body weight gains were reduced by 26, 24 and 52% for animals consuming diets containing AF_s, OTA, or both, respectively. Additional symptoms in pig ochratoxicosis were anorexia, fainting, uncoordinated movements, and increased water consumption and urination. Pigs also are susceptible to other mycotoxins, such as fumonsins and ergot alkaloids. Fumonisin B_1, for example, has been shown to cause pulmonary oedema and heart and respiratory dysfunction. The symptoms of swine pulmonary oedema included dyspnoea, cyanosis, and death [89, 90]. Mycotoxic porcine nephropathy is a serious disease, often associated with pigs consuming feed contaminated with OTA, especially in Scandinavian region. In addition to the enlarged and pale kidneys (with vascular lesions and white spots), morphological changes include a proximal tubular injury, epithelial atrophy, fibrosis and hyalinization of renal glomerular [80, 81]. Negative effects of ZEA on pigs' reproductive function have also been demonstrated [91]. Oestrogenic effects of ZEA on gilts and sows include oedematous uterus and ovarian cysts, increased maturation of follicles, more numerous litters or decreased fertility [92].

Aflatoxins affect the quality of the milk produced by dairy cows and result in a carry-over of AFM_1 with AFB_1-contaminated feed. Ten ruminally-canulated Holstein cows received AFB_1 (13 mg per cow daily) through a hole in the rumen for 7 days. The AFM_1 levels in the milk of the treated cows ranged from 1.05 to 10.58 ng/L. The carry-over rate was higher in early lactation (2-4 weeks) compared to late lactation (34 -36 weeks) [75, 93]. The T-2 toxin causes necrosis of the lymphoid tissues. Bovine infertility and natural abortion in the last trimester of pregnancy also result from consumption of feed contaminated with T-2 toxin. Calves consuming T-2 toxin in the amount of 10-50 mg/kg of feed showed abomasal ulcers and sloughing of papillae in the rumen [75, 94, 95].

6. Current EU regulations concerning mycotoxins

Since the discovery of aflatoxins in the 1960s, regulations have been established in many countries to protect consumers from harmful mycotoxins that can contaminate foods. Maximum levels of mycotoxins have been established by the European Commission after consultations with the Scientific Committee for Food, based on the analysis of scientific data collected by EFSA and the Codex Alimentarius.

These data include [73, 96]:

- toxicological properties of mycotoxins,
- mycotoxin dietary exposure,
- distribution of concentrations of mycotoxins in raw materials or a product batch
- availability of analytical methods,
- regulations in other countries with which trade contacts exist.

The first two factors provide the information necessary for risk assessment and exposure assessment, respectively. Risk assessment is the scientific evaluation of the likelihood of known or potential adverse health effects resulting from human exposure to food-borne hazards. It is a fundamental scientific basis for the notification of regulations. The third and fourth factors are important factors in enabling the practical enforcement of mycotoxins, through appropriate procedures as regards sampling and analysis. The last factor is the only one economic in nature, but it is equally important in decision-making to establish reasonable rules and restrictions for mycotoxins in foods and feeds [96].

According to the Commission Regulations, the maximum levels should be set at a strict level, which is reasonably achievable by following good agricultural and manufacturing practices and taking into account the risk related to the consumption of food. Health protection of infants and young children requires establishing the lowest maximum levels, which is achievable through the selection of raw materials used for the manufacturing of foods for this vulnerable group of consumers. Development of international trade, progress in research focused on mycotoxin food contamination and their toxicological properties cause changes in the mycotoxin-related legislationacross the European Union. The Commission Regulation 466/2001 [97] setting the maximum levels for certain contaminants in foodstuffs has been substantially amended many times. Te current maximum levels for mycotoxins in food are specified by the Commission Regulation EU 1881/2006 and the Commission Regulation EU 105/2010 as regards OTA, the Commission Regulation EU 165/2010 as regards aflatoxins, and the Commission Regulation EU 1126/2007 as regards *Fusarium* toxins [62, 65, 98, 99]. There have also been established maximum levels for aflatoxins, ochratoxin A, patulin, and *Fusarium* toxin (fumonisin, deoxynivalenol, zearalenone) in different products: nuts, cereals, dried fruit, unprocessed cereals, processed cereal-based food, coffee, wine, spices, and liquorices [62, 65, 97-99].

The number of countries that have regulations concerning mycotoxins is continuously increasing, and at least 100 countries are known to have founded specific limits for different combinations of mycotoxins and commodities, often accompanied by the prescribed or recommended procedures for sampling and analysis [100]. Specific regulations for food in different world regions were summarized by [101].

As for feeds, the legal situation is somewhat different and only aflatoxin B_1 is regulated by the Directive 2002/32/EC on undesirable substances in animal food amended by the Commission Directive (EC) 100/2003 [102, 103]. For other mycotoxins, such as deoxynivalenol, zearalenone, ochratoxin A and fumonisin B_1 and B_2 - only non-binding recommendation values in the Commission Recommendation 2006/57/EC [104] are determined for feeds (Table 6). This results from the fact that with the exception of aflatoxin-contaminated feed which either directly or indirectly affects human health, there is only a slight transfer to animal products [104, 105].

Table 5 presents the current maximum levels of mycotoxin content as regards cereals and cereal-based foods and feeds.

Mycotoxins in agricultural commodities are distributed heterogeneously. Therefore, sampling plays a crucial role in making the estimation of the levels of mycotoxin presence more precise. In order to obtain representative samples, sampling procedures, and particularly homogenisation, for different matrix types have been regulated. The EU Commission Regulation (EC) 401/2006 established the methods of sampling and analysis for the official control of mycotoxins in foodstuffs [106]. Official sampling plans for aflatoxins in dry figs, groundnuts, peanuts, oilseeds, apricot kernels and tree nuts and for ochratoxins in coffee and liquorice root are provided in the Commission Regulation (EU) No 178/2010 [107]. The sampling frequency and the method of sampling for cereals and cereal products for lots >50 tonnes and <50 tonnes, as well as for retail packed products were presented. Moreover, the procedures of subdivision of lots into sublots depending on the product and lot weight were also summarised [106, 107].

According to the current regulations where no specific methods for the determination of mycotoxin levels in food are required by the EU regulations, laboratories may select any method provided that they meet the relevant criteria presented in [106, 107]. These criteria are different in relation to individual mycotoxins, and the limit of detection, precision, and recovery depends on the concentration range. The analytical results must be submitted corrected or uncorrected for recovery and the level of recovery expressed in % must be reported too.

The main analytical procedures for the determination of the major mycotoxins from complex biological matrices consist of the following steps: sampling, extraction, purification, detection, quantification, and finally confirmation. The current development in mycotoxin estimation was reviewed by [108-110].

Regulation	Matrix	Maximum levels [ppb]				
		AFB$_1$	OTA	DON	ZEA	F
	FOOD					
Commission Regulation (EU) 165/2010	All cereals and all products derived from cereals	2.0	-	-	-	-
	Maize and rice	5.0	-	-	-	-
	Processed cereal-based foods for infants and young children	0.10	-	-	-	-
Commission Regulation (EC) 1126/2007	Unprocessed cereals	-	-	1250	100	-
	Unprocessed durum wheat and oats	-	-	1750	-	-
	Pasta (dry)	-	-	750	-	-
	Bread (including small bakery wares), pastries, biscuits, cereal snacks and breakfast cereals	-	-	500	50	-
	Maize-based breakfast cereals and maize-based snacks	-	-	-	-	800
	Unprocessed maize with the exception of unprocessed maize intended to be processed by wet milling	-	-	1750	350	4000
	Cereals intended for direct human consumption, cereal flour, bran and germ as an end product marketed for direct human consumption	-	-	750	75	-

Regulation	Matrix	Maximum levels [ppb]				
		AFB$_1$	OTA	DON	ZEA	F
	Milling fractions of maize and milling products with particle size "/> 500 micron not used for direct human consumption	-	-	750	200	1400
	Milling fractions of maize and maize milling products with particle size ≤ 500 micron not used for direct human consumption	-	-	1250	300	2000
	Processed cereal-based foods for infants and young children	-	-	200	20	200
	Processed maize-based foods for infants and young children	-	-	-	20	-
Commission Regulation (EC) 1881/2006	Unprocessed cereals	-	5.0	-	-	-
	All products derived from unprocessed cereals, including processed cereal products and cereals intended for direct human consumption	-	3.0	-	-	-
	Processed cereal-based foods for infants and young children	-	0.50	-	-	-
FEED						
Commission Recommendation (EC) 576/2006	Cereals and cereal products with the exception of maize by-products	-	250	8000	2000	-
	Maize by-products	-	-	12000	3000	-
	Complementary and complete feedingstuffs for pigs	-	50	900	250	-
	Complementary and complete feedingstuffs for calves, lambs and kids	-	-	2000	500	-
	Complementary and complete feedingstuffs for poultry	-	100	-	-	-
Commission Directive (EC) 100/2003	All feed materials	20	-	-	-	-
	Complete feedingstuffs for dairy animals	5	-	-	-	-
	Complete feedingstuffs for calves and lambs	10	-	-	-	-
	Complete feedingstuffs for pigs, poultry, cattle, sheep and goats	20	-	-	-	-

(-) limit not established; AFB$_1$ – aflatoxin B$_1$; OTA – ochratoxin A; ZEA – zearalenone; DON – deoxynivalenol; F – fumonisins

Table 5. Legislation on mycotoxins as regards cereals and cereal-based foods and feeds

7. Prevention strategies of exposure to mycotoxins

Several codes of practice have been developed by Codex Alimentarius for the prevention and reduction of mycotoxins in cereals, peanuts, apple products, and other raw materials. In order for this practice to be effective, it will be necessary for the producers in each country to consider the general principles given in the Code, taking into account their local crops, climate, and agronomic practices, before attempting to implement the provisions specified in the Code. The recommendations for the reduction of various mycotoxins in cereals are divided into two parts: recommended practices based on Good Agricultural Practice (GAP) and Good Manufacturing Practice (GMP); a complementary management system to consider in the future is the use of Hazard Analysis Critical Control Point (HACCP) [111].

Recommendations to be taken into account before the harvest in order to reduce the risk of mould contamination and mycotoxin production include [112]:

- use certified seed or ensure it is free from fungal infections;
- avoid drought stress – irrigate if possible;
- sow the seed as early as possible, so that crop matures early;
- when practising minimum or zero tillage, remove crop residues;
- weed regularly;
- control insect and bird pests;
- rotate crops;
- avoid nutrient stress – apply the appropriate amount of organic or inorganic fertiliser;
- plant resistant varieties where these are available

The main mycotoxin hazards associated with pre-harvest in Europe are the toxins that are produced by fungi belonging to the genus *Fusarium* in the growing crops. It is important to note that although *Fusarium* infection is generally considered to be a pre-harvest problem, it is certainly possible for poor drying practices to lead to crops' susceptibility in storage and mycotoxin contamination [113]. This part of the book will discuss some pre-harvest strategies appropriate to reduce the prevalence of fungi belonging to the genus *Fusarium* and their mycotoxins.

7.1. Resistance

There are inherent differences in the susceptibility of cereal species to *Fusarium* infections. The differences between crop species appear to vary between countries. This is probably due to the differences in the genetic pool within each country's breeding program and the diverse environmental and agronomic conditions in which crops are cultivated [114, 115]. It was observed that oats had higher levels of DON than barley and wheat in Norway from 1996 to 1999, whereas the DON levels in wheat, barley and oats were similar when grown under the same field conditions in Western Canada in 2001 [116].

7.2. Field management

Crop rotation

Numerous studies have shown that fumonisins or DON contamination in wheat is affected by the previous crop. It was shown that a higher incidence of F_s occurred in wheat after maize and, in particular, in wheat after a succession of two maize crops and in wheat following grain maize compared to silage maize. In Ontario, Canada, in 1983, the fields where maize was the previous crop had a significantly higher incidence of fumonisins than the fields where the previous crop was a small grain cereal or soybean [117]. In a repeated study, the following year, the fields where maize was the previous crop had a 10-fold DON content than the fields following a crop other than maize [118]. The research of [119] found higher levels of fumonisins in wheat following wheat rather than wheat following fallow.

An observational study performed using commercial fields in Canada [120] identified significantly lower DON content in wheat following soybean or wheat, compared to wheat following maize. In New Zealand, an observational study determined that higher levels of DON occurred in wheat grown after maize (mean = 600 ppb) and after grass (mean = 250 ppb), compared to small grain cereals (mean = 90 ppb) and other crops (mean = 70 ppb). The highest levels were recorded in wheat-maize rotations [121].

Codex recommends that crops such as potatoes, other vegetables, clover and alfalfa that are not hosts to *Fusarium* species should be used in rotation to reduce the inoculum in the field [122].

7.3. Soil cultivation

Soil cultivation can be divided into ploughing, where the top 10-30 cm of soil are inverted; minimum tillage, where the crop debris is mixed with the top 10-20 cm of soil; and no till, where seed is directly drilled into the previous crop stubble with minimum disturbance to the soil structure [111]. In the 1990s, a large observational study of F_s and DON was conducted in Germany (n=1600). The DON concentration of wheat crops after maize was ten-times higher in the field that was min-tilled compared to the ploughed one [123]. In wheat the DON concentration after min-till was 1300 ppb, after no-till it was 700 ppb and after ploughing it was 500 ppb [120]. Studies in France have determined that crop debris management can have a large impact on the DON concentration at harvest, particularly after maize. The highest DON concentration was found after no-till, followed by min-till, whereas the lowest DON levels were recorded after ploughing. The reduction in DON has been linked to the reduction in crop residue on the soil surface [124]. Large replicated field trials in Germany identified that there was a significant interaction between the previous crop and the cultivation technique [125]. Following sugar beet, there was no significant difference in the DON concentration between wheat plots receiving different methods of cultivation; however, following a wheat crop without straw removal, direct drilled wheat had a significantly higher DON level compared to wheat from plots which were either ploughed or min-tilled [125].

In accordance with the guidelines contained in the Codex Alimentarius, soil should be tested to determine if there is need to apply a fertilizer and/or soil conditioners to assure adequate soil pH and plant nutrition to avoid plant stress, especially during seed development [122].

Research of [126] showed that supplementary nitrogen and a plant growth regulator increased, by up to 125%, the incidence of infection by *Fusarium* species in the seed of wheat, barley and triticale. Similarly, in the studies of [127], a significant increase in fumonisins and deoxynivalenol contamination in the grain of wheat and kernels was observed with increasing N fertilizer from 0 to 80 kg/ha. That research concluded that in practical crop husbandry, F_s cannot be sufficiently controlled by only manipulating the N input [111]. The study of [128] showed that the use of six different combinations of agricultural practices (sowing time, plant density, N fertilization and European corn borer (ECB) control with insecticide) can effectively lead to good control of fumonisins and deoxynivalenol in maize kernels.

7.4. Use of chemical and biological agents

In accordance with the guidelines contained in the Codex Alimentarius [122], farmers should minimize insect damage and fungal infections of the crop by proper use of registered insecticides, fungicides and other appropriate practices within an integrated pest management program.

Some studies have been conducted to examine the effectiveness of the fungicides which are applied during flowering can reduce *Fusarium* infections and subsequent DON in the harvested grains. The results of [129] provided that azoles, tebuconazole, metconazole and prothioconazole significantly reduced the *Fusarium* disease symptoms and *Fusarium* mycotoxin concentrations. The greatest reduction in the DON concentration occurred with prothioconazole (10-fold). Azoxystrobin had little impact on the mycotoxin concentration in the harvested grain infected by *Fusarium* species, but could increasing the mycotoxin concentration in grains when *F. nivale* was the predominant species present [130, 131]. Fungicide mixtures of azoxystrobin and azole resulted in a lower reduction of DON, compared to azole alone [120, 132]. A number of trials in Germany have indicated that some strobilurin fungicides applied before anthesis can also result in increased DON compared to unsprayed plots [133]. Reductions in DON observed in field experiments using fungicides against natural infections of *Fusarium* are lower and inconsistent [134]. This is probably due to the fact that during a natural infection, the infection occurs over a longer period of time.

Alternatively, a limited number of biocompetitive microorganisms have been shown useful for the management of *Fusarium* infections [111]. Research has demonstrated the successful use of bacteria in biocontrol of mycotoxigenic fungi. One bacterium, *Enterobacter cloacae* was discovered as an endophytic symbiont of corn [135]. Corn plants with roots endophytically colonized by these bacteria were observed to be fungus-free and *in vitro* control of *F.verticillioides* and other fungi with this bacterium was demonstrated. An endophytic bacterium, *Bacillus subtilis* showed promising for reducing the mycotoxin contamination with *F.verticillioides* during the endophytic growth phase [136]. Yeast antagonists such as *Cryptococcus nodaensis* were isolated from wheat anthers. The antagonists reduced *Fusarium* head blight severity by up to 93% in greenhouse and by 56% in field trials when sprayed onto flowering wheat heads [137]. The most successful antagonists reduced the DON content of grain more than 10-fold in greenhouse studies [138].

Actions to be taken during harvest in order to reduce the risk of mould contamination and mycotoxin production include [112]:

- harvest as quickly as possible

- avoid field drying

- transport the crop to the homestead as soon as possible

- if lack of labour force or time prevents removal from the field, then dry the crops on platforms raised above ground (if climate is hot and the drying crop can be left to stay on the field on a platform or cut and tied into stooks) to dry

- bundles of stover should also be placed on platforms to dry and not left lying on the soil

The post-harvest strategies include improving the drying and storage conditions together with the use of chemical, physical or biological methods.

8. Methods of removing mycotoxins from cereals

When mycotoxin prevention is not satisfactory, some decontamination methods are needed. The use of detoxification methods is allowed only in the case of feed and feed components. Foodstuffs containing contaminants exceeding the maximum levels should not be placed on the market either as such, in the form of a mixture with other foodstuffs or used as an ingredient in other foods. Food contaminated with mycotoxins is not safe for consumers and no decontamination methods can be used.

According to FAO [111, 139, 140] the feed decontamination process must:

- destroy, inactivate or remove mycotoxins

- not produce toxic, carcinogenic or mutagenic residues in decontaminated final products

- not decrease the nutritive value and organoleptic properties

- destroy all fungal morphological forms

- not significantly increase the cost of production

There are some physical methods of decontamination of feed components such as sorting grains, washing procedures, gamma radiation and UV treatment and also extraction with organic solvents. These methods are summarized by [140]. Physical removal of damaged, mouldy or discoloured kernels significantly decreased the concentration of AF in peanuts. Sorting is not effective for maize and cottonseed. Washing with water or sodium carbonate solutions could decrease the concentration of DON, ZEA and fumonisins in wheat and maize.

High temperature is not used for decontamination of agricultural products, due to thermostability of mycotoxins. Different types of radiation were tested for mycotoxin detoxification, but the results were not effective enough.

Chemical compounds such as organic acids, ammonium, sodium hydroxide, hydrogen peroxide, ozone, chloride and bisulphite were tested for their efficacy in mycotoxin decontamination [141, 142]. Chemical decontamination is very effective, but these methods are expensive and affect the feedstuff quality. Among the chemical methods, only peroxide and ammonia are mostly used for aflatoxin removal from feed. Ammoniation works by irreversibly converting AFB_1 to less toxic products such as AFD_1 [143]. Data show that treatment of maize contaminated with 1000 or 2000 ppb aflatoxins with 1% of aqueous ammonia for 48 h removed 98% of the aflatoxins. There was no significant change in the dietary intake, body weight gain, and feed conversion ratio in chickens fed with ammonia-treated aflatoxin-contaminated maize, whereas these parameters were suppressed in birds fed with aflatoxin-containing diet [142]. Atmospheric ammoniation of corn does not appear to be an effective method for the detoxification of *F.moniliforme*–contaminated material. In the research of [144], the levels of fumonisin B_1 in naturally contaminated corn were reduced by about 45% due to the ammonia treatment. Despite this, the toxicity of the culture material in rats was not altered by ammoniation.

A recent and promising approach to protect animals against the harmful effects of mycotoxin-contaminated feed is the use of mycotoxin binders (MB). They are added to the diet in order to reduce the absorption of mycotoxins from the gastrointestinal tract and their distribution to blood and target organs. These feed additives may act either by binding mycotoxins to their surface (adsorption), or by degrading or transforming them into less toxic metabolites (biotransformation). Various inorganic adsorbents, such as hydrated sodium calcium aluminosilicate, zeolites, bentonites, clays, and activated carbons, have been used as mycotoxin binders. The use of mycotoxin binders is discussed in some review articles [145-147]. The best aflatoxin adsorbent seems to be HSCAS (hydrated sodium calcium aluminosilicate), which rapidly and preferentially binds aflatoxins in the gastrointestinal tract [148-150]. The prevention of aflatoxicosis in broiler folders was examined by [150]. HSCAS and activated charcoal were incorporated into the diets for broilers containing purified aflatoxin B_1 (7.5 ppm), or natural aflatoxin produced by *Aspergillus parasiticus* on rice (5 ppm). The authors showed that HSCAS significantly decreased the growth-inhibitory effects of AFB_1 or AF_s on the growing chicks, namely by 50 to 67%. The authors suggest that HSCAS can modulate the toxicity of aflatoxins in chickens; however, adding activated charcoal to the diet did not appear to have protective properties against mycotoxicosis [150].

Physical and chemical methods have a lot of disadvantages; in many cases they do not meet the FAO requirements. Therefore, the use of other methods is considered. Biological methods, involving decontamination with microorganisms or enzymes, give promising results. Recently, an increase in the research connected with mycotoxin detoxification by microorganisms has been observed. Several studies have shown that some bacteria, moulds and yeasts such as *Flavobacterium auriantiacum*, *Corynebacterium rubrum*, lactic acid bacteria (*Lactobacillus acidophilus*, *L.rhamnosus*, *L.bulgaricus*), *Aspergillus niger*, *Rhizopus nigricans*, *Candida sp.*, *Kluyveromyces sp.*, etc. are able to conduct detoxification of mycotoxins (Tab. 6). Unfortunately, few of these findings have practical application.

Already in 1966, a review of microorganisms was conducted by [151] as for their capability of degrading aflatoxins. It was found that yeasts, actinomycetes and algae did not show this trait, but some moulds, such as *Aspergillus niger*, *A. parasiticus*, *A. terreus*, *A. luchuensis*, and

Penicillium reistrickii, partially transformed aflatoxin B_1 to a new product. Among them, only the bacteria *Flavobacterium aurantiacum* (now *Nocardia corynebacterioides*) is able to remove aflatoxin, both from the media and from the natural environments such as milk, oil, cocoa butter and grain. It was shown that to obtain the apparent loss of the toxin, it was necessary to use the bacterial population with the density of more than 10^{10} CFU/ml [154, 188].

Mycotoxin	Microorganism	References
Aflatoxin B_1	*Flavobacterium aurantiacum (Nocardia corynebacterioides), Lactobacillus acidophilus, L.johnsonii, L.salivarius, L.crispatus, L.gasseri, L.rhamnosus, Lactococcus lactis, Bifidobacterium longum, B.lactis, Mycobacterium luoranthenivorans, Rhodococcus erythropolis, Bacillus megaterium, Corynebacterium rubrum, Kluyveromyces marxianus, Saccharomyces cerevisiae, Aspergillus niger, A. terreus, A.luchuensis, Penicillium reistrickii, Trichoderma viride*	[151-165]
Ochratoxin A	*Lactococcus salivarius subsp. thermophilus, Lactobacillus delbrueckii subsp. Bulgaricus, L. acidophilus, Bifidobacterium animalis, B. bifidum, Lactobacillus plantarum, L. brevis, L. sanfranciscensis, L.acidophilus, Acinetobacter calcoaceticus, Rhodococcus erythropolis, Oenococcus oeni, Saccharomyces cerevisiae, Kluyveromyces marxianus, Rhodotorula rubra, Phaffia rhodozyna, Xanthophyllomyces dendrorhous, Metschnikowia pulcherrima, Pichia guilliermondii, Trichosporon mycotoxinivorans, Rhizopus sp., Aureobasidium pullulans, Aspergillus niger, A.carbonarius, A. fumigatus, A. versicolor*	[166-183]
Fumonisin B_1	*Lactobacillus rhamnosus, Lactococcus lactis, Leuconostoc mesenteroides, Saccharomyces cerevisiae, Kluyveromyces marxianus, Rhodotorula rubra*	[176, 184]
Trichotecenes	Ruminant bacteria, chicken intestinal microflora, *Saccharomyces cerevisiae, Kluyveromyces marxianus, Rhodotorula rubra*	[176, 185, 186]
Zearalenone	Soil bacteria, *Propionibacterium fraudenreichii, Rhizopus sp., Trichosporon mycotoxinivorans*	[179, 183, 187]

Table 6. Decontamination abilities of microorganisms

It was observed that cultures of toxinogenic *Aspergillus flavus* and *Aspergillus parasiticus* were able to reduce aflatoxin contamination. Aflatoxins were degraded by the strains that produce them, but only after the fragmentation of the mycelium. The cause of this phenomenon was absorption into the cell wall of mycelium [165]. In the research of [176], 10 yeast strains of the *Saccharomyces, Kluyveromyces* and *Rhodotorula* genera were studied for their ability to perform biodegradation of fumonisin B_1, ochratoxin A and trichothecenes. Significant differences were demonstrated between the strains, but there were no preferences as to the types of mycotoxins. Fumonisins were removed by the majority of the strains in 100%, the removal rate for deoxynivalenol ranged from 63 to 100%, and for ochratoxin A from 69 to 100%. The possibility of using moulds to remove ochratoxin A was studied by [179, 182]. The authors selected two out of 70 isolates of the *Aspergillus* species - *Aspergillus fumigatus* and *Aspergillus niger*, which transformed ochratoxin A to ochratoxin α and phenylalanine within 7 days of incubation on both liquid and solid media.

In vitro studies conducted by [186] demonstrated the degradation of 12 trichothecene mycotoxins conducted by bacteria isolated from the digestive tract of chickens. The transformation of the toxin led to their partial or total deacylation and de-epoxidation. Similarly, it was shown, that the strains of anaerobic bacteria - isolated from the rumen, Gram positive, pre-classified to the genus *Eubacterium* - are able to perform the transformation of type A trichothecenes to non-toxic forms [185].

The above-presented examples of microbial activity aimed at removal of mycotoxins are mainly of scientific nature, allowing for a better understanding of the strains, their properties and the mechanisms of the processes. Their limited practical application made that research turned in the direction of such organisms, which can be used in biotechnological processes during production, such as fermented food production, where the raw material may be contaminated with mycotoxins. The most important among them are lactic acid bacteria and yeasts *Saccharomyces cerevisiae* [163].

Literature data indicate the existence of strains of lactic acid bacteria with different abilities to remove mycotoxins, as demonstrated both in *in vitro* and *in vivo* studies conducted by various authors with the use of some strains of probiotic *Lactobacillus rhamnosus, Lactobacillus acidophilus, Bifidobacterium bifidum, B. longum,* and *Streptococcus* spp., *Lactococcus salivarius, Lactobacillus delbrueckii* subsp. *bulgaricus* [155, 156, 158, 160, 169, 189, 190]. According to [191], the decontamination process is very fast; after 4h the toxin concentration was reduced from 50 to 77%. It was observed that heat-inactivated cells were more effective than living cells, which results from the changes in the surface properties of cells, which occur under high temperature [191]. The capacity to reduce the content of ochratoxin A in milk by lactic acid bacteria belonging to the species *Lactococcus salivarius, Lactobacillus delbrueckii* subsp. *bulgaricus* and *Bifidobacterium bifidum* was confirmed in [167]. The content of patulin in the medium decreased in the level from 10 to 82% under the influence of bacteria belonging to the genus *Lactobacillus* and *Bifidobacterium*. The decontamination process depends on the inoculum density, pH and the concentration of toxins. Among the studied strains, *L.acidophilus*, removes up to 96% of the toxin added to the medium in an amount of 1ppm [166].

Our *in vivo* experiments indicate that the use of probiotics as feed additives limited the effects of mycotoxins in animals, as well as reduced the accumulation of toxins in the tissues, thus reducing the contamination of food of animal origin with the toxins [192]. It was shown that *Lactobacillus rhamnosus* bacteria limited by 75% the adsorption of aflatoxin B_1 in the digestive tract of chickens [189].

The second group of organisms with a potential application in detoxification is constituted by *Saccharomyces cerevisiae* yeasts. Our own research demonstrated that these organisms are capable of eliminating ochratoxin A from the plant raw material during fermentation and chromatographic analysis did not show any products of OTA metabolism, which proves that it was not the case of biodegradation. The amount of ochratoxin A removed by bakery yeasts after 24-

hour contact equalled from 29% to 75% for 5 mg d.m/ml and 50 mg d.m./ml, respectively. The process of adsorption proved to be very fast; immediately after mixing the cells with the toxin its amount significantly decreased, and lengthening the contact up to 24 hours did not bring further notable changes. The presence of physiologically active cells is not necessary in order to remove the toxin; the dead biomass also removed OTA from the buffer and the amount of the toxin removed was much bigger than in the case of the active biomass. In the case of the 5 mg/ml density, 54% of the toxin was adsorbed, i.e. twice more than in the case of the active biomass [171]. The reason for OTA removal was adsorption of the toxin to the yeast cell wall. This mechanism was independent of the type of toxin, as demonstrated in relation to aflatoxin B_1, zearalenone and T-2 toxin and patulin. The compounds of the cell wall that are involved in the binding process are probably β-D-glucan and its esterified form [193, 194]. Yeasts and their cell wall components are also used as feed additives for animals, and as adsorbents, which effectively limits mycotoxicosis in farm animals [195, 196].

The potential application of yeasts as adsorbents for foods and feeds depends on the stability of the toxin binding to the cells in the conditions of the gastrointestinal tract. According to [194], zearalenone adsorption is most effective at a pH close to neutral and acidic, and therefore those which prevail in some regions of the gastrointestinal tract. The result of the use of yeasts to remove ochratoxin A is detoxification of the environment, as demonstrated in the cytotoxicity and genotoxicity tests using pig kidney cell lines [197]. Some yeasts also exhibit features of probiotic activity, which is an additional argument for the use of these organisms

The use of microorganisms or their cell components for decontamination of foods and feeds has raised high hopes, but also the controversy from the perspective of the consumer. There are no legal regulations devoted to this issue, and the data referring to the stability of the microorganism-toxin connection in the gastrointestinal tract, as well as toxicological data are still incomplete. The only group of microorganisms, which in addition to other advantageous features of health promotion has the ability to remove toxins, is probably that of probiotic lactic acid bacteria. Also, *Saccharomyces cerevisiae* yeast and its cell wall component - glucan can be used for this purpose. These factors can be applied both as human dietary supplements and ingredients in animal nutrition, as well as during biotechnological processes.

Author details

Małgorzata Piotrowska[1*], Katarzyna Śliżewska[1] and Joanna Biernasiak[2]

*Address all correspondence to: malgorzata.piotrowska@p.lodz.pl

1 Technical University of Lodz, Faculty of Biotechnology and Food Sciences, Institute of Fermentation Technology and Microbiology, Lodz, Poland

2 Technical University of Lodz, Faculty of Biotechnology and Food Sciences, Institute of Chemical Technology of Food, Lodz, Poland

References

[1] Legan JD. Cereals and cereal products. In: Lund BM, Baird-Parker TC, Gould GW. (eds.) The microbiological safety and quality of food. Gaithersburg: Aspen Publishers Inc; 2000. p. 759-783.

[2] Miller JD. Fungi and mycotoxins in grain: implications for stored product research. Journal of Stored Product Research 1995;31 1-16.

[3] Jeleń HH, Majcher M, Zawirska-Wojtasik R, Wiewiórowska M, Wąsowicz E. Determination of geosmin, 2-methylisoborneol, and a musty-earthy odor in wheat grain by SPME-GC-MS, profiling volatiles, and sensory analysis. Journal of Agriculture and Food Chemistry 2003;51 7079-7085. DOI:10.1021/jf030228g.

[4] Bhattacharya K, Raha S. Deteriorative changes of maize, groundnut and soybean seeds by fungi in storage. Mycopathologia 2002;155 135–141. DOI:10.1023/A: 1020475411125

[5] Filtenborg O, Frisvad JC, Thrane U. Moulds in food spoilage. International Journal of Food Microbiology 1996;33 85-102.

[6] Pacin AM, Gonzalez HHL, Etcheverry M, Resnik SL, Vivas L, Espin S. Fungi associated with food and feed commodities from Ecuador. Mycopathologia 2002;156 87-92. DOI:10.1023/A:1022941304447

[7] Tabuc C, Stefan G. Assessment of mycologic and mycotoxicologic contamination of soybean, sunflower and rape seeds and meals during 2002 – 2004. Archiva Zootechnica 2005;8 51-56.

[8] Miller W A, Roy KW. Mycoflora of soybean leaves, pods, and seeds in Mississippi. Canadian Journal of Botany 1982;60(12) 2716-2723.

[9] Villarroel DA, Baird RE, Trevathan LE, Watson CE, Scruggs ML. Pod and seed mycoflora on transgenic and conventional soybean [Glycine max (L.) Merrill] cultivars in Mississippi. Mycopathologia 2004;157 207-215. DOI:10.1023/B:MYCO. 0000020591.71894.48.

[10] Leslie JF, Pearson CAS, Nelson PE, Toussoun TA. Fusarium spp. from corn, sorghum, and soybean fields in the central and eastern United States. Phytopathology 1990;80 343-350.

[11] Ivić D, Domijan A-M, Peraica M, Miličević T, Cvjetković B. Fusarium spp. contamination of wheat, maize, soybean, and pea in Croatia. Archives of Industrial Hygiene and Toxicology 2009;60 435-442. DOI:10.2478/10004-1254-60-2009-196.

[12] Pereyra CM, Cavaglieri LR, Chiacchiera SM, Dalcero AM. Mycobiota and mycotoxins contamination in raw materials and finished feed intended for fattening pigs production in eastern Argentina. Veterinary Research Communication 2011;35 367-379. DOI:10.1007/s11259-011-9483-9.

[13] Barros GG, Oviedo MS, Ramirez ML, Chulze SN. Safety aspects in soybean food and feed chains: fungal and mycotoxins contamination. In: Tzi-Bun Ng (ed.) Soybean - Biochemistry, Chemistry and Physiology. InTech Open Access Company; 2011. p7-20.

[14] Weidenborner M, Wieczorek C, Appel S, Kunz B. Whole wheat and white wheat flour – the mycobiota and potential mycotoxins. Food Microbiology 2000;17 103-107.

[15] Biernasiak J, Piotrowska M, Śliżewska K, Libudzisz Z. Microbiological and mycotoxin characterization of animal feeds components (in polish).Cereals and milling review. 2012 (in press).

[16] Pacin AM, Broggi LE, Resnik SL, Gonzalez HHL. Mycoflora and mycotoxins natural occurrence in corn from entre Rios Province, Argentina. Mycotoxin Research 2001;17 31-38.

[17] Torres AM, Ramirez ML, Chulze SN. *Fusarium* and fumonisins in maize in South America. In: Rai M, Vatma A. (eds.) Mycotoxins in food, feed and bioweapons, Springer-Verlag Berlin Haidelberg; 2010, p179-200.

[18] Błajet-Kosicka A, Grajewski J, Twarużek M, Kosicki R, Rychlewska J. Mycotoxins and fungal contamination of conventional and ecological oats. In: Proceedings of the 9th International Confernece Mycotoxins and Moulds, 2010, Bydgoszcz, Poland, p59.

[19] Piotrowska M. Contamination with moulds and mycotoxins of cereal products for direct consumption. In: Proceedings of the 3rd International Conference on Quality and Safety in Food Production Chain, 2007, Wrocław, Poland, p380.

[20] Roy KW, Baird RE, Abney TS. A review of soybean (*Glycine max*) seed, pod, and flower mycofloras in North America, with methods and a key for identification of selected fungi. Mycopathologia 2000;150 15-27. DOI:10.1023/A:1010805224993.

[21] Yang XB, Feng F. Ranges and diversity of soybean fungal diseases in North America. Phytopathology 2001;91(8) 769-775.

[22] Aoki T, O'Donnell K, Scandiani MM. Sudden death syndrome of soybean in South America is caused by four species of *Fusarium*: *Fusarium brasiliense* sp. nov., *F.cuneirostrum* sp. nov., *F.tucumaniae*, and *F.virguliforme*. Mycoscience 2005;46 162-183. DOI: 10.1007/s10267-005-0235-y.

[23] Chelkowski J. Fungal pathogens influencing cereal seed quality at harvest. In: Chelkowski J. (ed.). Cereal Grain. Mycotoxins, fungi and quality in drying and storage. Amsterdam: Elsevier. 1991. p67-80.

[24] Bakutis B, Baliukonienė V, Lugauskas A. Factors predetermining the abundance of fungi and mycotoxins in grain from organic and conventional farms Ekologija 2006; (3) 122-127.

[25] Lacey J, Magan N. Fungi in cereal grains: their occurrence and water and temperature relationships In: Chelkowski J. (ed.). Cereal Grain. Mycotoxins, fungi and quality in drying and storage. Amsterdam: Elsevier, 1991. p77-112.

[26] Oviedo MS, Ramirez ML, Barros GG, Chulze SN. Effect of environmental factors on tenuazonic acid production by Alternaria alternata on soybean-based media. Journal of Applied Microbiology 2009;107 1186-1192. DOI:10.1111/j.1365-2672.2009.04301.x.

[27] Ostry V. *Alternaria* mycotoxins: an overview of chemical characterization, producers, toxicity, analysis and occurrence in foodstuffs. World Mycotoxin Journal 2008;1(2) 175-188. DOI:10.3920/WMJ2008.x013.

[28] Sweeney MJ, Dobson ADW, Mycotoxin production by *Aspergillus, Fusarium* and *Penicillium* species. International Journal of Food Microbiology 1998;43 141-158.

[29] Northold MD, Frisvad JC, Samson RA. Occurrence of food-borne fungi and factors for growth. In: Samson RA, Hoekstra ES, Frisvad JC, Filtenborg O. (eds.) Introduction to food-borne fungi. CBC, Baarn, The Netherland, 1996. p243-250.

[30] Versilovskis A, De Saeger S. Sterigmatocystin: Occurrence in foodstuffs and analytical methods – An overview. Molecular Nutrition and Food Research 2010;54 136-147. DOI:10.1002/mnfr.200900345.

[31] Garcia D, Barros G, Chulze S, Ramos AJ, Sanchis V, Marin S. Impact of cycling temperatures on *Fusarium verticillioides* and *Fusarium graminearum* growth and mycotoxins production in soybean. Journal of the Science of Food and Agriculture 2012; (wileyonlinelibrary.com) DOI:10.1002/jsfa.5707.

[32] Taniwaki MH, Hocking AD, Pitt JI, Fleet GH. Growth and mycotoxin production by food spoilage fungi under high carbon dioxide and low oxygen atmospheres. International Journal of Food Microbiology 2009;132 100-108. DOI:10.1016/j.ijfoodmicro. 2009.04.005

[33] Guzman-de-Peña D, Aguirre J Ruiz-Herrera J. Correlation between the regulation of sterigmatocystin biosynthesis and asexual and sexual sporulation in *Emericella nidulans*. Antonie van Leeuwenhoek 1998;73 199-205.

[34] Škrinjar M, Danev M, Dimic G. Interactive effects of propionic acid and temperature on growth and ochratoxin A production by *Penicillium aurantiogriseum*. Folia Microbiologica 1995;40(3) 253-256.

[35] Montville TJ, Shih P-L. Inhibition of mycotoxigenic fungi in corn by ammonium and sodium bicarbonate. Journal of Food Protection 1991;54(4) 295-297.

[36] Lee HB, Magan N. Environment factors influence in vitro interspecific interaction between *A.ochraceus* and other maize spoilage fungi, growth and ochratoxin production. Mycopathologia1999;146 43-47.

[37] Green H, Larsen J, Olsson PA, Jensen DF. Suppression of the biocontrol agent *Trichoderma harzianum* of the arbuscular mycorrhizal fungus *Glomus intraradices*. Soil Applied Environmental Microbiology 1999;65(4) 1428–1434.

[38] Wiśniewska H, Basiński T, Chełkowski J, Perkowski J. *Fusarium sporotrichioides* Sherb. toxins evaluated in cereal grain with *Trichoderma harzianum.* Journal of Plant Protection Research 2011;51(2) 134-139. DOI:10.2478/v10045-011-0023-y

[39] Vaamonde G, Bonera N. Zearalenone production by *Fusarium* species isolated from soybeans. International Journal of Food Microbiology 1987;4 129–133.

[40] Song DK, Karr AL. Soybean phytoalexin, glyceollin, prevent accumulation of aflatoxin B_1 in cultures of *Aspergillus flavus.* Journal of Chemical Ecology 1993;19(6) 1183-1194.

[41] Glenn AE. Mycotoxigenic *Fusarium* species in animal feed. Animal Feed Science and Technology 2007;137 213-240. DOI:10.1016/j.anifeedsci.2007.06.003.

[42] Desjardins AE, Proctor RH. Molecular biology of *Fusarium mycotoxins.* International Journal of Food Microbiology 2007;119 47-50. DOI:10.1016/j.ijfoodmicro.2007.07.024.

[43] Bullerman LB, Bianchini A. Stability of mycotoxins during food processing. International Journal of Food Microbiology 2007;119 140-146. DOI:10.1016/j.ijfoodmicro. 2007.07.035

[44] IARC (International Agency for Research on Cancer). Traditional herbal medicines, some mycotoxins, napthalene, and styrene. Monographs on the Evaluation of Carcinogenic Risks to Humans. IARC 2002;82-171.

[45] Moss MO. Risk assessment for aflatoxins in foodstuffs. International Biodeterioration & Biodegradation 2002;50 137-142.

[46] Peterson SW, Ito Y, Horn BW, Goto T. *Aspergillus bombycis,* a new aflatoxigenic species and genetic variation in its sibling species, *A. nomius.* Mycologia 2001;93 689-703.

[47] Ito Y, Peterson SW, Wicklow DT, Goto T. *Aspergillus pseudotamarii,* a new aflatoxin producing species in *Aspergillus* section *Flavi.* Mycological Research 2001;105 233-239.

[48] Larsen TO, Svendsen A, Smedsgaard J. Biochemical characterization of ochratoxin A producing strains of genus *Penicillium.* Applied of Environmental Microbiology 2001;8 3630-3635.

[49] IARC, International Agency for Research on Cancer. IARC monographs on the evaluation of carcinogenic risk to humans. IARC Lyon, France. 1993;56 445-466.

[50] EFSA Panel on Contaminants in the Food Chain (CONTAM); Scientific Opinion on the risks for public and animal health related to the presence of citrinin in food and feed1 EFSA Journal 2012;10(3) 2605 [82 pp.]. DOI:10.2903/j.efsa.2012.2605

[51] Scott PM, Recent research on fumonisins: a review. Food Additives & Contaminants: Part A: Chemistry, Analysis, Control, Exposure & Risk Assessment, 2012;29(2) 242-248. DOI:10.1080/19440049.2010.546000.

[52] Fumonisin B_1 IARC monographs. Monographs on the Evaluation of Carcinogenic Risks to Humans 2002;82 301-366.

[53] Carvet S, Lecoeur S. Fusariotoxin transfer in animal. Food and Chemical Toxicology, 2006;44 444-453. DOI:10.1016/j.fct.2005.08.021

[54] Opinion of the Scientific Committee on Food on *Fusarium* toxins part 3: Fumonisin B1 (2000).   SCF/CS/CNTM/MYC/   24   FINAL.   http://ec.europa.eu/food/fs/sc/scf/ out73_en.pdf (accesed 7 June 2012).

[55] Yazar S, Omrtag GZ. Fumonisins, trichothecenes and zearalenone in cereals. International Journal of Molecular sciences, 2008;9 2062-2090. DOI:10.3390/ijms9112062.

[56] EFSA Panel on Contaminants in the Food Chain (CONTAM); Scientific Opinion on the risks for animal and public health related to the presence of T-2 and HT-2 toxin in food and feed. EFSA Journal 2011;9(12) 2481, 1-187. DOI:10.2903/j.efsa.2011.2481.

[57] Rank C, Nielsen KF, Larsen TO, Varga J, Samson RA, Frisvad JC. Distribution of sterigmatocystin in filamentous fungi. Fungal Biology 2011;115 406-420. DOI:10.1016/ j.funbio.2011.02.013.

[58] Regulation (EC) 178/2002 of the European Parliament and of the Council of 28 January 2002 laying down the general principles and requirements of food law, establishing the European Food Safety Authority and laying down procedures in matters of food safety. Official Journal of the European Union 2002;L31 1-24.

[59] European Commission Health and Consumers Food: Rapid Alert System for Food and Feed. http://ec.europa.eu/food/food/rapidalert/ (accessed 3 June 2012)

[60] Lutfullah G, Hussain A. Studies on contamination level of aflatoxins in some cereals and beans of Pakistan. Food Control 2012;23 32-36. DOI:10.1016/j.foodcont. 2011.06.004

[61] Hajjaji A, Otmani ME, Bouya D, Bouseta A, Mathieu F, Collin S, Lebrihi A. Occurrence of mycotoxins (ochratoxin A, deoxynivalenol) and toxigenic fungi in Moroccan wheat grains: impact of ecological factors on the growth and ochratoxin A production. Molecular Nutrition and Food Research 2006;50 494-499. DOI:10.1002/mnfr. 200500196

[62] Commission Regulation (EU) No 165/2010 amending Regulation (EC) No 1881/2006 setting maximum levels for certain contaminants in foodstuffs as regards aflatoxins. Official Journal of the European Communities 2010;L50 8-12.

[63] Giray B, Girgin G, Engin AB, Aydin S, Sahin G. Aflatoxin levels in wheat samples consumed in some regions of Turkey. Food Control 2007;18 23-29. DOI:10.1016/ j.foodcont.2005.08.002.

[64] Pettersson H, Holmberg T, Lavsson K, Kaspersson A. Aflatoxins in acid-threated grain in Sweden and occurrence of aflatoxin M_1 in milk. Journal of the Science of Food and Agriculture 1989;48 411-420.

[65] Commission Regulation (EU) 105/2010 amending Regulation (EC) No 1881/2006 setting maximum levels for certain contaminants in foodstuffs as regards ochratoxin A. Official Journal of the European Union 2010;L35 7-8.

[66] Zinedine A, Brera C. Elakhdari S, Cantano C, Debegnach F, Angelini S, De Santis B, Faid M, Benlemlih M, Minordi V, Miraglia M. Natural occurrence of mycotoxins in cereals and spices commercialized in Marocco. Food Control 2006;17 868-874. DOI: 10.1016/j.foodcont.2005.06.001

[67] Pittet A. Natural occurrence of mycotoxins in food and feeds - an update review. Veterinary Medical Review 1998;149 479-492.

[68] Bancewicz E, Jędryczko R, Jarczyk A, Szymańska A. Contamination of feed components and established their mycotoxins in 2002-2003. Proceedings of VII International Scientific Conference on Mycotoxins and pathogenic molds in the environment. 2004. Bydgoszcz, Poland. 201-204.

[69] Abramson D, Hulasare R, White NDG, Jayas DS, Marquardt RK. Mycotoxin formation in hulless barley during granary storage at 15 and 19% moisture content. Journal of Stored Products Research 1999;35(3) 297-305.

[70] Nesheim S, Wood GE. Regulatory aspects of mycotoxins in soybean and soybean products. Journal of the American Oil Chemists' Society 1995;72(12) 1421-1423.

[71] Valenta H, Dänicke S, Blüthgen A. Mycotoxins in soybean feedstuffs used in Germany. Mycotoxins Research 2002;18(2) 208-2011.

[72] Binder EM, Tan LM, Chin LJ, Handl J, Richard J. Worldwide occurence of mycotoxins in commodities, feeds and feed ingredients. Animal Feed Science and Technology 2007;137 265-282. DOI:10.1016/j.anifeedsci.2007.06.005.

[73] CAST. Mycotoxins: Risks in plant, animal and human systems. Report No. 139. Council for Agricultural Science and Technology. USA: Ames, Iowa; 2003.

[74] James B, Adda C, Cardwell K, Annang D, Hell K, Korie S, Edorh M, Gbeassor F, Nagatey K, Houenou G. Public information campaign on aflatoxin contamination of maize grains in market stores in Benin, Ghana and Togo. Food Additives and Contaminants 2007;24(11) 1283-1291. DOI: 10.1080/02652030701416558

[75] Zain ME. Impact of mycotoxins on humans and animals. Journal of Saudi Chemical Society 2011;15 129-144. DOI:10.1016/j.jscs.2010.06.006

[76] Lewis L, Onsongo M, Njapau H, Schurz-Rogers H, Luber G, Kieszak S, Nyamongo J, Backer L, Dahiye AM, Misore A, DeCock K, Rubin C. Aflatoxin contamination of commercial maize products during an outbreak of acute aflatoxicosis in Eastern and Central Kenya. Environmental Health Perspectives 2005;113(12) 1763-1767. DOI: 10.1289/ehp.7998.

[77] Turner PC, Moore SE, Hall AJ, Prentice AM, Wild CP. Modification of immune function through exposure to dietary aflatoxin in Gambian children. Environmental Health Perspectives 2003;111 217-220.

[78] Jiang YI, Jolly PE, Ellis WO, Wang JS, Phillips TD, Williams J.H. AflatoxinB$_1$ albumin adduct levels and cellular immune status in Ghanaians. International Immunology 2005;17 807-814. DOI: 10.1093/intimm/dxh262.

[79] Montville TJ, Matthews KR. Food Microbiology, An Introduction. Washington: AMS Press; 2008.

[80] Krogh P. Porcine nephropathy associated with ochratoxin A. In: Smith JE, Anderson RA. (eds.), Mycotoxins and Animal Foods. Boca Raton CRC Press; 1991. p627-645.

[81] Hussein S, Brasel JM. Toxicity, metabolism, and impact of mycotoxins on humans and animals. Toxicology 2001;167 101-134.

[82] Fink-Gremmels J. Mycotoxins: their implications for human and animal health. Veterinary Questions 1999;21 115-120.

[83] Marasas WFO, Riley RT, Hendricks KA, Stevens VL, Sadler TW, Gelineau-van Waes J, Missmer SA, Cabrera J, Torres O, Gelderblom WCA, Allegood J, Martinez C, Maddox J, Miller JD, Starr L, Sullards MC, Roman A, Voss KA, Wang E, Merrill AH. Fumonisins disrupt sphingolipid metabolism, folate transport, and neural tube development in embryo culture and in vivo: a potential risk factor for human neural tube defects among populations consuming fumonisin contaminated maize. Journal of Nutrition 2004;134 711-716.

[84] Peraica M, Radic B, Lucic A, Pavlovic M. Toxic effects of mycotoxins in humans. Bulletin of the World Health Organization 1999;77 754-763.

[85] Pestka JJ, Zhou RR, Moon Y, Chung YJ. Cellular and molecular mechanisms for immune modulation by deoxynivalenol and other tricothecenes; unraveling a paradox. Toxicology Letters 2004;153 61-73. DOI:10.1016/j.toxlet.2004.04.023

[86] Smith EE, Kubena LF, Braithwaite RB, Harvey RB, Phillips TD, Reine AH. Toxicological evaluation of aflatoxin and cyclopiazonic acid in broiler chickens. Poultry Science 1992;71 1136-1144.

[87] Brake J, Hamilton PB, Kittrell RS. Effects of the trichothecene mycotoxin diacetoxyscirpenol on feed consumption, body weight, and oral lesions of broiler breeders. Poultry Science 2000;79 856-863.

[88] Huff WE, Kubena LF, Harvey RB, Doerr JA. Mycotoxin interactions in poultry and swine. Journal of Animal Science 1988;66 2351-2355.

[89] Osweiler GD, Ross PF, Wilson TM, Nelson PE, Witte ST, Carson TL, Rice LG, Nelson HA. Characterizations of an epizootic of pulmonary edema in swine associated with fumonisin in corn screenings. Journal of Veterinary Diagnostic Investigation 1992;4 53-59.

[90] Diaz FJ, Boermans HJ. Fumonisin toxicosis in domestic animals: a review. Veterinary and Human Toxicology 1994;36 548-555.

[91] Diekman MA, Green ML. Mycotoxins and reproduction in domestic livestock. Journal of Animal Science 1992;70 1615-1627.

[92] Glavitis R, Vanyi A. More important mycotoxicosis in pigs. Magyar Allatorvosak Lapja 1995;50 407-420.

[93] Veldman AJ, Meijs AC, Borggreve GJ, Heeres van der Tol JJ. Carry-over of aflatoxin from cows' food to milk. Animal Production 1992;55 163-168.

[94] Placinta CM, D'Mello JPF, MacDonald AMC. A review of worldwide contamination of cereal grains and animal feed with *Fusariam* mycotoxins. Animal Feed Science and Technology 1999;78 21-37.

[95] Cheeke PR. Mycotoxins associated with forages. In: Cheeke PR. (ed.) Natural Toxicants in Feeds, Forages, and Poisonous Plants. Danville: Interstate Publishers; 1998. p243-274.

[96] Van Egmond HP, Schothorst RC, Jonker MA. Regulations relating to mycotoxins in food. Analytical and Bioanalytical Chemistry 2007;389 147-157.

[97] Commission Regulation (EC) 466/2001 setting maximum levels for certain contaminants in foodstuffs. Official Journal of the European Union 2001;L77 1-13.

[98] Commission Regulation (EC) 1881/2006 setting maximum levels for certain contaminants in foodstuffs. Official Journal of the European Union 2006;L364 5-24.

[99] Commission Regulation (EC) 1126/2007 amending Regulation (EC) No 1881/2006 setting maximum levels for certain contaminants in foodstuffs as regards *Fusarium* toxins in maize and maize products Official Journal of the European Communities, 2007;L255 14-17.

[100] Fellinger A. Worldwide mycotoxin regulations and analytical challenges. World Grain Summit: Foods and Beverages, San Francisco, California, USA. 2006.

[101] Worldwide regulations for mycotoxins in food and feed in 2003. http://www.fao.org/docrep/007/y5499e/y5499e07.htm (accessed 7 June 2012)

[102] Directive (EC) 2002/32 on undesirable substances in animal feed. Official Journal of the European Union 2002;L140 10-22.

[103] Commission Directive 2003/100/EC Amending Annex I to Directive 2002/32/EC of the European Parliament and of the Council on undesirable substances in animal feed. Official Journal of the European Union 2003;L285 33-37.

[104] Commission Recommendation (EC) 2006/576 on the presence of deoxynivalenol, zearalenone, ochratoxin A, T-2 and HT-2 and fumonisins in products intended for animal feeding.Official Journal of the European Union 2006;L229 7-9.

[105] Siegel D, Babuscio T. Mycotoxin management in the European cereal trading sector. Food Control 2011;22 1115-1153.

[106] Commission Regulation (EC) No 401/2006 laying down the methods of sampling and analysis for the official control of the levels of mycotoxins in foodstuffs. Official Journal of the European Communities, 2006;L70 12-34.

[107] Commission Regulation (EU) No 178/2010 amending Regulation (EC) No 401/2006 as regards groundnuts (peanuts), other oilseeds, tree nuts, apricot kernels, liquorice and vegetable oil. Official Journal of the European Communities, 2010;L52 32-43.

[108] Turner NW, Subrahmanyam S, Piletsky SA. Analytical methods for determination of mycotoxins: A review. Analytica Chimica Acta 2009;632 168-180. DOI:10.1016/j.aca. 2008.11.010.

[109] Sheppard GS. Determination of mycotoxins in human foods. Chemical Society Review 2008;37(11) 2468-2477. DOI: 10.1039/B713084H

[110] Scott PM. Official methods of analysis. 15th ed., Vol. II, Association of Official Analytical Chemists, Arlington; 1990. p1184-1212.

[111] Kabak B, Dobson AD, Var I. Strategies to prevent mycotoxin contamination of food and animal feed: a review. Critical Reviews in Food Science and Nutrition 2006;46 593-619. DOI:10.1080/10408390500436185

[112] Golob P. Good practices for animal feed and livestock training manual on – farm mycotoxin control in food and feed grain. Food and Agriculture Organization of the United Nations, Rome 2007.

[113] Aldred D, Magan N. Prevention strategies for trichothecenes. Toxicology Letters 2004;153 165-171. DOI:10.1016/j.toxlet.2004.04.031

[114] Edwards SG. Influence of agricultural practices on *Fusarium* infection of cereals and subsequent contamination of grain by trichothecene mycotoxins. Toxicology Letters 2004;153 29-35. DOI:10.1016/j.toxlet.2004.04.022

[115] Langseth W, Rundberget T. The occurrence of HT-2 toxin and other trichothecenes in Norwegian cereals. Mycophatologia 1999;147 157-165.

[116] Tekauz A. *Fusarium* head blight of oat in western Canada – preliminary studies on *Fusarium* species involved and level of mycotoxins in grain. Journal of Applied Genetics 2002;43A 197-206.

[117] Teich AH, Nelson K. Survey of *Fusarium* Head Blight and Possible Effects of Cultural-Practices in Wheat Fields in Lambton County in 1983. Canadian Plant Disease Survey 1984;64 11-13.

[118] Teich AH, Hamilton JR. Effect of cultural-practices, soil-phosphorus, potassium and pH on the incidence of *Fusarium* head blight and deoxynivalenol levels in wheat. Applied and Environmental Microbiology 1985;44 1429-1431.

[119] Sturz AV, Johnston HW. Characterization of *Fusarium* colonization of spring barley and wheat produced on stubble or fallow soil. Canadian Journal of Plant Pathology 1985;7 270-276.

[120] Schaafsma AW, Tamburic-Ilinic L, Miller JD, Hooker DC. Agronomic considerations for reducing deoxynivalenol in wheat grain. Canadian Journal of Plant Pathology 2001;23 279-285.

[121] Cromey MG, Shorter SC, Lauren DR,Sinclair KI. Cultivar and crop management influences on *Fusarium* head blight and mycotoxins in spring wheat (*Triticum aestivum*) in New Zealand. New Zealand Journal of Crop and Horticultural Science 2002;30 235-247.

[122] Codex Alimentarius Commission Proposed draft code of practice for the prevention (reduction) of mycotoxin contamination in cereals, including annexes on ochratoxin A, zearalenone, fumonisins and trichothecenes, CX/FAC 02/21, Joint FAO/WHO Food Standards Programme, Rotterdam, The Netherlands. 2002.

[123] Obst A, Gleissenthall JL, Beck R. On the etiology of Fusarium head blight of wheat in south Germany - Preceding crops, weather conditions for inoculum production and head infection, proneness of the crop to infection and mycotoxin production. Cereal Research Communications 1997;25 699-703.

[124] Labreuche J, Maumene C, Caron D. Wheat after maize – Mycotoxin risk management. Selected papers from Arvalis 2005;2 14-16.

[125] Koch HJ, Pringas C, Maerlaender B. Evaluation of environmental and management effects on *Fusarium* head blight infection and deoxynivalenol concentration in the grain of winter wheat. European Journal of Agronomy 2006; 24 357-366. DOI10.1016/j.eja.2006.01.006

[126] Martin RA, Macleod JA, Caldwell C. Influences of production Inputs on incidence of infection by *Fusarium* species on cereal seed. Plant Disease 1991; 75 784-788.

[127] Lemmens M, Haim K, Lew H, Ruckenbauer P. The effect of nitrogen fertilization on *Fusarium* head blight development and deoxynivalenol contamination in wheat. Journal of Phytopathology 2004;152 1-8. DOI:10.1046/j.1439-0434.2003.00791.x.

[128] Blandino M, Reyneri A, Vanava F, Tamietti G, Pietri A. Influence of agricultural practices on *Fusarium* infection, fumonisin and deoxynivalenol contamination of maize kernels. World Mycotoxin Journal 2009;2 409-418. DOI: 10.3920/WMJ2008.1098.

[129] Nicholson P, Turner JA, Jenkinson P, Jennings P, Stonehouse J, Nuttall M, Dring D, Weston G, Thomsett M. Maximising control with fungicides of *Fusarium* ear blight (FEB) in order to reduce toxin contamination of wheat. HGCA Project Report 2003;297.

[130] Mesterhazy A, Bartok T, Lamper C. Influence of wheat cultivar, species of *Fusarium* and isolate aggressiveness on the efficacy of fungicides for control of *Fusarium* head blight. Plant Disease 2003;87 1107-1115.

[131] Ioos R, Belhadj A, Menez M, Faure A. The effects of fungicides on *Fusarium* spp. and *Microdochium nivale* and their associated trichothecene mycotoxins in French naturally-infected cereal grains. Crop Protection 2005;24 894-902. DOI:10.1016/j.cropro. 2005.01.014.

[132] Edwards SG, Pirgozliev SR, Hare MC, Jenkinson P. Quantification of trichothecene-producing *Fusarium* species in harvested grain by competitive PCR to determine efficacies of fungicides against *Fusarium* head blight of winter wheat. Applied and Environmental Microbiology 2001; 67 1575-1580. DOI:10.1128/AEM. 67.4.1575-1580.2001.

[133] Ellner FM. Results of long-term field studies into the effect of strobilurin containing fungicides on the production of mycotoxins in several winter wheat varieties. Mycotoxin Research 2006;21 112-115. DOI: 10.1007/BF02954432.

[134] Simpson DR, Weston GE, Turner JA, Jennings P, Nicholson P. Differential control of head blight pathogens of wheat by fungicides and consequences for mycotoxin contamination of grain. European Journal of Plant Pathology 2001;107 421-431. DOI: 10.1023/A:1011225817707.

[135] Hinton DM, Bacon CW. *Enterobacter cloacae* is an endophytic symbiont of corn. Mycophatologia 1995;129 117-125.

[136] Bacon CW, Yates II, Hinton DM, Mevedith F. Biological control *Fusarium moniliforme* in maize. Environmental Health Perspectives 2001;109 325-332.

[137] Khan NI, Schisler DA, Boehm MJ, Slininger PJ, Bothast RJ. Selection and evaluation of microorganisms for biocontrol of *Fusarium* head blight of wheat incited by *Gibberella zeae*. Plant Disease 2001;85 1253-1258.

[138] Schisler DA, Khan NI, Boehm MJ. Biological control of *Fusarium* head blight of wheat and deoxynivalenol levels in grain via use of microbial antagonists. In DeVries JW, Trucksess MW, Jackson LS. (eds.) Mycotoxins and Food Safety Kluwer Academic/ Plenum Publishers, New York, 2002 p53-59.

[139] Doyle MP, Applebaum RS, Brackett RE, Marth EH. Physical, chemical and biological degradation of mycotoxins in foods and agricultural commodities. Journal of Food Protection 1982;45 964-971.

[140] Sinha K K. Decontamination of mycotoxins and food safety. Toxicology Letters 2001;122 179-188.

[141] Kim J, Yousef A, Dave S. Application of ozone for enhancing the microbiological safety and quality of foods: A Review. Journal Food Protection 1999;62 1071-1087.

[142] Allameha A, Safamehrb A, Mirhadic SA, Shivazadd M, Razzaghi-Abyanehe M, Afshar-Naderia A. Evaluation of biochemical and production parameters of broiler chicks fed ammonia treated aflatoxin contaminated maize grains. Animal Feed Science and Technology 2005;122 289-301.

[143] Scott PM, Industrial and farm detoxification processes for mycotoxins. Revue de Mé-
decine Vétérinaire 1998;149 543-548.

[144] Norred WP, Voss KA, Bacon CW, Riley RT. Effectiveness of ammonia treatment in
detoxification of fumonisin-contaminated corn. Food and Chemical Toxicology
1991;29: 815-819.

[145] Kolosova A, Stroka J.Substances for reduction of the contamination of feed by mycotoxins: a review World Mycotoxin Journal 2001;4(3) 225-256. DOI:10.3920/
WMJ2011.1288

[146] Huwig A, Freimund S, Kappeli O, Dutler H. Mycotoxin detoxication of animal feed
by different adsorbents. Toxicology Letters 2001;122 179-188.

[147] Galvano F, Piva A, Ritieni A, Galvano G. Dietary strategies to counterackt the effects
of mycotoxins: a review. Journal of Food Protection 2001;64(1) 120-131.

[148] Diaz D, Hagler W, Hopkins B, Whitlow L. Aflatoxin binders: In vitro binding assay
for aflatoxin B_1 by several potential sequestering agents. Mycopathologia 2003;156
223-226. DOI: 10.1023/A:1023388321713

[149] Phillips TD, Lemke SL, Grant PG Characterization of clay-based enterosorbents for
the prevention of aflatoxicosis Advances in Experimental Medicine and Biology
2002;504 157-71.

[150] Manafi M. Aflatoxicosis and mycotoxin binders in commercial broilers: effects on
performance and biochemical parameters and immune status. Advances in Environmental Biology 2011;5(13) 3866-3870.

[151] Ciegler A, Lillehoj EB, Peterson RE, Hall HH. Microbial detoxification of aflatoxin.
Applied Microbiology 1996;14 934-938.

[152] D'Souza DH, Brackett E. Aflatoxin B_1 degradation by *Flavobacterium aurantiacum* in
the presence of reducing conditions and seryl and sulfhydryl group inhibitors. Journal of Food Protection 2001;64(2) 268-271.

[153] Hao YY, Brackett RE. Removal of aflatoxin B_1 from peanut milk inoculated with *Flavobacterium aurantiacum*. Journal of Food Science 1998;53(5) 1384-1386.

[154] Smiley RD, Draughon FA. Preliminary evidence that degradation of aflatoxin B_1 by
Flavobacterium aurantiacum is enzymatic. Journal of Food Protection 2000;63(3)
415-418.

[155] Pierides M, El-Nezami H, Peltonen K, Salminen S, Ahocas JP. Ability of dairy strains
of lactic acid bacteria to bind aflatoxin M_1 in a food model. Journal of Food Protection
2000;63(5) 645-650.

[156] El-Nezami H, Kankaanpaa P, Salminen S, Ahokas J. Ability of dairy strains of lactic
acid bacteria to bind a common food carcinogen, aflatoxin B_1. Food and Chemical
Toxicology 1998;36 321-326.

[157] Turbic A, Ahokas JT, Haskard CA. Selective in vitro binding of dietary mutagens, individually or in combination, by lactic acid bacteria. Food Additives and Contaminants 2002;19(2) 144-152.

[158] Megalla SM, Mohran MA. Fate of aflatoxin B₁ in fermented dairy products. Mycopathologia 1984;88 27–29.

[159] Teniola OD, Addo PA, Brost IM, Farber P, Jany KD, Alberts JF, Vanzyl WH, Steyn PS, Holzapfel WH. Degradation of AFB₁ by cell free extracts of *Rhodococcus erythropolis* and *Mycobacterium fluoranthenivorans* sp. International Journal of Food Microbiology 2005;105 111-117. DOI:10.1016/j.ijfoodmicro.2005.05.004

[160] Alberts JF, Engelbrecht Y, Steyn PS, Holzapfel WH, Vanzyl WH. Biological degradation of AFB₁ by *Rhodococcus erythropolis* cultures. International Journal of Food Microbiology 2006;109 121-126.

[161] Peltonen KD, El-Nezami H, Salminen S, Ahokas JT. Binding of aflatoxin B₁ by probiotic bacteria. Journal of the Science of Food and Agriculture 2000;80 1942-1945.

[162] Engler KH, Coker RD, Evans IH. Uptake of aflatoxin B₁ and T-2 toxin by two mycotoxin bioassay microorganisms: *Kluyveromyces marxianus* and *Bacillus megaterium*. Archives of Microbiology 2000;174(6) 381-385.

[163] Shetty PH, Hald B, Jespersen L. Surface binding of aflatoxin B₁ by *Saccharomyces cerevisiae* strains with potential decontaminating abilities in indigenous fermented foods. International Journal of Food Microbiology 2007;113 41-46. DOI:10.1016/j.ijfoodmicro.2006.07.013

[164] Mann R, Rehm HJ. Degradation products from aflatoxin B₁ by *Corynebacterium rubrum, Aspergillus niger, Trichoderma viride, Mucor ambignus*. European Journal of Applied Microbiology 1976;2 297-306.

[165] Brown RL, Cotty PJ, Cleveland TE. Reduction in aflatoxin content of maize by atoxigenic strains of *Aspergillus flavus*. Journal of Food Protection 1991;54(8) 623-628.

[166] Fuchs S, Sontag G, Stidl R, Ehrlich V, Kundi M, Knasmuller S. Detoxification of patulin and ochratoxin A, two abundant mycotoxins, by lactic acid bacteria. Food and Chemical Toxicology 2008;46 1398-1407. DOI:10.1016/j.fct.2007.10.008

[167] Škinjar M, Raaic JL, Stojicic V. Lowering of ochratoxin A level in milk by yoghurt bacteria and *Bifidobacteria*. Folia Microbiologica 1996;41(1) 26-28.

[168] Cheng-An H, Draughon FA. Degradation of ochratoxin A by *Acinetobacter calcoaceticus*. Journal of Food Protection 1994;57 410-414.

[169] Piotrowska M, Żakowska Z. The elimination of ochratoxin A by lactic acid bacteria. Polish Journal of Microbiology 2005;54(4) 279-286.

[170] Bejaoui H, Mathieu F, Taillandier P, Lebrihi A. Ochratoxin A removal in synthetic and natural grape juices by selected oenological *Saccharomyces strains*. Journal of Applied Microbiology 2004;97 1038-1044. DOI: 10.1111/j.1365-2672.2004.02385.x.

[171] Piotrowska M. Adsorption of ochratoxin A by *Saccharomces cerevisiae* living and non-living celles. Acta Alimentaria 2012;41 1-7. DOI:10.1556/AAlim.2011.0006.

[172] Piotrowska M, Żakowska Z. The biodegradation of ochratoxin A in food products by lactic acid bacteria and baker's yeast. Progress in Biotechnology. Food Biotechnology 2000;17 307-310.

[173] Mateo EM, Medina A, Mateo F, Valle-Algarra FM, Pardo I. Ochratoxin A removal in synthetic media by living and heat-inactivated cells of *Oenococcus oeni* isolated from wines. Food Control 2010;21 23-28. DOI:10.1016/j.foodcont.2009.03.012

[174] Angioni A, Caboni P, Garau A, Harris A, Orro D, Budroni M, Cabras P. In vitro interaction between ochratoxin A and different strains of *Saccharomyces cerevisiae* and *Kloeckera apiculata*. Journal of Agricultural Chemistry 2007;55 2043-2048. DOI:10.1021/jf062768u

[175] Bejaoui H, Mathieu F, Taillandier P,Lebrihi A. Conidia of black Aspergilli as new biological adsorbents for ochratoxin A in grape juices and musts. Journal of Agricultural Chemistry 2005;53 8224-8229.

[176] Štyriak I, Čonková E, Kmet V, Böhm J, Razzazi E. The use of yeast for microbial degradation of some selected mycotoxins. Mycotoxin Research 2001;17A(1) 24-27.

[177] Peteri Z, Teren J, Vagvolgyi C, Varga J. Ochratoxin degradation and adsorption caused by astaxanthin-producing yeasts. Food Microbiology 2007;24 205-210. DOI: 10.1016/j.ijfoodmicro.2004.10.034

[178] Patharajan S, Reddy KRN, Karthikeyan V, Spadaro D, Lore A, Gullino ML, Garibaldi A. Potential of yeast antagonists on in vitro biodegradation of ochratoxin A. Food Control 2011;22 290-296.

[179] Varga J, Peteri Z, Tabori K, Teren J, Vagvolgyi C. Degradation of ochratoxin A and other mycotoxins by *Rhizopus* isolates. International Journal of Food Microbiology 2005;99 321-328. DOI:10.1016/j.ijfoodmicro.2004.10.034.

[180] De Felice DV, Solfrizzo M, De Curtis F, Lima G, Visconti A, Castoria R. Strains of *Aureobasidium pullulans* can lower ochratoxin A contamination in wine grapes. Phytopathology 2008;98(12) 1261-1270. DOI:10.1094/PHYTO-98-12-1261

[181] Abrunhosa L, Santos L, Venancio A. Degradation of ochratoxin A by proteases and by a crude enzyme of *Aspergillus niger*. Food Biotechnology 2006;20 231-242. DOI: 10.1080/08905430600904369

[182] Varga J, Rigo K, Teren J. Degradation of ochratoxin A by *Aspergillus* species. International Journal of Food Microbiology 2000;59 1-7. DOI:10.1078/0723202042369947

[183] Molnar O, Schatzmayr G, Fuchs E, Prillinger H. *Trichosporon mycotoxinivorans* sp. nov. a new yeast species useful in biological detoxification of various mycotoxins. Systematic and Applied Microbiology 2004;27 661-671.

[184] Niderkorn V, Boudra H, Morgavi DP. Binding of *Fusarium* mycotoxins by fermenta-
tive bacteria *in vitro*. Journal of Applied Microbiology 2006;101 849-856. DOI:
10.1111/j.1365-2672.2006.02958.x

[185] Fusch E, Binder EM, Heidler D, Krska R. Structural characterization of metabolites
after the microbial degradation of type A trichothecenes by the bacterial strain BBSH
797. Food Additives and Contaminants 2002;19(4) 379-386.

[186] Young JC, Zhou T, Yu H, Zhu H, Gong J. Degradation of trichotecene mycotoxins by
chicken intestinal microbes. Food and Chemical Toxicology 2007;45 136-143. DOI:
10.1016/j.fct.2006.07.028

[187] Megharaj M, Garthwaite I, Thiele JH. Total biodegradation on the oestrogenic mycotoxin zearalenone by a bacterial culture. Letters in Applied Microbiology 1997;24
329-333.

[188] Line JE, Brackett RE. Factors affecting aflatoxin B_1 removal by *Flavobacterium aurantiacum*. Journal of Food Protection 1995;58 91-94.

[189] El-Nezami H, Mykkänen H, Kankaanpaa P, Salminen S, Ahokas J. Ability of *Lactobacillus* and *Propionibacterium* strains to remove aflatoxin B_1 from chicken duodenum.
Journal of Food Protection 2000;63(4) 549-552.

[190] Raaić JL, `kinjar M, Markov S. Decrease of aflatoxin B_1 in yoghurt and acidified
milks. Mycopathologia 1991;113 117-119.

[191] El-Nezami H, Kankaanpaa P, Salminen S, Ahokas J. Physicochemical alterations enhance the ability of dairy strains of lactic acid bacteria to remove aflatoxin from contaminated media. Journal of Food Science 1998;61(4) 466-468.

[192] Śliżewska K, Piotrowska M, Libudzisz Z. Influence of a probiotic preparation on the
aflatoxin B_1 content in feaces, blood serum, kidneys and liver of chickens fed on aflatoxin B_1 contaminated fodder (*in vivo* tests). Procceedings of 30th Mycotoxin Workshop, Utrecht, The Netherlands, 2008, p130.

[193] Raju MVLN, Devegowda G. Influence of esterified glucomannan on performance
and organ morphology, serum biochemistry and haematology in broilers exposed to
invidual and combined mycotoxicosis (aflatoxin, ochratoxin and T-2 toxin). British
Poultry Science 2000;41 640-650.

[194] Yiannikouris A, Francois J, Poughon L, Dussap CG, Bertin G, Jeminet G, Jouany JP.
Adsorption of zearalenone by beta-D-glucans in the *Saccharomyces cerevisiae* cell wall.
Journal of Food Protection 2004;67(6) 1195-1200.

[195] Baptista AS, Horii J, Calori-Domingues MA, Micotti da Gloria E.: The capacity of
manno-oligosaccharides, thermolysed yeast and active yeast to attenuate aflatoxicosis, World Journal of Microbiology and Biotechnology 2004;20 475-481. DOI: 10.1023/
B:WIBI.0000040397.48873.3b

[196] Richard JL. Some major mycotoxins and their mycotoxicoses - an overview. International Journal of Food Microbiology 2007;119 3-10. DOI:10.1016/j.ijfoodmicro. 2007.07.019.

[197] Piotrowska M, Arkusz J, Stańczyk M, Palus J, Dziubałtowska E, Stępnik M. The influence of lactic acid bacteria and yeasts on cytotoxicity and genotoxicity of ochratoxin A. Procceedings of 30th Mycotoxin Workshop, Utrecht, The Netherlands, 2008, p146.

Permissions

The contributors of this book come from diverse backgrounds, making this book a truly international effort. This book will bring forth new frontiers with its revolutionizing research information and detailed analysis of the nascent developments around the world.

We would like to thank Hany A. El-Shemy, for lending his expertise to make the book truly unique. He has played a crucial role in the development of this book. Without his invaluable contribution this book wouldn't have been possible. He has made vital efforts to compile up to date information on the varied aspects of this subject to make this book a valuable addition to the collection of many professionals and students.

This book was conceptualized with the vision of imparting up-to-date information and advanced data in this field. To ensure the same, a matchless editorial board was set up. Every individual on the board went through rigorous rounds of assessment to prove their worth. After which they invested a large part of their time researching and compiling the most relevant data for our readers. Conferences and sessions were held from time to time between the editorial board and the contributing authors to present the data in the most comprehensible form. The editorial team has worked tirelessly to provide valuable and valid information to help people across the globe.

Every chapter published in this book has been scrutinized by our experts. Their significance has been extensively debated. The topics covered herein carry significant findings which will fuel the growth of the discipline. They may even be implemented as practical applications or may be referred to as a beginning point for another development. Chapters in this book were first published by InTech; hereby published with permission under the Creative Commons Attribution License or equivalent.

The editorial board has been involved in producing this book since its inception. They have spent rigorous hours researching and exploring the diverse topics which have resulted in the successful publishing of this book. They have passed on their knowledge of decades through this book. To expedite this challenging task, the publisher supported the team at every step. A small team of assistant editors was also appointed to further simplify the editing procedure and attain best results for the readers.

Our editorial team has been hand-picked from every corner of the world. Their multi-ethnicity adds dynamic inputs to the discussions which result in innovative outcomes. These outcomes are then further discussed with the researchers and contributors who give their valuable feedback and opinion regarding the same. The feedback is then collaborated with the researches and they are edited in a comprehensive manner to aid the understanding of the subject.

Apart from the editorial board, the designing team has also invested a significant amount of their time in understanding the subject and creating the most relevant covers. They scrutinized every image to scout for the most suitable representation of the subject and create an appropriate cover for the book.

The publishing team has been involved in this book since its early stages. They were actively engaged in every process, be it collecting the data, connecting with the contributors or procuring relevant information. The team has been an ardent support to the editorial, designing and production team. Their endless efforts to recruit the best for this project, has resulted in the accomplishment of this book. They are a veteran in the field of academics and their pool of knowledge is as vast as their experience in printing. Their expertise and guidance has proved useful at every step. Their uncompromising quality standards have made this book an exceptional effort. Their encouragement from time to time has been an inspiration for everyone.

The publisher and the editorial board hope that this book will prove to be a valuable piece of knowledge for researchers, students, practitioners and scholars across the globe.

List of Contributors

Raman Bansal
Department of Entomology, Ohio Agricultural Research and Development Center, the Ohio State University, USA

Tae-Hwan Jun
Department of Entomology, Ohio Agricultural Research and Development Center, the Ohio State University, USA
Department of Horticulture and Crop Sciences, Ohio Agricultural Research and Development Center, the Ohio State University, USA

M. A. R. Mian
Department of Horticulture and Crop Sciences, Ohio Agricultural Research and Development Center, the Ohio State University, USA
USDA-ARS Corn and Soybean Research Unit, Madison Ave., Wooster, OH, USA

Andy P. Michel
Department of Entomology, Ohio Agricultural Research and Development Center, The Ohio State University, USA

Alexandre Ferreira da Silva
Embrapa Milho e Sorgo, Sete Lagoas-MG, Brazil

Leandro Galon
Universidad Federal da Fronteira Sul, Erechim-RS, Brazil

Ignacio Aspiazú
Universidad Estadual de Montes Claros-MG, Brazil

Evander Alves Ferreira
Universidad Federal dos Vales do Jequitinhonha e Mucuri, Diamantina-MG, Brazil

Germani Concenço
Embrapa Agropecuária Oeste, Dourados-MS, Brazil

Edison Ulisses Ramos Júnior
Embrapa Soja, Londrina-PR, Brazil

Paulo Roberto Ribeiro Rocha
Universidad Federal Rural do Semi-Árido, Mossoró-RN, Brazil

Steven C. Goheen
Pacific Northwest National Laboratory, Richland, Washington, USA

James A. Campbell
Pacific Northwest National Laboratory, Richland, Washington, USA

Patricia Donald
U. S. Department of Agriculture/ARS, Jacksonville, Tennessee, USA

Rafael Vivian
Brazilian Department of Agriculture, Agriculture Research Service – Embrapa Mid-North, Teresina, PI, Brazil

André Reis
Department of Civil and Environmental Engineering, Waseda University, Shinjuku-ku, Okubo, Tokyo, Japan

Pablo A. Kálnay
National University of Buenos Aires/Northwest, Pergamino, Buenos Aires, Argentina

Leandro Vargas
Brazilian Department of Agriculture, Agriculture Research Service – Embrapa Mid-North, Teresina, PI, Brazil

Ana Carolina Camara Ferreira
Federal University of Piauí, M.Sc. student – Soil conservation – Teresina – PI, Brazil

Franciele Mariani
Federal University of Pelotas, Ph.D. student – Weed management – Passo Fundo – RS, Brazil

Vincent P. Klink
Department of Biological Sciences, Mississippi State University, Mississippi State, MS, USA

Prachi D. Matsye
Department of Biological Sciences, Mississippi State University, Mississippi State, MS, USA

Katheryn S. Lawrence
Department of Biochemistry, Molecular Biology, Entomology and Plant Pathology, Mississippi State University, Mississippi State, MS,, U.S.A.

Gary W. Lawrence
Department of Entomology and Plant Pathology, Auburn University, Auburn, AL, USA

Renata Bažok, Maja Čačija, Ana Gajger and Tomislav Kos
University of Zagreb, Faculty of Agriculture, Zagreb, Croatia

M. H. Khan
Central Institute of Temperate Horticulture, Indian Council of Agriculture Research, Srinagar, India

S. D. Tyagi
Department of Plant Breeding and Genetics, Kisan (P.G) College, Simbhaoli, Ghaziabad, India

Z. A. Dar
S.K. University of Agricultural Sciences and Technology of Kashmir, Shalimar, Srinagar, India

Yaghoub Fathipour and Amin Sedaratian
Department of Entomology, Faculty of Agriculture, Tarbiat Modares University, Tehran, Iran

Małgorzata Piotrowska
Technical University of Lodz, Faculty of Biotechnology and Food Sciences, Institute of Fermentation Technology and Microbiology, Lodz, Poland

Katarzyna Śliżewska
Technical University of Lodz, Faculty of Biotechnology and Food Sciences, Institute of Fermentation Technology and Microbiology, Lodz, Poland

Joanna Biernasiak
Technical University of Lodz, Faculty of Biotechnology and Food Sciences, Institute of Chemical Technology of Food, Lodz, Poland

Printed in the USA
CPSIA information can be obtained
at www.ICGtesting.com
JSHW011457221024
72173JS00005B/1111

9 781632 393050